FINDING CONSCIOUSNESS IN THE BRAIN

ADVANCES IN CONSCIOUSNESS RESEARCH

ADVANCES IN CONSCIOUSNESS RESEARCH provides a forum for scholars from different scientific disciplines and fields of knowledge who study consciousness in its multifaceted aspects. Thus the Series will include (but not be limited to) the various areas of cognitive science, including cognitive psychology, linguistics, brain science and philosophy. The orientation of the Series is toward developing new interdisciplinary and integrative approaches for the investigation, description and theory of consciousness, as well as the practical consequences of this research for the individual and society.

Volume 8

Peter G. Grossenbacher (ed.)

Finding Consciousness in the Brain

A neurocognitive approach

FINDING CONSCIOUSNESS IN THE BRAIN

A NEUROCOGNITIVE APPROACH

Edited by

PETER G. GROSSENBACHER
National Institute of Mental Health, Bethesda

JOHN BENJAMINS PUBLISHING COMPANY
AMSTERDAM/PHILADELPHIA

∞^{TM} The paper used in this publication meets the minimum requirements of American National Standard for Information Sciences — Permanence of Paper for Printed Library Materials, ANSI Z39.48–1984.

Library of Congress Cataloging-in-Publication Data

Finding consciousness in the brain : a neurocognitive approach / edited by Peter G. Grossenbacher.
    p.   cm. -- (Advances in consciousness research, ISSN 1381-589X ; v. 8)
    Includes bibliographical references.
  1. Cognitive neuroscience. I. Grossenbacher, Peter G. II. Series.
QP360.5.F55   1998
612.8'2--dc21   98-16141
ISBN 90 272 5128 2 (Eur.) / 1 55619 188 x (US) (Pb; alk. paper)   CIP

John Benjamins Publishing Co. • P.O.Box 36224 • 1020 ME Amsterdam • The Netherlands
John Benjamins North America • P.O.Box 27519 • Philadelphia PA 19118-0519 • USA

Contents

Apology

When this collective volume finally goes to print it came to my mind to check back the correspondence relating to it and I found that the first arrangements with the Editor of this volume were made in 1995. And when I am writing this Preface, it is September 26, 2000. Five years passed — almost a millennium for the contemporary information society.

This is an apology to its contributors who accepted to take part in this project, who in time sent their articles and afterwards dutifully returned their proofs to the Editor of this volume expecting fast reaction on his side and an effort to get it published as soon as possible. Unfortunately this did not happen during all of the stages of the project. The manuscript was first delayed at the stage of compiling it. Some delay (a month or two) may sometimes happen, but this took much longer in the present case. The same trend followed again and again. Bertie Kaal, the Acquisition Editor for John Benjamins Publishers and I were expectant that this would not repeat itself, but we turned overly optimistic. At the final stage, Bertie accepted to check the final proofs and have the Index made (it was agreed that these were the volume editor's taks). I do not know what were the personal reasons for Dr. Grossenbacher to cumulatively delay the production procedure two or three times over its regular course. But I became once again keenly aware on this occasion of two of the duties of my editorship. The first is the commitment the author and the editor share — to write and/or compile something sensible and new. The second duty is related to accepting responsibility to a set of authors to get their word through to the professional audience. This is the midwifery service that editors accept. I hope I learnt something from this case and I undertake a further obligation to make every effort to detect and react in a radical way as early as possible the potential repetition of a case of delay like this in the future in the Series I am in charge of.

I would like to add several words about the topic of this volume. In my opinion, the cognitive neuroscience of consciousness is one of the most promising trends in the development of consciousness studies because it combines the interest in first-person experience, third-person conceptual analysis and application of modern brain imaging technology. The latter technology developed quite a lot in the last five years. Still the identification of a certain type of qualitative experience, its experimental study and conceptual analysis makes each article in this volume a unique contribution. I am very happy that this volume appears in our Series because it belongs to the class of 'sensible and new developments' in the field. Good luck to it!

Maxim Stamenov

Cognitive Neuroscience
and Consciousness

The problem of consciousness involves many difficult and overlapping issues (Block, 1995; Posner, 1994). Perhaps the most frequently discussed role of consciousness involves awareness of our sensory world but another equally important aspect of consciousness is the fact or illusion of voluntary control. In the course of development, a central issue is the awareness of one's self, and the form of voluntary control involved in self regulation. These functions develop within the dyad involving the child and the caregiver as a carrier of the culture's socialization process. It is possible, even likely, that brain mechanisms subserving these various forms of consciousness may cut across definitions in ways that defy the usual logical and philosophical distinctions.

Despite these difficulties, it is important and timely to devote a book to the task of finding consciousness in the brain. The task has been made somewhat easier by a generation of cognitive studies that have sought simple model systems to separate conscious from unconscious processes within normal subjects. These are summarized by Price in Chapter 2 for studies of visual masking.

In most of neuroscience, students have followed Crick's (1994) suggestion that sensory awareness is the aspect of consciousness most amenable to scientific analysis, placing awareness at the center of discussions of brain mechanisms of consciousness. Chapters 4 and 6 deal with this approach. Even with this more restricted definition, proposals published in recent years have ranged from the anatomical, for example, locating consciousness in the thalamus (Baars, 1997) to the physical, for example, the proposal that consciousness must rest on quantum principles (Penrose, 1989). Perhaps the most persuasively argued of these proposals is the effort of Weizkrantz (1996) to explore and localize the awareness of a moving visual signal that sometimes arises in cases of blindsight.

A quite different approach to consciousness stresses findings about attentional networks that lead to selection of sensory information, activate ideas stored in memory and maintain the alert state. This approach is taken in Chapter 1 and in the chapters by Whitehead & Schliebner (Ch. 6) and by Derryberry (Ch. 7) and with a somewhat more evolutionary and anatomical framework in Chapters 8 and 9.

Much of what is distinctive in this book and in the cognitive neuroscience approach rests upon our ability to image the anatomical location of mental operations in the human brain. The use of positron emission tomography and functional magnetic resonance imagery allow us to examine brain areas that are active during thought. When we can relate these anatomical areas to electrical or magnetic activity recorded from inside or outside the brain, we can also study the time course of mental activity in high level human skills (Posner & Raichle, 1994). These imaging studies have suggested particular brain areas of the frontal midline related to aspects of focal awareness and cognitive control (Posner & DiGirolamo, 1998; Posner & Rothbart, 1998).

A recent PET study has supported the mapping of different cognitive and emotional activations within the frontal midline. Derbyshire, Vogt & Jones (1998) used pain trials and Stroop trials in separate blocks. They found that both types of trials activated what are probably separate areas of the mid-cingulate. In a new set of ERP studies involving the Stroop task, DiGirolamo & Posner were able to show activation of responses over the cingulate at about 270 msec.

While this book presents a strong evolutionary perspective, except for the important analysis of autism by Baron-Cohen (Chapter 3) it does not stress development. In studies of normal infants and children, we have found specific events that mark the development of attention to sensory events (Posner & Rothbart, 1994) during the first year of life and these are earlier than equally dramatic changes that mark the development of self regulation in the second and third year of life (Posner & Rothbart, 1998). Developmental changes must eventually fit into the brain systems that are said to underlie consciousness.

The chapters in this book and efforts elsewhere do indicate progress in understanding mechanisms of sensory awareness, voluntary control and self regulation. There is no reason to suppose that the various philosophical distinctions concerning the unity or independence of various forms of consciousness will neatly divide anatomical or developmental systems discussed here or in future empirical studies. We still remain a long way from a locating consciousness, but we do have some new tools and judging from these chapters, a clear desire to explore this issue.

Michael Posner
Sackler Institute, New York City

References

Baars, B. 1997. *In the Theater of Consciousness*. New York: Oxford University Press

Block, N. 1995. On a confusion about a function of consciousness. *Behavioral and Brain Sciences,* 18:227-287.

Crick, F. 1994. *The Astonishing Hypothesis*. New York: Basic Books.

Derbyshire, S.W.G., Vogt, B.A., & Jones, A.K.P. 1998. Pain and Stroop interference tasks activate separate processing modules in anterior cingulate cortex. *Experimental Brain Research*, 118:52-60.

Penrose, R. 1989. *The Emperor's New Mind*. Oxford: Oxford University Press.

Posner, M.I. 1994. Attention: The mechanism of consciousness. *Proceedings of the National Academy of Sciences, U.S.A.*, 91 (16):7398-7402.

Posner, M.I. & DiGirolamo, G.J. 1998. Executive attention: Conflict, target detection and cognitive control. In R. Parasuraman (ed.), *The Attentive Brain*, Cambridge, MA: MIT Press (pp. 401-423).

Posner, M.I. & Raichle, M.E. 1994. *Images of Mind*. Scientific American Books.

Posner, M.I. & Rothbart, M.K. 1994. Attentional regulation: From mechanism to culture. In Bertelson, P., Eelen, & D'Ydavalle (eds.*), Current Advances in Psychological Science*. London: LEA Associates (pp. 41-56).

Posner, M.I. & Rothbart, M.K. 1998. Attention, self regulation and consciousness. *Transactions of the Philosophical Society of London B.*, 353: 1915–1927.

Weizkrantz, L. 1996. *Consciousness Lost and Found*. Oxford: Oxford University Press.

Acknowledgments

This book embodies a perspective which has been shaped by mentors to whom I owe great debt and gratitude. My undergraduate advisors Professors Eleanor Rosch and John Searle made the disciplines of psychology and philosophy come alive in the minds of impressionable students at U.C. Berkeley, and together they elucidated the compatibility between subjective and objective perspectives. As Advisor throughout my doctoral training at the University of Oregon, Professor Michael I. Posner ushered our cohort of fledgling researchers into the field of cognitive neuroscience with a limitless vision for the possibilities of this field which repeatedly challenged my own more restricted understanding. Professor Jon S. Driver (and U.K. Medical Research Council project grant G9123295) supported my work during the initial stages of this book at the University of Cambridge, where this volume was conceived amidst the stimulating buzz of experimentation on human attention across multiple sense modalities. For insightful tutorials in the consequences of brain pathology, I thank Dr. Allan F. Mirsky, Chief of the Section on Clinical and Experimental Neuropsychology at the National Institute of Mental Health.

My colleagues in the International Multisensory Research Forum have provided a wealth of advice and commentary, and to their collective expertise I am grateful. I also thank Tanner House and Lisa Korenman for help in preparing this manuscript.

I thank the reviewers who critically appraised one or more of the manuscripts which together comprise this volume: Matthew Beer, Scott Fischer, Mark Hallett, Loring Ingraham, Chris T. Lovelace, Allan F. Mirsky, Ewald Neumann, Mark Price, and Mike Wohl. My greatest thanks go to the contributors to this volume, having taken upon themselves the challenge of explaining how conscious experience depends on brain functions.

Preface

In intellectual discourse, consciousness often gets treated as an abstract thing. That approach is very misleading! In fact, *there is nothing more immediate and directly real than conscious experience.* It is the act of referring to our experience with the indirectness of concepts and words that renders consciousness into something abstract. For this reason, it makes sense to set aside many abiding philosophical and linguistic concerns in order to focus more directly on the conscious experience embodied in the sensitive and responsive nervous systems of human beings.

Why is there consciousness at all? Although ever-mounting evidence makes it clear that the brain's neural systems play a vital role in consciousness, it remains difficult to understand precisely why conscious experience subjectively feels ("from the inside") the way it does. For some reason, each human nervous system has a point of view from which many life events, large and small, are held in conscious regard. That is, there is an appreciation for, noticing of, or personal response to events transpiring within the brain. As to the question "Why is there consciousness?" — this is a poser on par with questions such as "Why is there matter?" Although such *Why* questions may be the first to come to mind, they are not easily answered. On the other hand, *How* questions are more easily addressed by the slow, piece-meal methods of science. Especially in the case of consciousness, it may be something of a mistake to ask "Why?" without first answering "How?"

How is it, then, that each of us is conscious? Conscious experience intimately depends on activity in the nervous system, an amazingly intricate universe of intercommunicating cells. We must take heed that an unstructured approach to studying the conscious brain would today encounter a bewildering array of facts about consciousness and facts about brains. Without the benefit of fundamental principles, an untutored survey might encounter only sparse links between these psychological and neurobiological findings. It is these mind-brain links which establish basic findings on the nature of conscious brain activity that constitute the backbone of this book.

When a mind wanders, where can it go? The entire mental landscape may be nothing more than the dynamically shifting patterns of activity in networks of living neurons in the central nervous system *as experienced from the inside.* This is the understanding presently emerging from several decades of increasingly detailed studies which link mental processes, including conscious phenomenology, with the corresponding activity of living neural systems.

Recognizing a new opportunity for better understanding the physical basis of mental activity, educational and research institutions have recently begun to train a new generation of scientists to think equally in terms of mind and brain. These cognitive neuroscientists mix two scientific strands, psychology and neurobiology, into a joint investigation into the neural basis of mind, which includes conscious experience. Exemplifying this new tradition, the contributors to this volume are researchers who have themselves conducted pioneering studies into mental functions of the brain.

With regard to the variety among the possible contents of conscious awareness, this book focuses primarily on awareness of the physical world, the bodily self and its surrounding environment. Without much concern for the conscious experience of recollecting past experience, we focus more on the sensory processes which establish the moment-to-moment experience of immediate physical and mental presence. Regardless of culture or language, every awake human being has this sort of conscious experience. Human sensory experience is thought to share much in common with other mammalian species, and in part because of this shared evolutionary heritage, sensory (and motor) systems are the best understood systems in the brain. Against this backdrop, the contributors to this book pursue the following questions:

1.   Which parts of the nervous system are accessible to consciousness? That is, which subsystems of the brain process the information pertaining to the contents of which we are aware?

2.   Which subsystems of the brain direct the mind to the next focus of attention, selecting from multiple potential contents the ones which will most occupy awareness?

3.   Which neural subsystems lend structure to the current contents of consciousness within a framework of energy and emotion?

4.   How has the human brain evolved to sustain conscious experience?

Peter G. Grossenbacher
Bethesda, Maryland

Contributors

Simon Baron-Cohen
Experimental Psychology
University of Cambridge
Downing Street
Cambridge CB2 3EB, UK

Douglas Derryberry
Department of Psychology
Oregon State University
Corvallis, OR 97331, USA

Peter G. Grossenbacher
National Institute of Mental Health
Building 15-K, Room 105A
15 North Drive MSC 2668
Bethesda, MD 20892-2668, USA
e-mail: Peter_Grossenbacher@nih.gov

John M. Kelley
Massachusetts General Hospital
and Harvard Medieval School
WACC 812
55 Fruit Street
Boston, Massachusetts 02114, USA
e-mail: jkelley3@partners.org

Stephen M. Kosslyn
Department of Psychology
Harvard University
Cambridge, MA 02138, USA

Daniel Levitin
Interval Research Corporation
1801 Page Mill Road, Bldg. C.
Palo Alto, CA 94304, USA

Phan Luu
Electrical Geodesics, Inc.
Riverfront Research Park
1900 Millrace Drive
Eugene, OR 97403, USA
e-mail: pluu@EGI.com

Michael J. Posner
The Sackler Institute
Dept. of Psychiatry, Box 140
Weill Medial College, Cornell University
1300 York Ave.
New York, NY 10021, USA

Mark Price
Department for Psychosocial Science
University of Bergen, Christiesgt. 12
N-5015 Bergen, Norway
e-mail: mark.price@psych.uib.no

Yves Rossetti
Espace et Action, INSERM 94
16, avenue doyen Lepine, case 13
F-69676 Bron Cedex, France

Scott D. Schliebner
Department of Educational Psychology
University of Utah
Salt Lake City, UT 84112, USA

Barry Stein
Wake Forest University
The Bowman Gray School of Medicine
Department of Neurobiology and Anatomy
Medical Center Boulevard
Winston-Salem, NC 27157-1010, USA

Mark Wallace
Wake Forest University
The Bowman Gray School of Medicine
Dept. of Neurobiology and Anatomy
Medical Center Boulevard
Winston-Salem, NC 27157-1010 USA

Roger Whitehead
Department of Psychology
Campus Box 173
University of Colorado at Denver
P.O. Box 173364
Denver, CO 80217-3364 USA

CHAPTER 1

A Phenomenological Introduction to the Cognitive Neuroscience of Consciousness

Peter G. Grossenbacher
National Institute of Mental Health

What is the nature of human consciousness? This question can be considered from either of two perspectives: (1) the subjective point of view — how things seem to you personally, from your first person perspective, and how the objective view — how things appear to a consensus of people.[1] Each perspective provides unique methods for understanding consciousness. Because neither on its own allows as complete an account of consciousness as they do in combination, this book attempts to bring subjective and objective perspectives together by explaining some of the neural mechanisms which underlie subjective aspects of conscious experience. You, the reader, will be asked to breath life into the subjective perspective embedded in this text. Authentic personal phenomenology, the observation of your own conscious experience, is the only way to ensure that the objective findings of cognitive neuroscience do in fact pertain to conscious experience.

The subjective perspective, a personal standpoint known directly "from the inside," has its limitations. Much of what goes on in a person's mind does so without being evident in conscious experience. Considered in this light, a subjective point of view implicitly distinguishes between conscious experience and unconscious mental processes not themselves apparent to the subject. Introspection, the subjective inspection of conscious phenomena, thus provides privileged but incomplete access to one's own mental activity. There is much more to a person than is met in their own conscious mind, and some of this unknown realm can become known through the objective lens of cognitive neuroscience.

Observing the behavior of other people, including what they say and how they act, enables us to infer something of their subjective experience. For purposes of controlled and replicated study, behavioral scientists restrict the conditions under which observations of somebody else's behavior are made. Methods for measuring neural activity in the human brain are becoming easier to combine with psychological methods for manipulating and measuring mental activity. Recent research into mental functions of the nervous system has been especially productive because of increasing collaboration between neuroscientists and psychologists, creating a cognitive neuroscience which studies the activity occurring throughout the human brain during conscious performance of mental tasks (Posner & Raichle, 1994).

1.1 *Subjective experience*

Immediate phenomenal quality, what it is like right now to be, for example, consciously perceiving objects in the world, seems to be as universal an aspect of human experience as one can find. Too many intellectual treatments of consciousness fail to fully include subjectivity. If we are concerned with what it is like to be conscious, we must draw explicitly upon the conscious experience that each author and reader makes available in the meeting of minds brought together through this text. To this end, throughout this book you will encounter suggested exercises marked by the symbol "‖••‖" that involve examining your own conscious awareness. The more that you participate in these exercises, the better basis you will have for judging the relevance of the objective findings reported by the contributors to this volume to subjective experience. Even when no specific exercise is suggested, you may wish to pause in order to test the ideas in this book against the subjective facts of your own consciousness.

Here is the first suggested exercise: look at the larger patterns made by the print on the two pages of this book now visible to you; notice the shapes formed by paragraph boundaries, and pay close attention to the branching rivulets which run vertically, crossing the lines of text, formed by the space between words. Do this for several seconds, and then read on; begin now.

‖••‖

After completing that exercise, take a moment to list what you experienced — what were you conscious of? If lighting conditions and your visual ability were adequate, and if you did in fact attend to the visible aspects of this book, then your conscious content included visual *percepts*, the consciously experienced representations of visible objects in the physical world. Perhaps you were aware of other things in addition to (or instead of) visual percepts. Did other sense modalities such as hearing or smell contribute percepts to your stream of consciousness? Were you aware of bodily movement or keeping still? Did you notice any emotions? Did you entertain thoughts about the exercise or about other matters? You may wish to repeat the exercise if you are unsure as to the details of your experience. Once you have listed the content of your conscious experience, examine the range of content in your list.

The concept "content of consciousness" is best defined inductively from *your* conscious experience, as given for example in the exercise above. Percepts, those contents arising most immediately from sensory stimulation, are one kind of conscious content, but there are others. What thoughts or feelings did you experience during the exercise? Were you imagining a familiar tune, or the face of a friend, or anything else that was not actually present? Content of consciousness includes thoughts, feelings and images in addition to percepts.[2] This subjective observation of content constitutes the first step in our project of finding consciousness in the brain.

1.2 *Subjective phenomena and cognitive neuroscience*

Science of the mind can be viewed in part as a search for links between conscious perceptual experience (involving private phenomenal content) and objectively observed bodily activity (including both overt behavior and neuro-physiological activity in the brain). Just as phenomenal content can be observed subjectively, it has now become possible to objectively observe the content of neural representations in the brain. Recording the activity of individual neurons has revealed that many neurons respond only to stimulation in a particular sense modality, such as vision, or touch. And more specifically, many sensory neurons respond only to stimulation of a particular subregion of the receptor sheet, such as the upper left quadrant of the retina, or the index finger of the right hand. Until recently, each neuron's receptive field, the portion of the receptor sheet which evokes responses in the neuron, was thought to be constant. Now we know that receptive fields can shift with changes in attention and body posture (Desimone & Duncan, 1995; Driver & Grossenbacher, 1996; Graziano & Gross, 1993).

There is much yet to discover concerning the neural basis of consciousness, and the search for conscious mind in the living brain is unlikely to be completed anytime soon. However, numerous links between mind and brain have already been discovered, and some of these links can help us to understand a great deal about how your nervous system, for example, is conscious. A variety of methods have been brought to bear by the contributors to this volume.

It is not clear how the objectivity of science can best embrace the subjectivity of personal experience, the subjectively known stream of feelings, percepts and concepts. In the first three sections of this book, recent results of cognitive neuroscience research are organized so as to emphasize their relation to a few fundamental aspects of conscious experience: (1) Conscious awareness has content. (2) This conscious content undergoes rapid changes. (3) The feel of any conscious content depends on something else besides the content itself. These three subjective observations are each elaborated, in turn, below.

2. Conscious experience

At the beginning of this chapter we inductively defined content of consciousness as whatever *you* are aware of. You, the reader, must add flesh to this bare-bones definition by considering your own conscious experience. It is left for you to decide whether or not this section, intended to capture human phenomenology, accords in fact with your own experience.

This section establishes a framework for linking conscious experience to brain activity by dissecting the subjective nature of conscious experience. Aiming to isolate a few elementary aspects of conscious experience, we will start with the simple observation that there is content to consciousness. Next, this conscious content can be observed to rapidly vary over time. And finally, the mental frame within which content is affectively experienced can itself be observed to undergo changes.

2.1 *Conscious awareness has content*

Consider the range of content which can appear in your conscious mind: there is a rich domain of mental life that includes perception, imagery, thought, language, planning, and action. It may be difficult for us all to agree on how best to carve subjective experience into exclusive categories. Perhaps the most universal categories of conscious experience arise in perception and imagery: the different sense modalities (such as vision and touch) are distinguished from each other *qualitatively*, according to their subjective feel. Sights and touches always seem qualitatively different from each other—one type of percept is not likely to be confused with the other. Indeed, humans distinguish between sense modalities so easily that it is natural for us to take this ability for granted.[3]

The probably infinite range of conscious content is constrained by a maximum limit on the total content which can be held in consciousness at one time. This limit seems to vary over time, and on some occasions may shrink to one item: when you really take notice of something, that particular content may fill your awareness to the complete exclusion of other content. As yet there is no clear link between this single-content focus and brain activity, although awareness of perceived objects can be *pathologically* limited to one single object at a time. This inability to simultaneously perceive multiple objects (simultanagnosia) has been observed in people with bilateral damage to parietal cortex (Berti, Papagno & Vallar, 1986).

The current content of consciousness consists in those objects of aware-ness which you experience right now. To the extent that multiple items are simultaneously conscious, these may seem to separate into foreground and background, but the nature of this distinction is a contentious issue.[4] Suffice it to say that the experience of a particular content usually extends briefly in time, from a fraction of a second up to several seconds. What then determines the content in any conscious moment? As described next, a subjective per-spective provides two answers: the particulars of conscious content can be influenced by very rapid changes in attention, and also by much slower changes in affective state (frame of mind).

2.2 *Conscious content frequently shifts*

The mind has a tendency to wander. To observe this yourself, try to restrict awareness to a single phenomenal content, whatever is most salient for you, without distraction. Try to keep the same focus of attention throughout one full minute.

‖••‖

Did you experience more than a single thought (or percept, or image, or whatever content you were aware of) during that exercise? Without practice, most people find it difficult or impossible to restrict conscious awareness to one single thing for more than a few seconds. Why does the conscious mind wander so easily? What are the winds that push the weather vane of attention to maintain so diverse a stream of content in conscious awareness?

An automatic tendency to shift attention from any representation in which there is little change may be a useful adaptive trait for acquiring more information about current events. For a neurobiological account of automatic spatial attention, see Chapter 5 (Stein and Wallace).

Every conscious content can be attended to a greater or lesser degree, and the focus of awareness seems at least partly under voluntary control. Use one of your hands to grasp and lift this book, and attend to the *object in your hand*.

||••||

Are you aware of its weight, size, or other attributes such as hardness, texture and temperature? Set the book down and pick it up again. This time attend to the *sensations in your fingers, hand and arm*.

||••||

Are you aware of pressure, or how different muscles feel? Compare your conscious experience during focus-book and focus-hand conditions; it may help to repeat each observation. Does the content of your conscious experience differ between these two conditions? This exercise illustrates the voluntary nature of conscious content.

2.3 The feel of consciousness depends on frame of mind

In addition to the relatively frequent changes in conscious content, we can identify another kind of change in consciousness. The experience of the content of consciousness is influenced by a discernible frame of mind, a mental state which may change from one day to the next (and even more frequently). Consider your experience on two separate but similar occasions, for example, when greeting an unfriendly neighbor. Encountering the neighbor's predictable grump might lead to your having very different experiences depending on what mood you are in to begin with. If you happen to start in a positive frame of mind, this encounter might be of only minimal annoyance. But if it has already been a bad day, you are more likely to experience strong displeasure. How is it that two episodes of essentially the same activity (greeting a grumpy neighbor) can yield such noticeably different subjective experiences?

The content of consciousness is experienced within the complex context of current degree of wakefulness and emotional state. The changes in subjective perspective produced by arousal and emotion can last from minutes to hours, regardless of the particular content which appears in conscious experience during this time. These factors are fully addressed in Section III of this book (Frame of Mind). If you have ever experienced the effects of caffeine, alcohol, nicotine, or other drugs, then you already know first-hand that brain chemistry can markedly influence your frame of mind. So it may not be surprising that several chemical neurotransmitters provide the keys to understanding the brain systems that control emotional tone and mental energy. The science of emotion is not very advanced yet, but important findings are already appearing for a few emotions such as anxiety (see Chapter 8 by Derryberry). In comparison, the brain mechanisms of arousal are well understood. Subcortical circuits deliver the neurotransmitters that control arousal in cortex (see Chapter 7 by Whitehead & Schliebner).

3.   Systems neuroscience

Several chapters in this book provide detailed analyses of how neural systems function to sustain conscious experience. Although all authors have thoroughly introduced their material, readers with absolutely no background in neuroscience would have difficulty understanding some of the finer points made in some chapters. This section provides a brief tutorial intended to offer such readers a sufficient knowledge base to fully enjoy the entire book. Neuroscience experts (only!) are invited to skip this section.

The relation between brain structure and function hinges on *connectivity*, the physical contact between neurons which transports high fidelity signals from one neuron to the next. This section explains brain function in terms of communication between neurons and in terms of the dynamic aspects of information processing which derive from this intercellular communication. This brief review hits only some of the highlights, and readers with little background in neurobiology are urged to read more thorough reviews of this fascinating field (Gazzaniga, 1995; Kandel & Schwartz, 1995).

3.1 *The nervous system*

Among the different kinds of cells which constitute the body, neurons are uniquely shaped, having long axons extending from the cell body to convey signals to other neurons. Each living human body contains a vast system of intricately networked neurons which communicate with each other via chemical neurotransmitters. As a connected ensemble, this nervous system conducts the internal orchestra of physiological functions in the living body, and also governs how the body moves through the world. The brain's ability to control the entire body depends on skilled processing of information about what goes on outside the brain.

Sensory receptors are the cells which transduce physical changes in energy occurring outside the nervous system into neural activity (see Figure 1.1). Sensory receptors are distributed throughout nearly all the tissue of the body, making available to the nervous system *interoceptive* information concerning internal bodily states and physiological processes. Sensory receptors suffuse the living body with sensitivity, especially the skin and other sensory organs such as tongue, nose, ears and eyes, which are specialized for picking up *exteroceptive* information concerning objects and events in the environment outside the body. This interoceptive and exteroceptive sensory information is always being processed within the brain, and has profound effects on both conscious and unconscious thought.

Within brain tissue, the physical changes most relevant to the person comprise the neural activity occurring there, the signaling from each neuron to its target neurons. Because neurons express themselves directly to the other parts of the nervous system, no transduction from non-neural physical energy into neural signals is required for the nervous system to become informed of what happens in this part of the body. As suggested by the many converging findings presented in this book, it is the living activity of these brain cells which embody the mind.

Cortex, the large and thickly wrinkled outermost portion of the human brain, is a laminated structure having parallel layers of neuronal cell bodies. These cortical layers comprise the brain's so-called "gray matter." The underlying "white matter" is made of densely packed axon fibers which project from neurons in one cortical area to reach neurons located in other areas of cortex, and also in subcortical brain areas. The cortex and its connections are critically important for consciously representing images and thoughts.

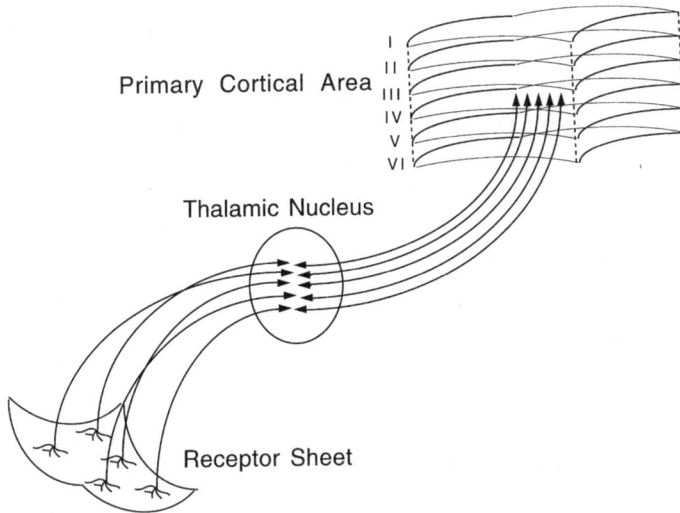

Figure 1.1. In each sense modality, afferent sensory information is typically conveyed from a peripheral sheet of receptors to a primary sensory cortical area by way of a thalamic nucleus. The thalamus is a subcortical structure containing nuclei (clumps of neurons) which send and receive signals from cortical areas. This organization of sensory pathways applies equally well to vision, audition, and somatosensation.

The complex nature of neural systems inside the human brain can be understood according to a few key principles: (1) Interconnected neurons communicate with each other. (2) Everyday experience affects this functional connectivity in the brain. (3) Systems of neurons which process information are organized into neural pathways. (4) The dynamic activity within these pathways occurs in neural circuits which convey both excitatory and inhibitory signals. Each of these principles will now be explained in some detail.

3.2 *Connectivity: The cornerstone of functional architecture*

The principle function of a neuron is to influence the activity of other cells by its signalling. The signaling activity of a neuron is influenced by the signals it receives from other neurons across *synapses*, the physical junctions between neurons. A vast number of synapses are found on dendrites, the small, branching elongations of the cell bodies, so a single neuron can receive signals from thousands of other neurons. Human brains contain many neurons which are not connected directly with either sensory inputs or outputs for controlling the body, but are instead devoted purely to communicating with other neurons.

We can distinguish between two types of neural networks. Anatomically *distributed* neural networks are comprised of interconnected neurons located in more than one part of the brain. In contrast, an anatomically *localized* neural network is made of interconnected neurons which are neighbors in the same part of the brain. In the cortex, there are localized networks constituting *cortical areas* (see Figure 1.1 for the example of a generic primary sensory area). Different kinds of network organization are found in different cortical areas. *Cytoarchitecture*, the anatomical distribution of neurons within a cortical area, concerns the number and types of neurons found in each cortical layer. In primate cortex, primary visual, auditory and somatosensory areas (V1, A1 and S1) have highly similar cytoarchitecture, all containing six clearly distinguishable layers, including a thick fourth layer containing a dense distribution of small neurons.

The information represented within a cortical area is determined by which neurons provide the input. Each network performs a particular kind of computation on the information conveyed from other networks. Different networks in the brain handle a variety of computational processes which transform information by analyzing into components, integrating into wholes, encoding into different formats, comparing with other information, and so on. Cortical areas dedicated to a particular sense modality are *content-specific:* only a specific kind of information can be represented by activity in the network. In humans, for example, primary somatosensory cortex (S1) processes sensory information from touch and bodily senses, but never handles information regarding smell or color.

Primary sensory areas of cortex V1, A1 and S1 (primary visual, primary auditory, and primary somatosensory) are among the best understood of cortical areas. Although the primary sensory areas are located in different regions of cortex, they share a similar pattern of input connectivity: neurons in the fourth layer receive inputs from afferent fibers projecting from thalamus in each of these three sense modalities. They are each organized as a *topographic map*, in which columns of neighboring neurons represent neighboring portions of the receptor surface. V1 is organized retinotopically, with adjacent portions of the retina being represented in adjacent portions of this cortical area. S1 is somatotopic, so that adjacent portions of the skin are represented adjacently. The map in A1 is organized tonotopically (by pitch), matching the organization of receptor cells in the cochlea, the acoustic sensing organ within the inner ear. Peripheral topography, the spatial arrangement of receptor cells within the receptor sheet, is reflected in the neural topography at both thalamic and cortical levels.

3.3 Brain connectivity changes with experience

The most general aspects of human brain organization develop during the first few years of life. During this time, brain circuits become tuned by *synaptic pruning,* the selective disappearance of those synapses which contribute least to the neuronal computations of each circuit. These changes in connectivity happen because the system is fundamentally restructuring itself in response to experience. By establishing new connections or eliminating old connections between neurons, the functional architecture of the brain irreversibly changes. A recent blood flow study found that cortical area V1 was activated by blind subjects while reading Braille (raised dots sensed through contact with finger-tips) whereas similar tactile perception in sighted subjects entailed deactivation of this typically visual area (Sadato et al., 1996). After early childhood, it may be impossible for new connections to form between neurons that are not already adjacent.

When a developing brain is surgically rewired so that a primary cortical area receives afferents from a sense modality other than the customary one, the internal connectivity of that cortical area becomes organized according to the nature of the rewired inputs (Sur, Pallas & Roe, 1990). But the external connectivity which links this cortical area to other localized networks persists in the customary arrangement. This suggests that the large-scale connections between localized networks are determined very early, whereas the nature of the representations and computations localized within a cortical area depend more on the information received as input.

Our brains have anatomically localized neural networks that support particular mental operations. For example, a cortical area localized in left occipital lobe is active[5] during perception of visually presented words, but this network is so content-specific that it does not appear active during perception of character strings which do not closely resemble words (Petersen et al., 1990). That study measured cortical blood flow using Positron Emission Tomography (PET), and found a sizable difference in this area between non-word letter strings that looked word-like and those strings which did not obey the orthography of words. Since neither of these types of stimuli have previously associated meanings, as real words do, the difference in blood flow activation must be entirely due to their visual appearance.

This localization of word orthography in occipital cortex demonstrates to what degree a cortical network can be dedicated to highly specific information content. This visual area, obviously geared for discriminating among a variety of similar visual shapes, may have developed in our evolution for some ancestral task such as identification of useful plants, but today can become specialized for word recognition. Children do not recognize words until they are several years of age. Because *learning* is necessary for the visual word area to develop its function, it is clear that individual experience strongly affects functional brain organization.

The unique organizational details of each person's brain continue to be refined as a result of ongoing experience throughout life. How does this unique pattern of connectivity within each individual's brain affect mental activity? Even when sensing the same physical stimulus, different people can have very distinct conscious experiences. Two people, one a musician and the other a painter, are likely to have brains which differ in the way auditory or visual sensory information is reflected in the content of conscious awareness. As yet, relatively little is known regarding the relation between idiosyncratic variation in neuroconnectivity on the one hand, and individual differences in phenomenal experience on the other.

3.4 *Neural pathways process sensory information*

A *neural pathway* is a type of circuit having neurons located in multiple parts of the brain which are interconnected to form a chain in which information flows from one end of the pathway to the other in a feedforward direction. Within a single sense modality, incoming (afferent) signals proceed through a sequence of distinct cortical areas comprising a modality-specific pathway. The organization of sensory representations changes according to the neural computations made along the pathway. Sensory cortical pathways transform "early" representations of sensory information in "upstream" parts of the pathway into "late" representations "downstream." For example, the visual areas of cortex which receive direct inputs from primary visual cortex share the retinotopic organization of V1, whereas areas positioned downstream are organized in other ways.

Neural pathways are not *sequential* in the sense of each processing stage feeding information forward to the next stage without any influence from the activity occurring further along (downstream) in the pathway. On the contrary, the feedforward connections are reciprocated by *feedbackward* connections which pass signals back from neurons in the downstream areas (Cynader et al., 1988). Because cortical areas within a pathway interconnect with reciprocal feed-forward and feed-backward connections, incoming sensory signals can be modified by top-down control signals. For example, feedbackward connections allow modality-specific processing to be influenced by multisensory representations computed downstream.

Consider a neural pathway that assembles sensory information into the representation of objects. Visual aspects of objects are represented in a connected series of cortical areas in the *inferior* (lower) temporal lobe (Baylis, Rolls & Leonard, 1987). This pathway is purely visual: it contains cells which are selectively responsive to particular views of hands, heads and other objects (Perret, Mistlin & Chitty, 1987). Damage to this pathway in human brains results in visual *agnosia*, a syndrome in which objects cannot be recognized by sight, despite unhindered perception of elementary visual attributes such as color (Farah et al., 1990). This early visual information is presumably available to consciousness in these brain-damaged patients because the cortical areas that normally feed inputs forward to the damaged object recognition pathway are still intact.

Parallel to the visual object recognition pathway, along the *superior* (upper) temporal lobe, runs a neural pathway having similar internal connectional architecture (see Figure 1.2). The superior temporal sulcus contains a series of interconnected cortical areas, each containing cells that respond to multiple sense modalities (Bruce, Desimone & Gross, 1981; Seltzer & Pandya, 1989). This strip of adjacent, connected cortical areas constitutes a neural pathway which may compute the representation of *multisensory* object attributes. During the perception of your neighbor's dog, this is the pathway in which the neural signals arising from the sound of a dog's bark and the sight of a swishing tail may come together (Grossenbacher, 1996).

The evolutionary history of cortical development helps to explain the organization of neural pathways in cortex (Pandya & Yeterian, 1985; Sanides, 1969; Sanides, 1972). Over the course of evolution, newly specialized forms of cortical tissue have developed to pre-process sensory signals, computing output representations adaptively suited as input to the cortical area to which

PETER G. GROSSENBACHER

Auditory
Area

Visual
Area

Adjacent Multisensory Areas in
Temporal Cortex
Somatosensory
Area

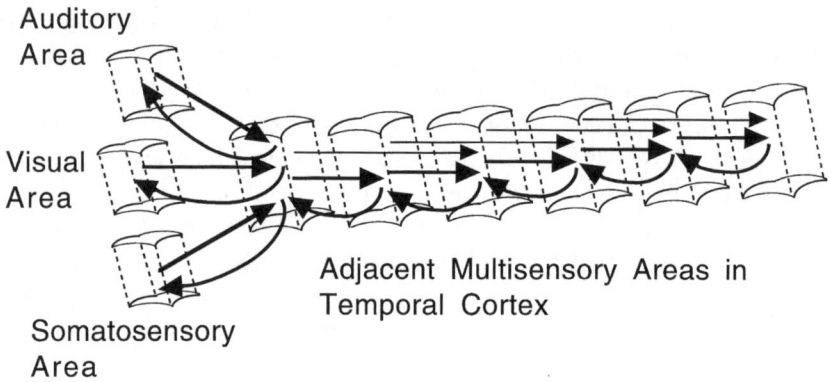

Figure 1.2. This figure shows a neural pathway located in superior temporal sulcus of primates. Inputs to this pathway come from widely separated areas of cortex which feed sensory information from distinct modalities forward to multisensory neural circuits in temporal cortex. Within this pathway, forward projections from each area reach to the adjacent downstream area (rightward thick arrows) as well as the area located one step further downstream (rightward thin arrows). Each cortical area feeds backward only to the preceding area (leftward curved arrows).

they project in a neural pathway. In primates, primary sensory cortical areas share a recently evolved cytoarchitecture which is unique to these areas. This input stage of cortical processing sends output signals to a series of downstream modality-specific cortical areas which together form a cortical sensory pathway. As signals are fed forward through a cortical pathway, they are conveyed through a series of cortical areas that gradually transition through progressively older forms of cytoarchitecture (Barbas, 1995; Pandya & Seltzer, 1982).

Visual, auditory, and tactile pathways lead to multimodal areas of cortex that receive information from all three senses (Pandya & Seltzer, 1982). These multimodal areas of cortex have a different cytoarchitecture than the primary sensory areas. The cytoarchitecture of multisensory cortical areas resembles that found in mammals with small, primitive-looking cortices. Paradoxically, this old type of cortical tissue handles some of the most evolutionarily advanced cognitive functions (Braak, 1978; Pandya & Yeterian, 1985). Chapter 10 of this book presents one possible resolution of this apparent paradox guided by the following observation. Because the older cytoarchitecture is preserved in multimodal association areas of cortex, any evolutionary accommodation in the most central networks appears not to require modification of the basic tissue structure. Perhaps the evolutionarily ancient forms of cortical tissue are ideally suited for neural networks which must handle complex or evolutionarily new patterns of information across a wide range of content domains.

3.5 *Circuit dynamics and neural computation*

The incredibly dynamic nature of living neural systems is evident in the ever-present signaling between neurons. Each synapse of a receiving neuron receives signals at a rate which reflects the activity level (firing rate) of the source neuron. Some synapses receive *inhibitory* signals which tend to quell activation of the receiving neuron. Other synapses receive *excitatory* signals which boost the firing rate of the receiving neuron. The moment-to-moment pattern of firing across these inhibitory and excitatory synapses controls the rate that the neuron will fire signals to its own output targets. This book does not delve into the intricacies of intracellular neurophysiology by which synaptic inputs cause a neuron to change its output firing rate. It is enough to know that neurons signal other neurons at rates that can rapidly change. These changes can alter the balance between inhibitory and excitatory inputs to receiving neurons, thereby affecting the output signaling rates of these cells.

A *neural circuit* is a set of synaptically connected neurons which transform one set of neural signals (received by the circuit's input neurons) into another set of signals (conveyed from the circuit's output neurons as input to other circuits). This process of neural computation involves millisecond-by-millisecond alterations in the signaling of neurons throughout the circuit. The specific computations made by a neural circuit are determined by the organizational architecture of inhibitory and excitatory connections between the neurons within the circuit.

Exactly how does neural activity relate to conscious experience? One set of clues has been found in the patterns of electrical changes measured in and around the brain. Electrical changes in neurons reflect their signaling activity, and can be measured with needle electrodes inserted into the brain or, without damaging any tissue, with disk electrodes placed on the scalp (Tucker, 1993). In humans, awareness of physical stimulation is reflected by the activity of cortical neurons in their response to afferent signals (Parasuraman & Beatty,

1980). In that study, as early as 100 milliseconds after the onset of a quiet sound, the brain's electrical activity reflects whether the person is aware of the sound or not.

Not only does the natural electrical activity of the brain reflect conscious experience, but electrical stimulation of brain tissue can evoke sudden changes in the content of consciousness. Stimulation of cortex in the brains of people with chronic seizures alters ongoing perception or triggers experiential recall of a previous autobiographical experience (Penfield, 1970). This stimulation technique helps in determining the kind of experience that might be lost if that part of the brain were to be surgically removed. Recently, a much less invasive method has been developed for cortical stimulation. Without even touching the head at all, Transcranial Magnetic Stimulation (TMS) can alter brain activity (Barker, Jalinous & Freeston, 1985), and a remarkable range of applications are likely to develop from this technique. For example, magnetic stimulation of visual cortex prevents conscious awareness of visual stimuli presented 100 msec earlier (Amassian et al., 1993).

4. Book preview

This chapter has discussed three aspects of conscious experience which are to be explained in terms of neural activity in the brain. And a brief review of systems neuroscience has introduced fundamental concepts necessary for the reader to understand the rest of this book.

Section I of this book aims to elucidate the nature of human consciousness by discussing boundary conditions. Section II describes how neural systems in the brain contribute to the representation of phenomenal content and the rapid changes in this content. Section III reveals the neural basis of the mental frame which colors the experience of conscious content. Section IV takes a much longer view and considers the impact of brain evolution on human consciousness.

In an attempt to directly tap into subjectivity, this chapter was heavily salted with exercises in subjective observation. Using your own mind as a "laboratory" for observation, these exercises involve pausing from the task of reading in order to observe your own conscious experience. It may help for you to continue this phenomenological habit when reading the rest of this book, regardless of the frequency with which explicit exercises or examples come along. Your own conscious experience helps form the basis for understanding the links between brain activity and consciousness explored in this book. I hope that you find the results from objective laboratory experiments in cognitive neuroscience to be equally relevant to understanding consciousness.

Notes

1. Objectivity is here operationalized as that which is held in common among multiple subjective perspectives.

2. No assumption of exclusive divisions between these kinds of content are implied, only that consciousness does include at least these kinds of content.

3. Many other species of mammals have brains in which it may be much easier to confuse two sense modalities (see Chapter 10).

4. This book does not grapple with the subtle distinction between foreground and background in awareness. Differences among introspective reports pertaining to this distinction may reflect a variety of ways in which information represented by neural activity may contribute to current conscious content.

5. When neurons do more work, they consume more chemical fuel, and this metabolic consumption can be observed with techniques such as positron emission tomography (PET). The brain's system of blood vessels adapts quickly to changes in neural activity in order to bring fuel more quickly to where it is needed. Localized changes in blood flow can be measured by PET or functional magnetic resonance imagery (fMRI). Functional brain activation studies of metabolism and blood flow are now offering objective access to the levels of activity occurring in different pathways, and in different portions of a single pathway.

References

Amassian, V. E., Maccabee, P. J., Cracco, R. Q., Cracco, J. B., Rudell, A. P., & Eberle, L. 1993. Measurement of information processing delays in human visual cortex with repetitive magnetic coil stimulation. *Brain Research, 605*, 317-321.
Barbas, H. 1995. Anatomic basis of cognitive-emotional interactions in the primate prefrontal cortex. *Neuroscience and Biobehavioral Reviews, 19*(3), 499-510.
Barker, A. T., Jalinous, R., & Freeston, I. L. 1985. Non-invasive magnetic stimulation of human motor cortex. *Lancet, 1*(8437), 1106-7.
Baylis, G. C., Rolls, E. T., & Leonard, C. M. 1987. Functional subdivisions of the temporal neocortex. *Journal of Neuroscience, 7*, 330-342.
Berti, A., Papagno, C., & Vallar, G. 1986. Balint syndrome: a case of simultanagnosia. *Italian Journal of Neurological Sciences, 7*(2), 261-4.

Braak, H. 1978. On magnopyramidal temporal fields in the human brain—probable morphological counterparts of Wernicke's sensory speech region. *Anatomy and Embryology, 152*, 141-169.

Bruce, C., Desimone, R., & Gross, C. G. 1981. Visual properties of neurons in a polysensory area in superior temporal sulcus of the macaque. *Journal of Neurophysiology, 46*(2), 369-84.

Cynader, M. S., Andersen, R. A., Bruce, C. J., Humphrey, D. R., Mountcastle, V. B., Niki, H., Palm, G., Rizzolatti, G., Strick, P., Suga, N., von Seelen, W., & Zeki, S. 1988. General principles of cortical operation. In P. Rakic & W. Singer (Eds.), *Neurobiology of Neocortex*, (pp. 353-371). New York: John Wiley & Sons.

Desimone, R., & Duncan, J. 1995. Neural mechanisms of selective visual attention. *Annual Review of Neuroscience, 18*, 193-222.

Driver, J. S., & Grossenbacher, P. G. 1996. Multimodal spatial constraints on tactile selective attention. In T. Inui & J. L. McClelland (Eds.), *Attention and Performance XVI: Information Integration in Perception and Communication*, (pp. 209-235). Cambridge, MA: MIT Press.

Farah, M. J., Brunn, J. L., Wong, A. B., Wallace, M. A., & Carpenter, P. A. 1990. Frames of reference for allocating attention to space: evidence from the neglect syndrome. *Neuropsychologia, 28*, 335-347.

Gazzaniga, M. S. (Ed.). 1995. *The Cognitive Neurosciences*. Cambridge, MA: MIT Press.

Graziano, M. S. A., & Gross, C. G. 1993. A bimodal map of space - Somatosensory receptive-fields in the macaque putamen with corresponding visual receptive-fields. *Experimental Brain Research, 97*, 96-109.

Grossenbacher, P. G. 1996. Consciousness and evolution in neocortex. In P. A. Mellars & K. R. Gibson (Eds.), *Modelling the Early Human Mind*, (pp. 119-130). Cambridge, England: McDonald Institute for Archaelogical Research.

Kandel, E. R., & Schwartz, J. H. 1995. *Essentials of Neural Science and Behavior*. Norwalk, CT: Appleton & Lange.

Pandya, D. N., & Seltzer, B. 1982. Association areas of the cerebral cortex. *Trends in Neurosciences, 5*, 386-390.

Pandya, D. N., & Yeterian, E. H. 1985. Architecture and connections of cortical association areas. In A. Peters & E. G. Jones (Eds.), *Cerebral Cortex. Volume 4. Association and Auditory Cortices*, (pp. 3-61). New York: Plenum Press.

Parasuraman, R., & Beatty, J. 1980. Brain events underlying detection and recognition of weak sensory signals. *Science, 210*(4465), 80-83.

Penfield, W. 1970. Memory and perception. *Research Publications - Association for Research in Nervous and Mental Disease, 48*, 108-122.

Perret, D. I., Mistlin, A. J., & Chitty, A. J. 1987. Visual neurones responsive to faces. *Trends in Neuroscience, 10*, 358-364.

Petersen, S. E., Fox, P. T., Snyder, A. Z., & Raichle, M. E. 1990. Activation of extrastriate and frontal cortical areas by visual words and word-like stimuli. *Science, 249*(4972), 1041-4.

Posner, M. I., & Raichle, M. E. 1994. *Images of Mind*. New York: Scientific American Press.

Sadato, N., Pascual-Leone, A., Grafman, J., Ibanez, V., Deiber, M. P., Dold, G., & Hallett, M. 1996. Activation of the primary visual cortex by Braille reading in blind subjects [see comments]. *Nature, 380*(6574), 526-8.

Sanides, F. 1969. Comparative architectonics of the neocortex of mammals and their evolutionary interpretation. *Annals of the New York Academy of Sciences, 167*, 404-423.

Sanides, F. 1972. Representation in the cerebral cortex and its areal lamination patterns. In G. F. Bourne (Ed.), *Structure and Function of Nervous Tissue*, (Vol. 5, pp. 329-453). New York: Academic Press.

Seltzer, B., & Pandya, D. N. 1989. Intrinsic connections and architectonics of the superior temporal sulcus in the rhesus monkey. *Journal of Comparative Neurology, 290*, 451-471.

Sur, M., Pallas, S. L., & Roe, A. W. 1990. Cross-modal plasticity in cortical development: differentiation and specification of sensory neocortex. *Trends in Neuroscience, 13*(6), 227-33.

Tucker, D. M. 1993. Spatial sampling of head electrical fields: the geodesic sensor net. *Electroencephalography and Clinical Neurophysiology, 87*(3), 154-63.

Section I

Edges of Consciousness

Introduction

During the last several years, several perceptual processes have been physiologically localized to particular areas of cortex. For example, visual area "V4" contains neural networks involved in processing color in both humans and monkeys (Schein & Desimone, 1990; Zeki et al., 1991). Because many mental representations are anatomically localized within discrete cortical areas, the death of brain tissue (lesion) can yield a specific loss of one kind of experience. For example, lesion of an appropriate area of cortex can produce loss of conscious color vision and color imagery and knowledge about the color of familiar objects. In this case it is clear that consciousness of color is only possible if the requisite brain tissue is alive and connected with other brain areas. Neural activity in sensory areas of cortex is necessary (but perhaps not sufficient) for the appearance of sensory content in consciousness.

The phenomenal content of awareness can be directly known subjectively. At least some aspects of phenomenal contents can even be communicated to other people. Although this content is limited in several ways, its boundaries may be very difficult to observe subjectively, oneself. In order to learn more about the limits of conscious content, each chapter in this section examines *lack of awareness* in one form or another. Each limitation on conscious awareness discussed in these chapters defines a facet of consciousness that can be observed objectively (that is, by other people).

In perception, conscious awareness requires adequate neural representation of the physical stimulus. Masking is a technique in which precise manipulation of the physical stimulus disrupts the conscious registration of sensory information. In Chapter 2, Price explains how this technique has provided unique insights into the nature and timing of conscious awareness. Price's experimental results reveal the complex nature of chronometric (timing) limits on sensory awareness.

How can an edge of consciousness be observed subjectively? Suppose that you are searching for something and are asked "What are you trying to find?" You might have one of two responses, depending on whether the thing you have in mind is fully available in that moment as content of present awareness or not. If it is adequately accessible, you might respond, "I can tell you exactly what I am looking for..." Or, if you cannot specify even to yourself the exact thing, only knowing at this moment that you are searching for *something*, you might say, "I will know it when I find it, but right now I forget the exact thing." In this latter case, the thing sought might be a familiar object in your home which you came into a room to find, only to realize that you have forgotten the specific identity of what you are looking for. Or you could be trying to recall a word; so-called "tip-of-the-tongue" phenomena are intriguing because the precise word is not at this moment conscious, but some of its aspects (for example, its approximate meaning) may be held in conscious awareness. In either case, the object of your search, whether physically present in the environment or existing only in your own mind, lacks fully conscious representation, and this incompleteness can itself be noticed by you.

Thankfully, human experience can include an awareness of one's own awareness. In addition to consciously perceiving, one can also be conscious of an ongoing mental activity such as that indicated by the statement "I realize that I am seeing a page of this book." We should be careful not to confuse this *experiential* meta-awareness with a logically similar (but phenomenologically different) awareness of an *abstraction* of awareness. For example, as you read this sentence you can become aware of the *concept* of awareness: your idea of what the term "awareness" means without specifying any particular moment of experience. But concepts are not the same as that to which they refer! In order to actually experience meta-awareness, you might pause in your reading and quietly introspect, focusing your mind on your own present awareness, to become explicitly *aware of your awareness as subjectively experienced.*

||••||

While performing this exercise, did some aspect of your own awareness appear in your phenomenological content? Conscious regard of your own conscious experience constitutes awareness of conscious experience, or meta-awareness.

This kind of meta-awareness seems to be within easy reach of most people. However, Baron-Cohen reveals in Chapter 3 that individuals with the syndrome of autism have a specific deficit for awareness of mental states. Very little is known about the neurobiology of meta-awareness, but Baron-Cohen finds that awareness of mental processes can depend on the functions of brain systems which are not important for awareness of external stimuli. On the contrary, the capacity for regarding the conscious experience of other people depends on the ability to reflect on one's own awareness.

Applying this result to our own project of finding consciousness in the brain, there are ramifications for the scientific study of consciousness from a "third-person" perspective. Each perceptual scientist, as an investigator of someone else's sensory awareness, necessarily operates from a perspective shaped by their own experience in meta-awareness. Once pointed out, this sounds almost obvious: Expertise in detecting, conceptualizing, and manipulating the conscious experience of other people may be partly determined by skill in being aware of one's own conscious experience.

References

Schein, S. J., & Desimone, R. 1990. Spectral properties of V4 neurons in the macaque. *Journal of Neuroscience, 10* (10), 3369-89.

Zeki, S., Watson, J. D. G., Lueck, C. J., Friston, K. J., Kennard, C., & Frackowiak, R. S. J. 1991. A direct demonstratrion of functional specialization in human visual cortex. *Journal of Neuroscience, 11* (3), 641-649.

CHAPTER 2

Now You See It, Now You Don't
Preventing consciousness with visual masking

Mark C. Price

University of Bergen

As a classroom demonstration, visual masking always gets a gasp. A visual stimulus such as a word is first presented for a few milliseconds on its own. Despite its brevity, the word is easy to identify. However when the word is presented for exactly the same period of time, but is followed after a blank delay by a random jumble of letters (Figure 2.1), most observers consciously perceive only the second stimulus. The preceding word seems to have completely disappeared. If a longer delay between the two stimuli is used, the masking effect of the second stimulus, or mask, is reduced. The first stimulus might now be perceived as a vague flash, even though none of the letters are

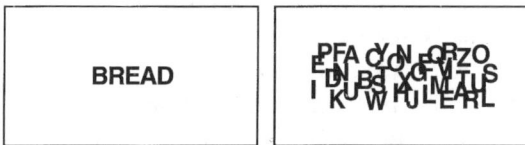

Figure 2.1. Examples of stimulus displays in a masking experiment.
The left panel depicts a target word, and the right panel depicts a mask composed of overlapping letters. Each display might appear in the same location for about 10 milliseconds. Depending on the exact conditions of presentation, and on individual variation between subjects, effective masking which precludes conscious awareness of the stimulus might typically be achieved with an interval of 20 - 50 milliseconds between displays.

discernible. A yet longer inter-stimulus delay will typically allow the observer to identify some letters. If a long enough delay is used, the entire word will again become visible.

Similar effects can be obtained using a variety of masks. These can simply be a light flash or a random dot pattern, in which case they are referred to as noise masks. Alternatively the mask can consist of letter fragments, letter strings or a real word. These are referred to as pattern masks because they are visually similar to the target. Although the example of masking described above is a backward effect of a mask on a preceding target, forward masking in which the mask interferes with a succeeding target can also occur. Masking also takes place if masks flank or surround the spatial location occupied by the target, rather than occurring in the same position. This is referred to as "paracontrast" masking when the masking is forwards, and "metacontrast" if backwards. An example is the masking of a small solid black disc by a slightly larger black circle; all the subject sees is an empty circle.

Whatever the exact details of the type of masking used, it is clear that the rapid succession of two stimuli in some sense oversteps the temporal resolution of the visual system. The mask appears to interfere with processing of the target somewhere between peripheral sensory registration of the target and its emergence into subjective awareness. Exactly how and where such interference takes place has been the subject of much debate. Perhaps the point on which there is most agreement is that masking embraces not one but many varieties of interaction at several levels of stimulus processing. Although part of the story appears merely to involve the temporal summation of the target and mask in peripheral processing channels between the retina and cortex, much higher level processes are also involved.

One of the most striking pieces of evidence for this comes from studies of nonconscious perception in which the meaning of a masked stimulus is shown to be processed, despite the fact that the stimulus is not perceived consciously. For example, an undetectable masked word (e.g., BREAD) can speed up reaction times to make a judgment about a subsequently presented word to which it is semantically associated (e.g., BUTTER). The fact that masking can prevent conscious awareness of sensory input, while leaving nonconscious perceptual and recognition processes at least partially intact, has been influential in leading some theorists (Bachmann, 1993; Marcel, 1983b) to propose that masking may interfere with brain processes that are critically involved in the emergence of consciousness itself.

In this chapter I discuss three ways in which masking contributes to the study of how neurocognitive events in the brain could give rise to conscious experience. Although masking occurs in the auditory and somatosensory modalities as well as in vision, this discussion will be confined to vision since it is here that the vast majority of empirical and theoretical work in masking has been conducted. Not all of the methodological and theoretical issues that will be raised are exclusive to masking, but few other phenomena in psychology bring one so readily and so comprehensively into contact with them.

The first section of the chapter describes the salutary lesson that masking provides for the way we operationalize and measure consciousness. In particular, masking studies highlight the difference between *subjective* introspective measures of what somebody is conscious of, and more *objective* behavioral measures. While the latter are commonly used to avoid the methodological pitfalls of introspective techniques, they may confound conscious and nonconscious processing. I argue that an introspective approach is possible, extremely informative, and at times necessary. Masking experiments also force us to distinguish subtly different ways in which various aspects of stimulus information can become manifest in consciousness, and therefore sharpen our definition of what it means to be conscious of something. For example, awareness of masked stimuli can range from awareness of stimulus presence without awareness of stimulus meaning, to counterintuitive instances of awareness of meaning without awareness of stimulus presence.

The second section examines what masking studies can tell us about the limits of stimulus processing *without* conscious awareness. In order to identify the brain processes which give rise to subjective awareness, we need to identify which processes can occur in the absence of consciousness, and which cannot (the approach labelled as "contrastive analysis" by Baars, 1997). One way to do this is to compare the qualitative differences between conscious and nonconscious perception (see Merikle & Daneman, 1998). Studies involving masking have formed a major part of such research, illustrating both the sophistication and the limitations of perception without awareness and exposing some complex but poorly understood interactions between conscious and nonconscious processes.

In the last section, I turn in more detail to theoretical accounts of the mechanisms that underlie masking. Some theories propose that masking prevents consciousness by disrupting early perceptual analysis. However theories of this kind are difficult to reconcile with the kind of nonconscious perception effects described in the second section, and I therefore concentrate on those theories which hold that masking directly disrupts higher processes involved in giving rise to consciousness.

Throughout the chapter a recurring and unifying theme will be the importance of phenomenological data, and I repeatedly argue that failure to address such data can impoverish or even mislead our theoretical understanding.

1. Measuring consciousness

1.1 Objective measures of consciousness

Anyone interested in the relationship between consciousness and brain processes will at some stage have to consider the problem of how to *measure* consciousness. I refer here to consciousness in its strong sense of subjective experience, rather than the purely functional sense used by some authors to refer to high-level cognitive processes of control and integration. Visual masking is, *par excellence*, an area where the issue of measuring the contents of consciousness has been impressively thrashed out. This is partly because of the role that masking has played in studies of nonconscious perception, where operationalizing the presence or absence of subjective awareness is of central importance. It is also partly because of the different varieties of stimulus awareness that can be induced by changes in the conditions and strength of masking.

The first experiments on nonconscious perception, conducted at the end of the last century, relied on subjects' introspective reports to determine whether stimuli had been presented nonconsciously. For example, Sidis (1898) showed that letter identification was above chance, even though stimuli were presented so far away that subjects claimed they were guessing their responses. However with the advent of the behaviorist movement, the reliability and objectivity of introspective techniques fell under question. It was correctly argued that introspective reports are both theory-laden and idiosyncratic because subjects (and experimenters) differ in the way they interpret questions, in the type of experience they consider relevant to their response, and in the way they translate their experience into spoken language. In particular, subjects are often unwilling to admit awareness under conditions of minimal stimulus information and low confidence. This makes it hard to distinguish whether they are truly unaware of stimulus information or are merely employing conservative response criteria which bias them against reporting awareness.

Recent studies of nonconscious perception have therefore tended to operationalize their definition of stimulus awareness in terms of *objective* psychophysical measures which assess perceptual sensitivity independently of observer bias. These objective indices of consciousness take the form of forced-choice tasks, such as forced identification of a masked stimulus, or forced detection judgments where a subject has to decide whether or not a stimulus has even occurred. On every trial subjects have to choose among a small set of alternative responses (e.g., "Yes" or "No") even if they lack confidence or are entirely guessing. Chance performance is taken to indicate the absence of stimulus awareness. Many authors refer to this type of task as a *direct* measure of stimulus processing: subjects are instructed to directly respond to a particular aspect of a stimulus, and the dependent variable is the effect of the specified stimulus information on their responses (Greenwald, Klinger & Schuh, 1995; Holender, 1986; Humphreys, 1981; Reingold & Merikle, 1988).

Having established in this way that a subject is unaware of a stimulus, nonconscious perception of the stimulus can then be demonstrated using an *indirect* measure of processing. These measure the indirect effect of stimulus information to which the subject has *not* been instructed to respond. A typical indirect measure is the ability of a stimulus to facilitate or inhibit (i.e., prime) the RT or accuracy of responses to another stimulus. Suppose for example that forced-choice detection of a masked word is at chance. It may nevertheless be shown that the semantic relationship between this word (the prime), and a second letter-string (the probe) presented perhaps 1000 msec after the mask, can affect RT to decide whether the second letter-string is a word or a non-word (the lexical decision task; e.g., Balota, 1983; Fowler et al., 1981; Marcel, 1980, 1983a). A semantic priming effect of this kind implies that the meaning of the prime must have been accessed, despite the fact that the subject was unaware even of its presence.

Of the many experiments on nonconscious perception that have been based on dissociations between direct and indirect measures, the majority have employed visual masking as the technique to prevent awareness of perceptual input. Masking has been popular because it appears to allow tight control over stimulus awareness. At the start of an experiment, the masking

conditions which prevent stimulus awareness can be established independently for each individual subject by gradually decreasing the interval between target and mask until performance the direct measure falls to chance.
The same masking conditions can then be used to prevent stimulus awareness
in the second part of the experiment which examines the influence of the
masked stimulus on an indirect measure of processing.

Other techniques to present stimuli nonconsciously do exist, such as
parafoveal presentation (presenting items in peripheral vision) or dichotic
listening (ignoring a message played to one ear while listening to the message
played to the other ear). However these rely on directing the focus of a
subject's attention away from the stimulus whose processing is being indirectly tested. It is therefore always difficult to entirely rule out momentary
switches of attention to the supposedly unattended stimuli. In contrast, a well
masked stimulus is hidden from a subject however hard they try to perceive it.

Although other forms of methodological criticism (Briand et al., 1988;
Doyle, 1990; Holender, 1986; Purcell et al., 1983; Reingold & Merikle, 1988)
have dogged many claimed demonstrations of nonconscious perception that
employ masking, a number of careful masking studies have now claimed to
strongly establish that indirect measures of processing can be more sensitive
than direct ones (e.g., Ansorge et al., 1998; Greenwald et al., 1988, 1995;
Humphreys, 1981; Kemp-Wheeler & Hill, 1988; Neumann & Klotz, 1994). In
the study by Greenwald et al. (1995), subjects were at chance in deciding
whether a word had been presented on the right or the left of a masked display
(direct measure), but the meaning of the letter stings "RIGH" or "LEFT"
nevertheless indirectly biased the outcome of these position judgements.

1.2 Limitations of objective measures

Despite the popularity of experiments that dissociate direct and indirect measures of processing, it is not clear that this is the most appropriate way to
characterize nonconscious perception. The problem has to do with the use of
objective forced-choice tasks as the criterion for conscious awareness. This
criterion rests on the assumption that direct measures of processing reflect
only conscious processing, i.e., that above-chance performance in such tasks
cannot be based on nonconscious processing of stimulus information. However it has been argued that this assumption should not be taken for granted. If
nonconscious perceptual processing can influence performance on indirect

measures, then why should it not also influence performance on direct measures (Reingold & Merikle, 1988)? Even though subjects feel they are guessing whether or not a stimulus has preceded the mask, their guesses might be biased by nonconscious information. Under these circumstances, increasing masking strength until performance on the direct measure is at chance may simply establish masking conditions under which nonconscious perception no longer takes place. Failure to find a dissociation between direct and indirect measures would therefore not necessarily argue against the occurrence of nonconscious perception.

To discover whether objective forced-choice performance on a direct measure of processing can indeed reflect nonconscious perception, we need to assess what subjects are aware of, *subjectively,* when they are making their forced-choice responses. This may seem a strange demand, given that the measurement of subjective awareness is precisely what the forced-choice tasks were designed for. The crucial point is that these tasks, which were adopted in the interests of methodological rigor, may not always be appropriate. If we want to assess subjective awareness, we need to do it introspectively after all.

This distinction between objective and subjective performance measures has been stressed and developed in the work of Cheesman and Merikle (1984, 1985; see also Henley, 1984). Cheesman and Merikle showed that two types of masking threshold could be obtained for forced identification. One was the objective threshold, defined as the inter-stimulus-interval (ISI) between a target and a mask at which accuracy on forced-choice identification among four different color-nouns fell to chance. The other was the subjective threshold, defined as the ISI where subjects no longer had any confidence in their responses being correct. The subjective threshold was higher; i.e., forced-choice guessing was above chance even when subjects had zero confidence in their guesses. Cheesman and Merikle showed that color words which were masked at the subjective threshold were able to prime RTs to name color patches (the Stroop effect), but had no effect on this indirect measure of processing when they were masked at the objective threshold. So although there was no dissociation between objective performance on the direct and indirect measures of processing, nonconscious perception was nevertheless demonstrated by the dissociation between the indirect measure and subjectively defined awareness.

An objective criterion for consciousness is too strict and potentially misleading. However, distrust of introspective methods is deep-seated and has contributed to continued attempts to demonstrate nonconscious perception in terms of the traditional dissociation between objectively defined direct and indirect measures. Using a subjective criterion for awareness may be desirable theoretically, but it will only be considered acceptable if it the methodological problems which led to its original abandonment can be overcome.

1.3 Validating subjective measures of consciousness

The technique used by Cheesman and Merikle (1984, 1985) to assess subjective awareness on their forced identification task was, first, to collect confidence ratings at the end of each block of trials. They recognized however that these ratings did not on their own guarantee unawareness of stimulus information. Subjects who gave ratings of zero confidence could have been momentarily aware but have forgotten by the time of reporting (after each block of trials). Also, remembered awareness on the occasional trial could have been ignored when subjects made an average assessment of the whole block. More fundamentally, claimed confidence does not depend just on awareness of stimulus information. It also reflects subjects' beliefs concerning the relevance of the information they are aware of, what it means to say one has no confidence, and so on.

In order to counter these objections, Cheesman and Merikle showed that the subjective threshold corresponded to a natural boundary across which there were qualitative differences in the effect of masked words on Stroop color-patch naming. Above the subjective threshold, increasing the proportion of trials on which the word matched the color patch led to larger priming effects on naming RT. Below the subjective threshold it did not. This suggested that subjective threshold, as defined by the confidence ratings, is not just an arbitrary cut-off at which subjects are no longer willing to indicate any confidence. Instead, it reflects an important transition between attentional priming effects which are mediated by conscious expectation, and nonconscious automatic effects.

An alternative approach to validating introspective reports of stimulus awareness was explored by Price (1991). Subjects were presented with single words that were backward masked by a pattern of letter fragments. The words were either animate (e.g., SHEEP) or inanimate (e.g., DIARY). In one experiment a word was not always present and the task was forced-choice detection of word presence. In a second experiment, a word was always present, and subjects had to make a forced-choice animacy categorization, replying "living" or "non-living." In both tasks the ISI between word and mask was progressively reduced, allowing objective categorization performance to be plotted against ISI. After making each categorization or detection judgment, subjects rated the extent to which their answer was a guess on a scale which ran from "absolutely sure" to "a complete guess."

As expected, reducing ISI led to a general decrease in objective performance, and an increase in the number of trials rated as complete guesses. However at very short ISIs, where detection judgments were at chance and nearly all trials were rated as complete guesses, categorization performance showed a small recovery, deviating once again from chance. This recovery was manifest even on just those trials rated as complete guesses. For both categorization and detection, performance also deviated from chance on the small number of "complete guess" trials which occurred at the longer ISIs (where average performance was well above chance). One interpretation of these results is that subjects were truly unaware of any task-relevant stimulus information when they claimed to be guessing, and that their categorization responses were influenced by nonconsciously perceived information. An alternative interpretation is that claimed guesses were not really guesses at all, and that categorization responses were in fact influenced by conscious stimulus information.

These two interpretations were tested in the following way. At the end of each block of trials, subjects were asked for further information about the contents of their awareness on trials rated as a "complete guess." Using a numerical scale, they rated the proportion of their "complete guesses" which had been based on something they had seen or felt (however seemingly irrelevant), and the proportion which had been completely "spontaneous" (i.e., on which they were aware of no reason for having chosen that response rather than any other). These subjective rating data from the post-block questionnaires were then statistically regressed against objective performance on "complete guesses" at different ISIs.

For detection judgments, performance on "complete guesses" was positively correlated with the extent to which subjects rated they were influenced by visual cues. This suggests that useful visual cues were consciously available after all, despite the claim to be guessing, and argues against nonconscious perception. For animacy categorization judgments, the only correlation was a positive one between objective performance and the extent to which "complete guesses" were rated as spontaneous; i.e., on trials rated as "complete guesses," subjects did better when their responses felt more spontaneous. Since stimulus awareness would most plausibly cause a decrease rather than an increase in spontaneity ratings, this correlation is the *inverse* of what we would expect if improvements in performance were due to awareness of stimulus information. It therefore provides strong evidence that nonconscious perception was taking place. The evidence does not depend on the absolute accuracy of introspective ratings, but only on the correlation between ratings and performance. And because the evidence depends on the presence rather than the mere absence of a correlation, it cannot be dismissed on the traditional grounds of insensitivity. In addition, it is unlikely that fluctuations in the spontaneity rating were themselves just the result of a nonconscious bias; if this were the case then one would expect other ratings to be similarly affected, which they were not.

This study demonstrates again that objective forced-choice measures do not necessarily reflect subjective awareness, and shows how introspective measures of awareness can be successfully applied without external validation from qualitative differences in indirect measures of processing. Instead the study relied on a detailed battery of introspective ratings which also provided informative data on the importance of spontaneity that would otherwise have been missed. This is not the first time that nonconscious perception has been found to be enhanced by a passive response strategy in which the answer is left to "pop out of the subject's head," as opposed to an active strategy of scouring the stimulus display for possible cues (Marcel, 1983a; Snodgrass et al, 1993). Reasons why spontaneity may enhance nonconscious perception are discussed later in this chapter.

1.4 *Shades of stimulus awareness*

So far I have shown that a subjective definition of consciousness is theoretically desirable, methodologically possible, and potentially very informative. However, the definition of "stimulus awareness" is in need of greater sophistication.

When a target stimulus is well masked, we may not even be aware of its presence. Without the mask, we are fully aware of the stimulus. But between these two extremes there is wide variation in our possible experience of stimulus information. Suppose an observer is trying to detect the presence or absence of a masked word. Even if he never perceives the stimulus, its presence can alter the perceived form of the mask in subtle ways such as by speeding the subjective onset of the mask (Bachmann, 1988a), or enhancing the apparent brightness of the mask (Bachmann, 1988b). An observer could use any of these conscious cues to infer the presence of the target which was itself phenomenally suppressed. At longer ISIs, the observer may be directly aware of a discrete target but be unable to recognize its form, or he may consciously perceive a few letters without identifying the whole word. If we are to talk about stimulus awareness, we must be clear which aspects of stimulus information we are referring to. To demonstrate nonconscious semantic priming from a masked stimulus, we would not have to insist that the prime was totally undetected; we need only show that *task-relevant* information (in this case the identity of the prime) is not consciously discernible. To talk merely of "stimulus awareness" is, from a methodological point of view, too simplistic. Rather, we must specify exactly *what* it is that we are aware of, or unaware of.

That we can be unaware of stimulus identity, while conscious of stimulus presence, seems unsurprising; the detection of a source of light energy requires less processing of stimulus information than recognition of form and meaning, and seems more likely to escape any interference caused by the mask. That we can process stimulus meaning nonconsciously, without being conscious of stimulus presence, is more surprising; it implies that high-level stimulus information can be processed without the mechanisms of consciousness being engaged even by more rudimentary stimulus information. It would be even more counter-intuitive to find that we could be *conscious* of stimulus identity, while still unaware of stimulus presence. Yet there is a sense in which conscious identification without conscious detection can take place, providing a further illustration of the care that is needed when describing awareness of stimulus information.

When stimuli such as words are backward pattern masked, subjects may report the percept of a word that has visual qualities but seems more like a mental image than a word on the display screen (Price, 1991). These imaged words sometimes appear to differ in size from the displayed words, or to appear after the mask rather than before it, but their most important characteristic is that they seem to have an internal rather than an external source. For example, the following descriptions were recorded from subjects during the forced-choice categorization of animate and inanimate words described earlier: The experience was *"not a visual representation of what it looks like when you see a word in the machine'* or was *"like an after-image"* (Price,

1991; p. 192). In these examples, subjects were aware of semantic aspects of stimulus information, but this information was not integrated with its correct episodic source. The semantic content was divorced from its spatial and temporal location, and was instead attributed to an internally generated image.

A still more extreme example from the same experiment shows that semantic information from a masked word can manifest itself in consciousness without being attached to the visual percept of a word at all. During a block of animacy categorization trials in which masking ISI was very short, one subject paused, without responding, at the end of a trial. He then began to laugh and was asked why. He replied: *"I wanted to say "camels" really loud and cracked up. I've got camels on my brain."* (Price, 1991; p. 193). The previous trial, correctly classified as "living" and rated as a "complete guess," had in fact been the word "CAMEL." The subject insisted that he had been guessing and expressed great surprise when told the identity of the first stimulus. According to him, the idea of "camel" had just popped up in his head. Here the meaning of the masked stimulus seems likely to have affected the contents of consciousness without being associated with any discrete percept at all. It would therefore be untrue to say that the subject was totally unaware of any stimulus information, but it would likewise be untrue to say he was aware of the stimulus in any conventional sense. To do justice to situations like this, subjective experience needs to be characterized on several dimensions. These should include not only a precise description of the *type* of stimulus information that is experienced, but also the perceived relevance and attributed source of the experience.

The shades of stimulus awareness revealed by masking studies have parallels with anomalous forms of stimulus awareness revealed by neuropsychological studies of brain-damaged patients. For example, patients with alexia (greatly impaired reading ability, despite intact ability to write, speak, or comprehend speech) may have awareness of the general semantic category of a word whose basic physical form they are unable to discern (Farah, 1994). In "blind-sight" patients with hemianopia (subjective blindness in part of the visual field following damage to primary visual cortex), preserved abilities to locate and reach for objects presented in the blind field may never the less sometimes be associated with "non-visual" awareness of the objects (Weiskrantz, 1986; 1997).

However, acknowledgement and classification of anomalous forms of stimulus awareness has usually been even less rigorous in patient studies than in studies of nonconscious perception in normal subjects. Often it is deemed sufficient to classify the behavioral discrimination response of a patient simply as a "guess." But in my own studies of nonconscious perception using masking in normals (Price, 1991), subjects applied the label "guess" to responses which on close analysis turned out to embrace a wide variety of possible experiences. These ranged from complete unawareness of any task-relevant stimulus information to the vivid "after-images" of masked words described above.

Whether one is studying patient groups or normal subjects, an overly simplistic dichotomization between consciously and nonconsciously mediated behavior will clearly impoverish investigation of the relation between consciousness and cognition. Attempts to understand the mechanisms of consciousness need to account not only for subjective experience itself, but also for what one can be conscious *of*. Such accounts will have either to accommodate the evident variety of possible contents of consciousness, or will have to carefully specify what types of content are being referred to.

2.   Exploring nonconscious perception

2.1 *Nonconscious perception and consciousness*

One way to explore the relationship between cognitive processing and consciousness is to discover which processes are inevitably associated with the emergence of consciousness and which are not. In studies of nonconscious perception, the particular focus of interest is the relationship between processing and awareness of external sensory input. Although awareness of external stimuli is only a subset of the possible contents of consciousness, it particularly lends itself to research for three reasons. First, the methodology of measuring stimulus awareness has undergone considerable progress. Second, processing of nonconsciously perceived stimulus information can be assessed indirectly in a variety of ways. Third, stimulus awareness can be manipulated using the technique of masking. In simple perceptual experiments the relationship between consciousness and cognition is therefore at least empirically tractable, and the possibility remains that the principles uncovered will have wider applicability.

Like any experimental technique used to prevent consciousness, masking can only help identify the processes underlying consciousness by a method of elimination. Showing that a particular aspect of stimulus processing is prevented by masking does not imply that it is important for consciousness; masking may interfere at a point in the chain of information processing events well prior to that at which processes integral to the emergence of consciousness occur. However if a particular aspect of stimulus processing is indirectly shown to occur despite masking, it can be eliminated as a sufficient condition for consciousness. Evidence has already been presented that masking does not prevent extensive visual analysis of a stimulus, nor lexical access, nor even recognition of its superordinate semantic category (Humphreys, 1981; Price, 1991). Given that masking shows basic semantic recognition to be independent of consciousness, what *are* the qualitative differences between conscious and nonconscious processing?

2.2 The flexibility of nonconscious perception: Evidence from indirect measures of processing

One simple view is that the transfer from preconscious to conscious perception is merely a matter of activation strength (Shallice, 1978). A way to increase the activation produced by a stimulus is to repeat it. Marcel (1983a) showed that if the pairing of a word plus mask was rapidly repeated several times on each trial, the activation produced by the word did indeed seem to increase since its priming effect on responses to semantically related probe words was enhanced. However the repetition had no effect on awareness of the masked word. Activation strength is therefore probably not sufficient for consciousness, even if it may sometimes be a necessary condition (e.g., see Farah, 1994 for discussion of the latter in relation to deficits of stimulus awareness after brain injury).

Another commonly made distinction between conscious and nonconscious perception is that nonconscious processing is relatively inflexible and insensitive to context, whereas consciously represented information can be flexibly used in response to changing context (e.g., Jacoby et al., 1997; Merikle & Daneman, 1998). An example is provided by Cheesman and Merikle's (1984) demonstration that the priming effect of consciously identifiable primes is affected by the proportion of trials on which primes and probes are related, whereas the priming effect of more heavily masked primes which are unidentifiable is unaffected.

A masking experiment by Marcel (1980) further supports this kind of division. Marcel presented trials consisting of three successive letter strings. The first was a word (e.g., HAND or TREE), intended to provide a context capable of biasing the interpretation of the middle string which was a polysemous word (e.g., PALM). Subjects were to decide as fast as possible if the last string (e.g., WRIST), which could be semantically related to one of the meanings of the middle word, was a word or a non-word. In one experimental condition all three words were consciously visible, and in another condition the middle word was masked sufficiently that subjects were unable to guess whether it had been presented or not.

In the unmasked condition, results were as follows. If the context word (HAND) specified an interpretation of the middle word (PALM) that was semantically related to the last word (WRIST), RTs to the last word were faster than with a completely unrelated middle word (RACE). If the context word (TREE) specified the interpretation of the middle word (PALM) that was not related to the last word (WRIST), RTs to the last word were slower than with an unrelated middle word. This was taken to show that conscious perception is selective and affected by prior context. Only the contextually appropriate meaning of the polysemous word was activated, and the inappropriate meaning appeared to be actively suppressed, leading to inhibitory priming. A different set of results was obtained when the polysemous word was backward pattern masked. The polysemous word now produced facilitatory priming of the last word regardless of whether its semantically related or unrelated meaning had been specified by the context word. Nonconscious perception therefore appeared unselective and insensitive to prior context, with all possible meanings of the masked word being automatically activated.

These results support the view that masking prevents higher integration of stimulus information, and that processes required for more flexible and contextually sensitive responding may be involved in the emergence of consciousness. Further examples of studies reaching similar conclusions are summarised by Merikle and Daneman (1998). However other experiments show that a masked prime *can* have an effect, this time on motor responses, that is highly flexible and responsive to changing response requirements.

Neumann and Klotz (1994) conducted a series of experiments in which subjects were presented simultaneously with two shapes on each trial (e.g., an outline square, and an identical square with a line above and below it). One of the two shapes was designated as the target, and the task was to indicate which side of the display it appeared on by pressing one of two keys as fast as possible. The two shapes also acted as metacontrast masks which obscured the presence of two, smaller, preceding "primes." The primes were either (1) both non-target shapes (neutral condition), (2) a target and a non-target in the same relative positions as the masking shapes (congruent condition), or (3) a target and a non-target in the opposite positions to the masking shapes (incongruent condition). Compared to neutral trials, responses in the congruent condition were faster, while responses in the incongruent condition were slower and less accurate. Subjects almost always realized when they had made a mistake, but could nevertheless not prevent themselves from occasionally pressing the wrong response key. The same priming effects were obtained when a less natural stimulus-response mapping was employed, with subjects having to press the left-hand key in response to a right-hand target and vice versa. The effects were even obtained when the response mapping was redefined from trial to trial, using a *pre-trial* cue. It therefore appears that undetectable masked stimuli can affect RT to another stimulus in a way which depends on the binding of form and location information in the masked stimuli, and in a way which reflects continually varying stimulus-response mappings.

Neumann and Klotz (1994) argued that the priming was not just due to an effect of the prime on sensory processing of the masks, but was at least partly a direct effect on motor processes that produce the subjects' responses. Two pieces of evidence supported this assertion. First, in the condition where response mapping was varied from trial to trial, priming was only obtained if the pre-trial cue was given at least 250 msec before prime onset. This timing requirement is best explained by a motor priming effect which only occurs if enough time has elapsed for the stimulus-response mapping specified by the cue to have been programmed. The second piece of evidence was derived from a further experiment in which the prime and mask displays both consisted of three shapes presented side by side. One shape was the target and the remaining two were non-targets. Subjects were to press the left key for a left-hand target and the right key for a right-hand target. If the target was in the middle, some subjects were to press the left key and some the right. The interesting trials were those on which a target shape appeared in the middle position of either the prime display or the mask display. If priming just affected sensory processing, we would expect a middle "target" prime to have an equal effect on right and left target masks, and we would expect a middle target mask to be equally affected by right and left primes. In fact priming on these trials depended on whether the prime and mask displays dictated the same motor response, rather than on the relative spatial positions of the target shape in the two displays. When the prime and mask displays dictated different responses, RT was slowed. However there was no facilitation of RT when the displays dictated the same response, indicating that facilitatory effects may well be sensory, even though inhibitory effects seem to be motor.

The experiments of Neumann and Klotz (see also Leuthold & Kopp, 1998; Taylor & McCloskey, 1996) claim to show that processing of nonconsciously perceived sensory information is not entirely inflexible and stimulus driven. Instead, it can directly prime motor responses in a flexible and task-dependent manner. This challenges convergence between two conventional criteria used to distinguish automatic from controlled processing. Automatic processes are usually characterized as both inflexible and preconscious. However processing clearly *can* be flexible without being conscious. (For further critical discussion of the mapping between consciousness and the inflexibility of processing, see for example Shiffrin, 1997)

The concept of automaticity is also qualified by another set of experiments which show the effect of context on nonconscious perception. Smith, Besner and Miyoshi (1994) have shown that facilitatory semantic priming (again of lexical decision RT) by well masked primes disappears when trials are mixed with others on which longer ISIs allow the primes to be easily identified. It could be argued that the context of easily identifiable primes merely raises the signal detection criterion for perceptual encoding. To test this, Smith et al. looked at the effect of context on "repetition priming" in which primes were either the very same words as the probes, or a different and unrelated word. Facilitatory priming by well masked primes was now unchanged by mixing with longer ISI trials, showing that the changed context did not affect encoding of prime identity. Moreover this encoding must have reached a lexical level since repetition priming was not obtained from nonword primes. Smith et al. argued that the effect of context therefore acted at a post-recognition level and directly modulated the spread of semantic activation. Again this refutes the view of nonconscious perception as always based on inflexible automatic processes that are immune to contextual influence.

Nervertheless, the degree of flexibility shown by the processing of masked stimuli in these experiments must be qualified. For example, the results of Neumann and Klotz (1994) discussed above appear to show that the processing of masked stimuli can vary in accordance with the response contingencies represented by a *conscious pre-trial cue*. So although one of two different stimulus-response mappings is being automatically triggered by a masked stimulus, the selection of the appropriate mapping is being consciously mediated and pre-set *before* the nonconscious stimulus is presented. The flexibility of processing would be much more impressive if it could be shown that *post-trial* cues could modulate the processing of a nonconscious stimulus that had already been presented. A result of the latter kind would be needed to more strongly challenge the notion (Baars, 1997, Merikle & Daneman, 1998) that ongoing nonconscious processing is indeed not globally accessible to behavioural response systems in a truly flexible manner.

2.3 *Guessing, spontaneity and passivity: Evidence from direct measures of processing*

One aspect of the way in which information is processed is the use to which it can be put. It is therefore of interest to ask whether there are differences between the direct and indirect effects of masked stimuli. (Direct effects were earlier defined as the effects of stimulus information to which a subject has been explicitly instructed to respond.) The extreme view that nonconsciously perceived information cannot be used directly at all has already been refuted by the described dissociations between objective forced-choice measures and subjective measures of awareness (Cheesman & Merikle, 1984; Price, 1991). The nature of these direct effects and the circumstances under which they are manifested now merit further discussion.

Let us return to the experiment by Price (1991) in which semantic categorization of masked words took place without subjective awareness of stimulus information. Another finding of this experiment was that the deviations of categorization performance from chance were *below* as well as above chance. The direction of the deviation did not just vary between subjects, but was liable to change even within subjects during the break between blocks of trials. Although the masked words were capable of biasing response choice on the basis of their semantic category, the mapping of stimulus information to the correct motor response seems to have been disrupted.

This contrasts markedly with Neumann and Klotz's (1994) claim that masked stimuli can prime motor responses defined by complex stimulus-response mappings. Although masked stimuli can affect responses to other stimuli that are consciously presented, response control therefore becomes problematic when subjects are required to respond directly to the masked stimulus (i.e., are instructed to guess). A possible reason is that the absence of a conscious rationale for response choice may engage alternative decision processes that interfere with the influence of nonconscious information.

Support for this view is provided by the relationship that Price (1991) observed between categorization performance and spontaneity ratings. The fact that deviations of performance from chance increased with self-rated spontaneity of responding does more than just provide good evidence for nonconscious perception. It suggests that the influence of nonconscious information is greater when subjects are not consciously thinking about which response to make. This is consistent with other observations that a passive mode of responding is conducive to nonconscious perception (Dixon, 1981; Marcel, 1983a; Snodgrass, Shevrin & Kopka, 1993). For example, Marcel (1983b) found that judgments of the semantic similarity of test words to unidentifiable masked words were above chance, but only for subjects who adopted a passive response strategy of letting answers pop spontaneously into their head. Subjects who tended to scrutinize the display for every last clue showed chance performance.

The importance of passivity or spontaneity is usually explained, using a signal detection analogy, by proposing that they are the subjective correlate of an attentional mode in which potentially distracting noise is ignored. This increases the likelihood that the weak signal from the nonconsciously activated stimulus can influence responding (Greenwald et al., 1995; Marcel, 1983a; Price, 1991; Snodgrass et al., 1993). Noise can consist of irrelevant information in the visual display, such as fragments of the mask that subjects mistakenly take to be clues to the identity of the target. It can also consist of subjects' thoughts about which response to make, given the lack of consciously represented criteria. For example, subjects will typically choose one response simply to alternate from the response given on the preceding trials. All these sources of noise are reduced by a passive approach.

This explanation of the relationship between passivity and performance can account for why Price (1991) found animacy categorization to be influenced by nonconscious perception at long masking ISIs and short masking ISIs, but not at intermediate ISIs: At long ISIs, where target words are usually visible and overall performance is above chance, lapses in attention lead subjects to miss the target stimulus on a small proportion of trials. On these trials subjects therefore have to guess out of the blue. These guesses will tend to be spontaneous and influenced by nonconscious perception. At intermediate ISIs the words are harder to see. The proportion of guessed responses increases and subjects therefore tend to adopt idiosyncratic response strategies which impair spontaneity and interfere with the effects of nonconscious perception. Additionally, awareness of display cues that are just sufficient to allow stimulus detection (but not identification or categorization) acts as further distracting noise. At very short ISIs the masked words are completely undetectable, subjects tend to give up trying to see anything in the display, and guesses are more spontaneous. In the words of one subject, he was *"just sort of placidly sitting there and the more relaxed I was the more the words just came"* (Price, 1991, p. 201). Under these conditions, a nonconscious influence of word identity on responses will again be found.

The importance of a passive response mode may also explain why Price (1991) found a dissociation between subjective awareness and forced-choice categorization, but not forced-choice detection (see also Van Selst & Merikle, 1993). With categorization, it is evident to subjects that no helpful stimulus information is consciously available, and a strategy of concentrating hard on the visual display is likely to be seen as futile. With detection it is much easier to believe that minute variations in the appearance of the mask can act as clues to stimulus presence, encouraging subjects to persist with an active rather than passive response strategy.

The apparent inability to obtain nonconscious effects of stimulus presence on forced detection contrasts with the robust effects of stimulus presence on other measures of processing. In experiments using metacontrast masking, it has been shown that if subjects have to make some kind of manual response (e.g., press a key) as soon as they detect the onset of the first stimulus in a visual display, RT is no slower when the first stimulus is masked and consciously undetectable than when it is unmasked and visible (Fehrer & Raab, 1962; Taylor & McCloskey, 1990). This RT finding, sometimes referred to as the "Fehrer-Raab effect", appears to hold even if the required response is one of 2 possible motor programs which are contingent on the location of the masked stimulus (e.g., flex the right arm for a target on the right and the left arm for a target on the left; Taylor & McCloskey, 1996). It therefore appears that motor responses are being *directly initiated* by the masked stimulus. In the masked-shape priming experiments by Neumann and Klotz (1994) that were discussed above, behavioural evidence was already used to argue that masked primes were *pre-activating* motor processes. This kind of pre-activation is corroborated by electrophysiological data from Leuthold and Kopp (1998) who measured Event Related Potentials across motor cortex during similar masking experiments. However in order for the presence of masked stimuli to show the Fehrer-Raab effect and directly initiate a motor response, rather than merely prime motor responses to subsequent stimuli via motor pre-activation, it is perhaps crucial that subjects are not trying to respond to a stimulus they cannot see, but instead *believe* that they are responding to the consciously visible mask. They therefore do not have to search for a rationale for their response, or become distracted by irrelevant clues. Instead responses are left to the unhindered guidance of fast motor systems which can be driven by nonconscious sensory input. The role played by motor systems in controlling behavioral responses to nonconsciously represented information is discussed at length in the chapter by Rossetti (this volume).

Another example of the effect of passivity is provided by Snodgrass et al. (1993) in a study where performance was again analyzed into positive and negative deviations from chance. Subjects had to guess which of four undetectable words preceded a backward mask. They were instructed either to look hard for clues in the displays, or to let the answers pop into their head. They were also asked which of these two strategies they preferred. Overall performance was at chance, but when subjects were split according to preferred strategy, significant deviations from chance were found. Under instructions to use the passive pop strategy, subjects who preferred this strategy did better than chance, but subjects who preferred the active strategy did worse than chance. As with Price's (1991) experiment, non-chance responding was therefore associated with passivity, but Snodgrass et al. offered a different explanation for below-chance responding. They noted that the effect of passivity is consistent with a fundamental tenet of psychoanalytical theory, that unconscious processes are more likely to manifest themselves when subjects freely associate as opposed to controlling their thought processes. They then suggested that the below-chance score of subjects who preferred the active strategy may have resulted from active suppression of stimulus information.

The relationship between passivity and performance clearly represents a set of complex interactions between many contextual variables which include experimental instructions and individual differences. The processes of nonconscious perception are not entirely automatic and inflexible. With the kind of work described above, the study of nonconscious perception has moved on from dogmatic methodological battles as to whether nonconscious perception exists in the first place, to consider the way in which nonconscious perception interacts with, and is modulated by, conscious attentional processes. More questions have been raised than answered, but this is often the inevitable cost to be paid for the excitement of exploring new territory.

3. Theories of masking

3.1 *Basic types of masking*

As mentioned in the introduction to this chapter, there are many types of masking and these reflect different processing interactions. The most informative masking techniques for studying conscious awareness will be those which disrupt the causal chain of information processing events leading to consciousness at as late a stage as possible. The later the locus of masking, the greater the number of processes that will be left intact under masking, and the greater the number of processes that can be eliminated as being causally sufficient for consciousness.

With this qualification in mind, the most basic distinction to make is between *integrative* masking, which occurs mainly at a peripheral level of the visual system, and *interruptive* masking which has a later, more central locus.

Integrative masking is thought to involve the summation of the neural signal of the target and mask due to the limited temporal resolution of processing channels. The resulting stimulus signal is a composite, much like two photographs exposed on the same emulsion. In this type of masking (Turvey, 1973): (a) The stimulus that dominates the conscious percept is the stimulus with the highest energy (where "energy" is approximately equivalent to the product of the duration and luminance of the display). If masking is only partial, the subjective appearance of the display is of a bold stimulus superimposed over a faint "ghost" stimulus. (b) The temporal order of the two stimuli is relatively unimportant; the most energetic stimulus masks the other, regardless of whether it comes first (forward masking) or second (backward masking). (c) Masking is obtained even from light flashes or random patterns of dots (*noise* masks). (d) Masking is relatively ineffective if the two stimuli are presented dichoptically (i.e., to separate eyes) rather than monoptically or binocularly. For this reason the main locus of integrative masking must be prior to binocular convergence in the visual cortex. (A small amount of integrative masking does however occur at more central levels; Michaels & Turvey, 1979.)

Although noise masks are often experimentally employed to prevent visual persistence, and thereby to specify how long stimulus information is available for processing, this method is based on the questionable assumption that the mask interrupts stimulus processing (see Schultz & Eriksen, 1978; Eriksen, 1980; Bridgeman, 1986). Under the integration account, the mask does not so much limit processing time as limit the duration and information content of the target signal itself by reducing its signal-to-noise ratio. In contrast, interruption of stimulus processing has usually been proposed as the main type of masking interaction at more central levels. It is these higher-level interactions that will be of most interest here.

Central interruptive masking has the following properties: (a) The relative energy of the target and mask does not greatly affect the degree of masking. It is the first stimulus, rather than the least energetic stimulus, that is phenomenally suppressed; i.e., only backward masking takes place. Because target energy and hence duration are unimportant, the temporal variable that determines the effectiveness of central pattern masking is often given as the stimulus-onset asynchrony (SOA) between target and mask (Turvey, 1973). (b) The effectiveness of masking is greatly increased by masks which are perceptually similar to the target (*pattern* masks). (c) Masking persists at longer ISIs than found with peripheral noise masking. (d) If masking is partial, the masked stimulus may subjectively appear as a brief unrecognizable flash which *precedes* the mask, but it does not seem temporally superimposed by the mask. (e) Masking is still effective under dichoptic conditions and so occurs at a cortical locus where information from the two eyes converges.

The interruptive effect of masks on higher processing of target stimuli has been directly observed in studies that have recorded neuronal responses from single cells in the cortex of monkeys (Kovács et al., 1995; Rolls & Tovee, 1994). Recordings were made from areas of temporal cortex, known to be involved in object recognition, in animals showing similarity to humans in their psychophysical responses to masked stimuli. Although neurons showed an initial response to backward masked shapes for which they were selective (e.g., faces), the onset of the mask appeared to drastically shorten the neuronal firing period (e.g. from 200-300ms to only 20-30ms, Rolls & Tovee, 1994).

Masking experiments on nonconscious perception usually aim to maximize central interruptive masking by employing backward pattern masks, and to minimize peripheral integrative masking by using a mask of lower energy than the target, or by presenting the target and mask dichoptically. If integrative masking is allowed to happen, early perceptual processing may be interfered with and nonconscious perception effects may not be obtained (Marcel, 1983b). Some studies have however reported higher processing of energy-masked stimuli (e.g., Snodgrass et al.,1992).

To understand theories of central masking it is necessary to introduce a technical distinction concerning the relationship between the effectiveness of masking and target-mask SOA. It has so far been implied that the effectiveness of masking increases as SOA is reduced. Often this is the case, as in integrative masking. However under some conditions the plot of masking against SOA can show a "U-shaped" function; as progressively shorter SOAs are used, phenomenal suppression of the target increases, but only until very short SOAs are reached at which point it decreases again. Sometimes recovery from masking at short SOAs is almost total. U-shaped functions of this kind, in which masking is maximal at intermediate SOAs, can be obtained with metacontrast or backward pattern masks, and are usually attributed to central interruptive mechanisms. Any adequate theory of central masking must be able to explain U-shaped functions, and masking theories can in fact be largely characterized by the way in these functions are addressed.

One of the most detailed explanations of U-shaped functions, and probably the best known, is the transient-on-sustained hypothesis which was first proposed by Breitmeyer and Ganz (1976; for more recent refinements see, e.g., Breitmeyer, 1984; Breitmeyer & Williams, 1990; Williams et al., 1991). Although most of the evidence for this hypothesis derives from experiments using metacontrast masking, it also plays a central role in the masking theory of Michaels and Turvey (1979) which was developed to account for non-metacontrast backward masking. The hypothesis is based on neurophysiological evidence for a division of the visual pathway from retina to cortex into *sustained* and *transient* channels. The two types of channel differ systematically along several dimensions; notably the sustained channel carries higher frequency spatial information which permits the fine analysis of visual form, and takes longer to respond after stimulus onset than the transient channel. Interruptive masking is thought to occur because sustained channel activity is inhibited by transient channel activity. When two stimuli are presented in rapid succession, the transient activity elicited by the second stimulus (the mask) coincides with, and inhibits, the longer latency sustained activity elicited by the preceding stimulus (the target). At shorter or longer SOAs, superimposition will be less than optimal so masking peaks at intermediate SOAs, producing the U-shaped function.

However, details of the neurophysiological evidence which underlies the transient-on-sustained hypothesis have been challenged on both empirical and theoretical grounds (e.g., Bridgeman, 1986). In addition, the hypothesis fails to easily accommodate nonconscious perception. According to the hypothesis, masking eliminates or at least severely reduces target information that is carried by the sustained channel. However this channel carries exactly the range of spatial frequencies that are crucial to detailed form recognition of the target. This makes it hard to account for nonconscious perception effects which depend on recognition of shapes or written words, such as semantic priming or forced-choice categorization. By postulating that masking precludes consciousness by interfering with basic perceptual analysis of stimuli, the theory cannot satisfactorily account for nonconscious perception effects obtained either with metacontrast masking or with non-metacontrast backward pattern masking.

Masking theories which can accommodate nonconscious perception also tend to be those in which masking is tied more directly to processes involved in giving rise to consciousness. Two interesting examples are the Recovery theory of Marcel (1983b) and the Perceptual Retouch theory of Bachmann (1993).

3.2 The Recovery Theory of masking

Marcel (1983b) proposed that all centrally masked sensory data are preconsciously processed to a highly abstract level. Central masking prevents conscious awareness of stimulus information by interfering with the "recovery" of information. "Recovery" consists of linking perceptual information to its spatio-temporal sensory source, and involves the synthesis of information from the different specialist processors that analyse different aspects of sensory information in parallel. According to Marcel (p.263), *"Without segmentable evidence of particular form or of particular location, the separate existence of an environmental event or aspect cannot be acknowledged or experienced."*

Despite interfering with integration of episodic and semantic information into a stable sensory record, masking does not however prevent the activation of partial representations of masked input, thus allowing masked information to bias brain activity and behaviour. It is notable that the studies of monkeys' neuronal responses to masked stimuli referred to in section 3.1 (Kovács et al., 1995; Rolls & Tovee, 1994), found that object recognition neurons do show a brief selective response to masked targets, even though this response is then interrupted. Similarly, Leuthold and Kopp (1998) have used Event Related Potentials in the human brain to indicate pre-activation of motor programs by masked stimuli. Masked stimuli therefore seem capable of influencing brain activity all the way from perceptual classification to response preparation.

Marcel claimed certain rules account for some of the data of central masking, proposing that candidate signals for recovery are determined by factors such as temporal and spatial proximity, expectancy, economy and level of description etc. This may account for why backward masking is stronger when the mask can be described at a higher and more economical level of description than the target; e.g., a non-word is more effectively masked by a word than by another non-word (Taylor & Chabot, 1978) because the word can activate a more unitary and abstract representation than a meaningless and unintegrated string of letters. It may also account for why target letters sometimes migrate to the conscious percept of a mask if this allows the letters from each stimulus to be recombined into a meaningful word.

U-shaped functions occur because, at longer SOAs, the relative recency of the mask gives it privileged status for recovery, whereas at short SOAs this advantage is reduced and out-weighed by the potential level of description of the target. Marcel supports this by asserting that most non-metacontrast studies showing U-shaped functions use targets with higher potential levels of description than the mask (e.g., a word target and a non-word mask). However this is simply not correct: e.g., Bachmann and Allik (1976) obtain recovery at short SOAs with targets and masks that are both geometrical shapes.

According to Marcel's theory, the behavioural biasing effect of masked stimuli should be expected to reflect a lack of stabilised integration in the internal representation of the masked stimulus. This is exactly what was shown by Marcel's (1980) finding that semantic priming effects are obtained from both interpretations of a backward masked polysemous word, despite the presence of a disambiguating contextual cue. When no mask was used, only the cued interpretation of the polysemous word gave a priming effect; the cue could now be integrated into the perceptual interpretation of the word (see section 2.2). However, as also argued in section 2.2, experiments such as those of Neuman and Klotz (1994) do show that some aspects of a metacontrast-masked stimulus (e.g., form and location) *are* sufficiently integrated to support indirect priming effects. This demands a greater clarification of the way that masking interferes with stimulus integration, as well as of any differences in the effects of different types of mask. Future studies of nonconscious perception may play an important role in this endeavor.

Marcel's account of central masking was intended as part of a broad approach to the relationship between nonconscious and conscious perception, rather than as a detailed theory. Nevertheless it is limited by a lack of convergent evidence and by counter-examples to some of its predictions. Also, in contrast to the next masking theory to be discussed, no basis for the mechanisms of central masking is given at a neurophysiological level of description.

3.3 *The Perceptual Retouch theory of masking*

Like Marcel, Bachmann (Bachmann & Allik, 1976; Bachmann, 1984, 1988a, 1988b, 1993, 1997) proposes that the perceptual trace derived from a stimulus is not in itself sufficient for conscious awareness of the stimulus to arise. For Bachmann, visual consciousness and U-shaped masking functions both depend on the interaction between "specific" and "non-specific" neurophysiological systems. These two systems are independently activated by sensory input but later reconverge at a cortical level.

The specific system, represented by the pathway from the retina, via the LGN to visual cortex, processes information specific to a particular stimulus and provides the contents of consciousness. The non-specific system is mediated by the subcortical midbrain structures of the brainstem reticular formation and the non-specific thalamic activating system, which together provide the arousal necessary for consciousness itself. Non-specific activation has two aspects, tonic and phasic (see chapter by Whitehead and Schliebner, this volume). The tonic aspect is due to the reticular formation and is important for general arousal and wakefulness. However Bachmann's theory is especially concerned with the dynamics of the interaction between specific activation and phasic non-specific activation from the thalamus. This interaction, referred to as "perceptual retouch" (PR), is essential for the specific stimulus activation to become consciously represented: *"PR refers to the psychological process which, being only partially under voluntary control of the subject, has functions of ... allotting the quality of introspective awareness to the formerly preconscious stimulus representation built up by specific neural activities...*" (Bachmann, 1984, p. 70). Perceptual retouch need not be an all or none process, allowing for *"smooth gradients in the introspective clarity of perceptual data"* (p. 70).

To account for masking, Bachmann proposes that perceptual retouch has the following two crucial properties. First, the non-specific signal elicited by a stimulus takes longer to arrive at the cortex than its specific signal; this is reflected by the form of scalp recorded electrical evoked potentials, whose later negative components correspond to non-specific input. Second, when two specific signals arrive at the cortex in rapid succession, the perceptual retouch initiated by arrival at the cortex of a non-specific signal operates mainly on the specific signal with the highest current signal-to-noise ratio.

At short SOAs the specific neural signals from a target and a backward mask both reach the cortex before either of their respective non-specific signals. When the non-specific retouch impulses arrive at the cortex, the signal-to-noise ratios of the specific signals from target and mask are roughly equivalent, and perceptual retouch operates on both. The conscious perceptual outcome is the "retouch" of a composite target-mask signal. Which signal perceptually dominates the composite percept depends on the relative energies and structures of the target and mask.

At intermediate SOAs (40-90 msec), specific activation from the target and mask again reaches the cortex before the respective non-specific signals. However the slightly longer delay between the two stimuli means that the first non-specific signal to arrive roughly coincides with the arrival of the specific signal from the mask. The mask is therefore "retouched" by the non-specific signal from the target, and enters consciousness in place of the target. U-shaped masking functions result because masking of the target will be maximal at intermediate SOAs where the non-specific signal from the target optimally coincides with the specific signal from the mask. Since perceptual retouch is a process which happens *after* perceptual analysis, the preconsciously represented stimulus that is not retouched can nevertheless continue to affect behavior on the basis of its form or meaning.

One of the main pieces of evidence claimed for Bachmann's account is that masking SOAs which lead to the strongest backward masking seem to have an average value of about 50 msec, which is similar to the delay of arrival at the cortex between specific and non-specific signals (Bachmann, 1984, 1993). Further evidence is provided by patients receiving electrostimulation as treatment for Parkinson's Disease (Bachmann, 1993). The backward masking functions of these patients were tested just after their non-specific thalamic system had been heavily stimulated by electrodes, and it was found that masking almost disappeared. Bachmann argues that the specific signal from the target is being retouched, as soon as it reaches the cortex, by the previously applied non-specific activation. Retouch of the target is therefore relatively immune from competition by the mask. Bachmann interprets the results of another study in a similar way. Lester et al. (1979) measured electrical evoked potentials during backward masking and observed that the biggest difference in brain activity, between trials on which the target was consciously perceived or not, occurred *before* stimulus onset. According to Bachmann, "hit" trials in which subjects did perceive the target were probably those occasions on which there was a build up of non-specific activation prior to target onset.

Bachmann proposes there is also an attentional component to masking at longer SOAs (90-250 msec) where subjects often report that a clear conscious percept precedes the mask but that it is too brief to identify. According to Bachmann, the target is retouched, but the sudden appearance of the mask results in a switch of attention away from the target in favor of the more recent stimulus. The possible existence of a late attentional component to masking would imply that perceptual retouch was not sufficient for consciousness, even if it was necessary.

Like other theories of masking, it is difficult for the PR theory to account for all aspects of masking data. For example, the theory predicts that repeated presentation of target and mask should lead to a build up of non-specific activation, permitting perceptual retouch of the target. However, as discussed, Marcel (1983a) has demonstrated that repeated cycling of target and mask does not decrease the power of masking. Another drawback of the PR theory in its current form is that, unlike Marcel's theory, it does not directly predict why masking should interfere qualitatively with processes such as binding of target information. Therefore the theory less readily predicts qualitative differences between masked and unmasked priming effects, such as Marcel's (1980) observation that both meanings of a contextually disambiguated polysemous word give priming if the word is masked, but not if it is consciously perceived.

The PR theory is close in some respects to a growing general literature on the importance of non-specific thalamo-cortical interactions as an essential ingredient in constructing conscious representations (e.g., Newman, 1997a, 1997b). However, in the PR theory, masking is modeled as a consequence of temporal interactions between primarily *feedforward* specific and non-specific signals. In contrast, a major emphasis in neurophysiological models of thalamo-cortical interactions has been the reciprocation of signals from thalamus to cortex and back *downwards* to thalamus in dynamic reentrant (iterative) loops which allow a functional integration of spatially distributed brain activity. These loops include non-specific thalamic signals which are hypothesised to modulate both (1) the binding together of cortical information, and (2) the selective inhibitory gating of emerging cortical representations so that, via a winner takes all competition, "only one of many competing streams of potential contents reaches consciousness" (Newman, 1997b, p. 112).

Bachmann (1997) is explicit in wishing his PR masking model to stand independent of the validity of this more complex dynamic picture of thalamo-cortical reentry. It is nevertheless tempting to ask whether his simple feedforward masking model, along with its temporal predictions, could be elaborated to fit this more complex picture. Studies of masking and nonconscious perception should in turn be able to contribute to the more global neuroscience approach to attention and consciousness outlined by Newman (1997b).

3.4 *Do U-shaped masking curves reflect conscious or nonconscious perception?*

I would now like to suggest that the full implications of nonconscious perception have not been acknowledged even by some of the masking theories which permit high-level processing of masked stimuli. This stems from a failure to pay sufficient attention to phenomenological data. Masking theories which address the origin of U-shaped masking functions tend to assume that the apparent decrease in the effectiveness of masking at short SOAs is due to a return of the *conscious* percept of the target. A possible alternative is that the target remains phenomenally suppressed, and that subjects' performance improves because nonconscious perception enhances forced-choice guessing.

Admittedly the latter possibility is not the case for metacontrast masking. One of the most salient features of the metacontrast literature is that U-shaped functions are specifically related to the *conscious* perceptibility of contour information, and are not obtained if criteria such as forced-choice detection are employed (Breitmeyer, 1984). This is because subjects can infer target presence from an illusory apparent motion of the mask away from the target location. (If the target is a ring that spatially surrounds a target disc, the ring looks as if it is expanding outwards.) This apparent motion is preserved even if the target is itself completely invisible.

For non-metacontrast masking the situation is quite different. To start with, a true U-shaped function is only obtained if the mask display is monoptically presented and also of lower energy than the target display. At short SOAs these conditions lead to peripheral integration of target and mask into a composite signal that is dominated by the target. A large component of any improvement in target perceptibility is therefore just due to forward integrative masking of the mask itself by the target. However if dichoptic presentation is used to eliminate integrative masking and obtain a purer measure of central interruptive masking, improvement in forced-choice performance at short SOAs is very small; i.e., the masking function becomes J-shaped rather than U-shaped. Whether this improvement is a conscious effect at all is questioned by three lines of evidence.

First, none of the relatively few studies on dichoptic, non-metacontrast, backward masking (e.g., Bachmann & Allik, 1976; Michaels & Turvey, 1979) provide any introspective data to show that the observed improvement was necessarily conscious. The performance measure in such studies has invariably been forced-choice identification of the target, and it may just not have been noticed that the small improvement was due to above-chance guessing. In this respect it is extremely notable that the J-shaped functions of Bachmann and Allik (1976) were reported to disappear with minor changes in the report task that subjects were asked to perform.

Second, no evidence for conscious recovery was found in the study by Price (1991) on forced detection and forced categorization of dichoptically masked words, described earlier in this chapter. Despite systematically measuring performance across many SOAs which descended well into the range where improvement is claimed to take place, the only recovery found was a *nonconscious* bias on the outcome of guesses. Furthermore, the best predictor of this nonconscious recovery seemed not to be SOA itself, but subjective "passivity" which was argued to reflect attentional strategies that modulate nonconscious perception.

Third, a *nonconscious* improvement in performance at short SOAs has also been noted when using indirect measures of stimulus processing (Dagenbach et al., 1989; Greenwald et al., 1995). For example, Dagenbach et al. (1989)

reported that masked priming of lexical decision gradually fell to zero as SOA was shortened, but subsequently reappeared at even shorter SOAs. The interpretation put on these results was the same as that discussed in section 2.3 for direct measures of processing. At intermediate SOAs nonconscious perception is hindered by distracting noise in displays which contain detectable but unidentifiable targets. At shorter SOAs no such target fragments are consciously visible and subjects are more likely to adopt a passive response mode conducive to nonconscious perception.

Thus the typical recovery from non-metacontrast central masking that is claimed at short SOAs may actually reflect an upturn in nonconscious perception rather than a release from masking. On this interpretation, reductions in SOA steadily increase the effectiveness of interruptive masking until, at intermediate SOAs, forced-choice measures of target perceptibility reach chance. At even shorter SOAs, conditions may be especially conducive to nonconscious perception and a small *nonconscious* improvement in direct or indirect measures of performance may occur. A large *conscious* improvement may also occur if peripheral masking is allowed to take place and if the target has higher energy than the mask (Turvey, 1973); interruptive masking, which depends on the interaction of two separate signals, is now prevented by the integration of target and mask into a composite signal, which is dominated by the target.

This represents a serious challenge to theories of masking that are in part based on the shape of masking functions. Unless a distinction is made between conscious and nonconscious effects, interpretation of the functions will be flawed. Important characteristics of their shape may reflect interactions between conscious and nonconscious processing whose complexity we are only beginning to appreciate. Just as masking studies can help to advance our neurophysiological or cognitive understanding of consciousness, so phenomenological data must be used to constrain and guide theories of masking. We must be prepared to recognize the importance of this interplay which all too often is still denied by the legacy of behaviorism. Masking can tell us about consciousness, but consciousness too can tell us about masking.

4.   Summary

Studying brain function by pushing its processing sub-systems to their limits is a common technique in psychology. For example, perception is often studied by presenting stimuli that are too fast or too degraded for easy recognition. Attention is often studied by giving subjects too many streams of information to monitor concurrently. And memory may be studied by giving subjects too much to remember. Consciousness is no exception. This chapter has described some of the ways in which masking can be used to study consciousness by observing the breakdown of normal awareness when the temporal resolution of conscious perception is exceeded. In particular, I have discussed how masking studies help us to measure and define consciousness, how they challenge assumptions about the causes of consciousness by revealing the sophistication of nonconscious perception, and how some of the masking theories they generate make claims about the direct involvement of certain cognitive processes or brain mechanisms in consciousness.

One of the goals of future theoretical development in the field of masking must be to achieve greater synthesis across different levels of description. For example, Bachmann's (1993) neurophysiological approach describes brain structures and events that may be important for consciousness, but does not satisfactorily explain their computational or cognitive role. On the other hand cognitive approaches such as that of Marcel (1983b) need to be more constrained by neurophysiological data. Even an integrated neurocognitive approach will be incomplete unless it takes on board the phenomenological level of description whose importance is well illustrated by the problems of interpreting U-shaped masking functions.

If masking studies tell us more about what consciousness *is not* than about what it *is,* this merely illustrates the inevitability of an eliminative approach to consciousness. As discussed, progress in mapping the relationship between neurocognitive events and consciousness can, logically, only proceed by a patient elimination of those processes which are independent of consciousness. Our current patchwork of empirical knowledge, however frustrating, must therefore be regarded as a necessary stepping stone to a convergent understanding.

Acknowledgment

I would like to thank Jason Mattingley and Peter Grossenbacher for their helpful comments on the first drafts of this chapter which were prepared while the author was working at the University of Cambridge in 1996.

58 MARK C. PRICE

References

Ansorge, U., Klotz, W., & Neumann, O. 1998. Manual and verbal responses to completely masked (unreportable) stimuli: Exploring some conditions for the metacontrast dissociation. *Perception* 27: 1177-1189.

Baars, B.J. 1997. Global workspace theory, a rigourous scientific theory of consciousness. *Journal of Consciouness Studies* 4(4): 292-309.

Bachmann, T. 1984. The process of perceptual retouch: Nonspecific afferent activation dynamics in explaining visual masking. *Perception and Psychophysics*, 35(1): 69-84.

Bachmann, T. 1988a. Microgenesis as traced by the transient paired-forms paradigm. *Acta Psychologia* 70: 3-17.

Bachmann, T. 1988b. Time course of the subjective contrast enhancement for a second stimulus in successively paired above-threshold transient forms: Perceptual retouch instead of forward masking. *Vision Research* 28(11): 1255-1261.

Bachmann, T. 1993. *The Psychophysiology of Visual Masking; The Fine Structure of Conscious Experience*. New York: Nova Science Publishers.

Bachmann, T. 1997. Visibility of brief images: The dual-process approach. *Consciousness and Cognition*, 6: 491-518.

Bachmann, T. and Allik, J. 1976. Integration and interruption in the masking of form by form. *Perception* 5(1):79-97.

Balota, D.A. 1983. Automatic semantic activation. *Journal of Verbal Learning and Behavior* 22: 88-104.

Breitmeyer, B.G. 1984. *Visual Masking: An Integrative Approach*. Oxford University Press.

Breitmeyer, B.G. and Ganz, L. 1976. Implications for sustained and transient channels for theories of visual pattern masking, saccadic suppression, and information processing. *Psychological Review* 83(1): 1-36.

Breitmeyer, B.G. and Williams, M.C. 1990. Effects of isoluminant-background colour on metacontrast and stroboscopic motion. *Vision Research* 33(7): 1069-1075.

Briand, K., Den Heyer, K., and Dannenberg, G.L. 1988. Retroactive semantic priming in a lexical decision task. *Quarterly Journal of Experimental Psychology* 40(A): 341-359.

Bridgeman, B. 1986. Theories of visual masking. *Behavioural and Brain Sciences,* 9(1): 25-26.

Cheesman, J., and Merikle, P.M. 1984. Priming with and without awareness. *Perception and Psychophysics* 36: 387-395.

Cheesman, J., and Merikle, P.M. 1985. Word recognition and consciousness. In D. Besner, T. Gary Waller, and G.E. MacKinnon (eds), *Reading Research: Advances in Theory and Practice*. Orlando, FL: Academic Press, 311-352.

Dagenbach, D. Carr, T.H. and Wilhelmsen, A. 1989. Task-induced strategies and near-threshold priming: Conscious effects on unconscious perception. *Journal of Memory and Language* 28:412-443.

Dixon, N. 1981. *Preconscious Processing*. London: Wiley.

Doyle, J.R. 1990. Detectionless processing with semantic activation? A footnote to Greenwald, Klinger & Liu (1989). *Memory and Cognition* 18(4): 428-429.

Eriksen, C.W. 1980. The use of a visual mask may seriously confound your experiment. *Perception and Psychophysics* 28(1): 89-92.

Farah, M.J., 1994. Visual perception and visual awareness after brain damage: A tutorial overview. *Attention and Performance XV*, 37-75.

Fehrer, E. and Raab, D. 1962. Reaction time to stimuli masked by metacontrast. *Journal of Experimental Psychology* 63: 143-147.

Fowler, C.A., Wolford, G., Slade, R., and Tassinary, L. 1981. Lexical access with and without awareness. *Journal of Experimental Psychology: General* 110: 341-362.

Greenwald, A. G. and Klinger, M.R. 1990. Visual masking and unconscious processing: Differences between backward and simultaneous masking? *Memory and Cognition* 18(4): 430-435.

Greenwald, A.G., Klinger, M.R and Schuh, E.S. 1995. Activation by marginally perceptible ("subliminal") stimuli: Dissociation of unconscious from conscious cognition. *Journal of Experimental Psychology: General* 124(1): 22-42.

Henley, S.H.A. 1984. Unconscious perception revisited: A comment on Merikle's (1982) paper. *Bulletin of the Psychonomic Society* 22: 121-124.

Holender, D. 1986. Semantic identification without conscious identification in dichotic listening, parafoveal vision, and visual masking: a survey and appraisal. *The Behavioral and Brain Sciences* 9:1-66.

Humphreys, G.W. 1981. Direct vs. indirect tests of the information available from masked displays: What visual masking does and does not prevent. *British Journal of Psychology* 72: 323-330.

Jacoby, L.L., Yonelinas, A.P., and Jennings, J.M. 1997. The relation between conscious and unconscious (automatic) influences: A declaration of independence. In J.D. Cohen & J.W. Schooler (eds) *Scientific Approaches to Consciousness*, Lawrence Erlbaum Associates: New Jersey, 13-47.

Kemp-Wheeler, S.M., and Hill, H.B. 1988. Semantic priming without awareness: some methodological considerations and replications. *Quarterly Journal of Experimental Psychology* 40(A): 671-692.

Kovács, G., Vogels, R., and Orban, G.A. 1995. Cortical correlate of pattern backward masking. *Proceedings of the National Academy of Sciences, USA* 92: 5587-5591.

Lester, M.L., Kitzman, M.J., Karmel, B.Z. and Crowe, G.J. 1979. Neurophysiological correlates of central masking. In H. Begleiter (ed), *Evoked brain potentials and behaviour,* New York: Plenum.

Leuthold, H., and Kopp, B. 1998. Mechanisms of priming by masked stimuli: Inferences from event-related brain potentials. *Psychological Research* 9(4): 263-269.

Marcel, A.J. 1980. Conscious and preconscious recognition of polysemous words: locating the selective effects of prior verbal context. In R.S Nickerson (ed), *Attention and Performance, Vol. VIII.* Erlbaum, Hillsdale, NJ.

Marcel, A.J. 1983a. Conscious and unconscious perception: Experiments on visual masking and word recognition. *Cognitive Psychology* 15: 197-237.

Marcel, A.J. 1983b. Conscious and unconscious perception: An approach to the relations between phenomenal experience and perceptual processes. *Cognitive Psychology* 15: 238-300.

Michaels, C.F. and Turvey, M.T. 1979. Central sources of masking: Indexing structures supporting seeing at a single, brief glance. *Psychological Research* 41: 1-61.

Merikle, P.M., and Daneman, M. 1998. Psychological investigations of unconscious perception. *Journal of Consciousness Studies*, 5(1), 5-18.

60   MARK C. PRICE

Neumann, O. and Klotz, W. 1994. Motor responses to nonreportable, masked stimuli: Where is the limit of direct parameter specification? *Attention and Performance* 15: 123-150.
Newman, J. 1997a, 1997b. Putting the puzzle together, Part 1 and Part II: Towards a general theory of the neural correlates of consciousness. *Journal of Consciousness Studies* 4(1): 47-66, and 4(2): 100-121.
Price, M.C. 1991. Processing and awareness of masked stimuli. *Unpublished Ph.D. dissertation, University of Cambridge.*
Purcell, D.G., Stewart, A.L., and Stanovich, K.K. 1983. Another look at semantic priming without awareness. *Perception and Psychophysics*, 34: 65-71.
Reingold, E.M., and Merikle, P.M. 1988. Using direct and indirect measures to study perception without awareness. *Perception and psychophysics* 44: 563-575.
Rolls, E.T., and Tovee, M.J. 1994. Processing speed in the cerebral cortex and the neurophysiology of visual masking. *Proceedings of the Royal Society of London, Series B, Biological Sciences* 257 (1348): 9-15.
Schultz, D.W. and Eriksen, C.W. 1977. Do noise masks terminate target processing? *Memory and Cognition* 5(1): 90-96.
Shallice, T. 1978. The dominant action system: An information-processing approach to consciousness. In K.S. Pope and J.L. Singer (eds), *The Stream of Consciousness: Scientific Investigations into the Flow of Human Experience.* NY: Plenum.
Shiffrin, R.M. 1997. Attention, automatism, and consciousness. In J.D. Cohen & J.W. Schooler (eds), *Scientific Approaches to Consciousness,* Mahwah, NJ: Lawrence Erlbaum Associates, 49-64.
Sidis, B. 1898. *The Psychology of Suggestion.* New York: Appleton.
Smith, M.C., Besner, D. and Miyoshi, H. 1994. New limits to automaticity: Context modulates semantic priming. *Journal of Experimental Psychology: Learning, Memory and Cognition* 20(1): 104-115.
Snodgrass, M., Shevrin, H. and Kopka, M. 1993. The mediation of intentional judgments by unconscious perceptions: The influences of task strategy, task preference, word meaning, and motivation. *Consciousness and Cognition* 2: 169-193.
Taylor, G.A and Chabot, R.J. 1978. Differential backward masking of words and letters by masks of varying orthographic structure. *Memory and Cognition* 6: 629-635.
Taylor, J.L., and McCloskey, D.L. 1990. Triggering of preprogrammed movements as reactions to masked stimuli. *Journal of Neurophysiology.* 63(3), 439-446.
Taylor, J.L., and McCloskey, D.L. 1996. Selection of motor responses on the basis of unperceived stimuli. *Experimental Brain Research* 110: 62-66.
Turvey, M.T. 1973. On peripheral and central processes in vision: Inferences from an information-processing analysis of masking with patterned stimuli. *Psychological Review* 80: 1-52.
Van Selst, M. and Merikle, P.M. 1993. Perception below the objective threshold? *Consciousness and Cognition* 2: 194-203.
Weiskrantz, L. 1986. *Blindsight: A case study and implication.* Oxford University Press: Oxford.
Weiskrantz, L. 1997. *Consciousness Lost and Found.* Oxford: Oxford University Press.
Williams, M.C., Breitmeyer, B.G., LovegroveW.J. and Gutierrez, C. 1991. Metacontrast with masks varying in spatial frequency and wavelength. *Vision Research* 31(11): 2017-2023.

CHAPTER 3

Consciousness
of the Physical and the Mental
Evidence from autism

Simon Baron-Cohen

University of Cambridge

At the close of the last millennium and in the dawn of the new one, scientists and philosophers of all hues (Crick, 1994; Dennett; 1993; Penrose, 1992; Gray, 1995; Bloch, 1995) are seizing on one big question: the explanation of consciousness. Presumably this recent surge of interest is because we think that if we can crack this one, we will finally have understood what it is that makes humans special. But by a strange irony, almost all accounts seem to focus on one type of consciousness that in all likelihood makes humans indistinguishable from many (if not all) animals: consciousness of the physical world. In Dennett's (1978) terminology, this is first-order consciousness. Thus, questions driving most accounts are along the lines of "Why do we 'see' something when our visual system is stimulated?", or "Why do we 'feel' something when our tactile system is stimulated?", or "Why are we 'aware' of some things, but unaware of others?" etc., These are important questions all right. But they all center on what I call our *consciousness of the physical world*. That side of consciousness will figure very small in this essay. The other side of consciousness — and let's for the moment assume there are only two aspects to this slippery thing — I call *consciousness of the mental world*. Exactly which entities count as mental needs a little more spelling out, which I do next.

1. The mental-physical distinction

I draw this distinction not because I am a dualist (I am not - there are in my world no mental events that do not also have a physical instantiation); but because, according to Brentano (1823), we can identify two sorts of (ultimately physical) entities in the universe: things with intentionality — which I shall call 'mental' — and things without intentionality — which I shall call 'physical'.

This begs the question as to how intentionality is to be defined. Brentano's definition is probably as good as one can get. Here it is, paraphrased. Those things that refer to (or are about) things other than themselves are intentional, and everything else is not. Contrast therefore, a rock, with a thought about a rock. The rock is not 'about' anything else. It is just a rock! In contrast, a thought about a rock necessarily is about something else: a rock. The thought is both something in its own right, and is about something other than itself. It is in this sense that intentional and non-intentional objects are distinct. I will argue that humans have two dissociable forms of consciousness for these two classes of entity, and the evidence for this claim comes largely from neuropsychological experiments with children with autism.

2. Are children with autism conscious of the physical?

This section can be relatively brief, since the answer to the above question is clearly "yes." We know this because of the following pieces of evidence. First, children with autism search to find occluded objects (Sigman, Ungerer, Mundy, and Sherman, 1987; Frith and Baron-Cohen, 1987). That is, they behave in ways which are intended to cause them to *see* something. Secondly, they are capable of mental rotation (Shah, 1988). This suggests that they have representations in their mind of how physical objects appear from different visual perspectives, in the same way that the rest of us do. Thirdly, they respond to the same range of sensory stimuli as other people (though they may be hypersensitive to some sounds and tactile stimuli) (Wing, 1976; Frith, 1989). Fourthly, as far as we know, their color perception is normal. They may attend to parts of objects in a different manner to others (Shah and Frith, 1983, 1993; Frith and Happe, 1994; Jolliffe and Baron-Cohen, 1997), but there is little doubt that they have conscious experiences of the physical world.

Perhaps the best evidence for their consciousness of the physical world is that, when questioned, they can *report* their awareness of this plane of existence. After all, verbal report is pretty much the only way we have of confirming that someone (other than ourselves) is conscious. Even this is not a fool-proof form of evidence, of course. A zombie could in principle produce words which apparently report conscious experience even though s/he was totally unaware of the physical world. But if we accept what people tell us, on face-value, then we find that when asked if they can see, hear, touch, smell, or taste something, children and adults with autism will affirm that they can. (At least, those who speak do so). Since this aspect of their consciousness is relatively non-contentious, we can pass swiftly on to the next question, of whether children with autism are also conscious of the mental world.

3.   Are children with autism conscious of the mental?

This section is necessarily longer than the previous one, since a long line of evidence collected over the last two decades suggests that children with autism are relatively *unaware of the mental*. Wimmer and Perner (1983) devised an elegant paradigm to test when normally developing children be-come conscious of the mental — specifically, when they are aware of another person's beliefs. The child was presented with a short story, with the simplest of plots. The story involves one character not being present when an object was moved, and therefore not *knowing* that the object was in a new location. The child being tested is asked where the character *thinks* the object is. Wimmer and Perner called this the False Belief test, since the focus was on the subject's ability to infer a story character's mistaken belief about a situation. These authors found that normal 4 year olds correctly infer that the character thinks the object is where the character last left it, rather than where it actually is. This is impressive evidence for the normal child's ability to distinguish between their own knowledge (about reality) and someone else's false belief (about reality).

When this test was given to a sample of children with autism, with mild degrees of mental handicap, a large majority of them 'failed' this test by indicating that the character thinks the object is where it actually is (Baron-Cohen, Leslie, and Frith, 1985). That is, they appeared to disregard the critical fact that, by virtue of being *absent* during the critical scene, the character's mental state would necessarily be different to the child's own mental state. In contrast, a control group of children with Down's Syndrome, with moderate degrees of mental handicap, passed this test as easily as the normal children. The implication was that the ability to infer mental states may be an aspect of social intelligence that is relatively independent of general intelligence (Cosmides, 1989), and that children with autism might be specifically impaired in their consciousness of the mental.

Of course, simply failing one test would not necessarily mean that children with autism lacked awareness of the mind. One swallow does not make a summer. There might be many reasons for failure on such a test. (Interestingly, control questions in the original procedure ruled out memory, or language difficulties, or inattention as possible causes of failure). The conclusion that children with autism are indeed impaired in the development of a normal awareness of the mind only becomes possible because of the convergence of results from widely differing experimental paradigms. These are reviewed in detail in two edited volumes (Baron-Cohen, Tager-Flusberg, and Cohen, 1993, 2000) and for that reason are only briefly summarized here, next.

3.1 *Summary of results suggesting that children with autism are impaired in their awareness of the mental*[1]

The majority of children with autism
i.   are at chance on tests of the *mental-physical distinction* (Baron-Cohen, 1989a). That is, they do not show a clear understanding of how physical objects differ from *thoughts* about objects. For example, when asked which can be touched: a biscuit, or a thought (about a biscuit), young normal 3 year olds rapidly identify the former, whereas most children with autism respond at chance levels.
ii.   They also have an appropriate understanding of the functions of the brain, but have a poor understanding of the functions of the mind (Baron-Cohen, 1989a). That is, they recognize that the brain's physical function is to make you move and do things, but they do not spontaneously mention *the mind's mental function* (in thinking, dreaming, wishing, deceiving, etc.,). Again, contrast this with normal 3 year old children who do spontaneously use such mental state terms in their descriptions of what the mind is for (Wellman and Estes, 1983).
iii.   Most children with autism also fail to make the *appearance-reality*

distinction (Baron-Cohen, 1989a), meaning that, in their description of misleading objects (like a red candle in the shape of an apple), they do not distinguish between what the object *looks* like, and what they *know* it really is. For example, the normal 4 year old child will say of an ambiguous object, when asked what it looks like, and what it really is, that "It *looks* like an apple, but *really* it's a candle made of wax" (Flavell, Flavell, and Green, 1983). In contrast, children with autism tend to refer to just one aspect of the object (e.g., saying "It looks like an apple, and it really is an apple."

iv.   Most children with autism fail a range of *first-order false belief* tasks, of the kind described in the previous section (Baron-Cohen et al., 1985, 1986; Perner, Frith, Leslie, and Leekam, 1989; Swettenham, 1996; Reed and Petersen, 1990; Leekam and Perner, 1991). That is, they show deficits in thinking about someone else's different beliefs.

v.   They also fail tests assessing if they understand the principle that *"seeing leads to knowing"* (Baron-Cohen and Goodhart, 1994; Leslie and Frith, 1988). For example, when presented with two dolls, one of whom touches a box, and the other of whom *looks inside* the box, and when asked "Which one *knows* what's inside the box?", they are at chance in their response. In contrast, normal children of 3-4 years of age correctly judge that it is the one who looked, who knows what's in the box. (This experimental procedure is schematically shown in Figure 3.1).

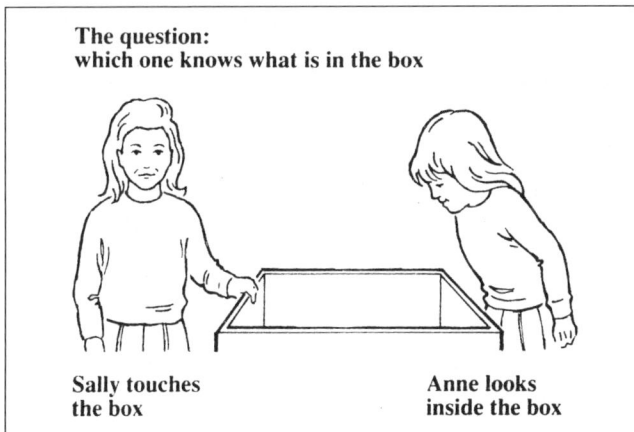

Figure 3.1. The seeing leads to knowing distinction. After Baron-Cohen and Goodhardt (1994).

We continue our survey of the evidence relevant to a 'consciousness of the mental' deficit in autism here:

vi. Whereas normally developing children are rather good at picking out *mental state words* (like "think", "know", and "imagine") in a wordlist that contains both mental state and non-mental state words, most children with autism are at chance (Baron-Cohen, Ring, Moriarty, Shmitz, Costa, and Ell, 1994). In contrast, they have no difficulty in picking out words describing physical states.

vii. Nor do most children with autism *produce* the same range of mental state words in their spontaneous speech (Tager-Flusberg, 1992; Baron-Cohen et al., 1986). Thus, from about 18-36 months of age, normally developing children spontaneously use words like "think", "know", "pretend", "imagine", "wish", "hope", etc., and use such terms appropriately (Wellman, 1990). In contrast, such words occur less frequently, and are often even absent, in the spontaneous speech of children with autism.

viii. They are also impaired in the production of *spontaneous pretend play* (Baron-Cohen, 1987; Wing, Gould, Yeates, and Brierley, 1977; Lewis and Boucher, 1988). Pretend play is relevant here simply because it involves understanding the mental state of *pretending*. The normal child of even 2 years old effortlessly distinguishes between when someone else is acting veridically, versus when they are "just pretending" (Leslie, 1987). Sometimes mommy is *actually* eating (putting a real spoon with real food into her mouth), whilst at other times mommy is just pretending to eat (holding a pen to her lips, and making funny slurping noises, in between her smiles).

Young normal children make rapid sense of such behavior, presumably because they can represent the latter case as driven by the mental state of "pretending." They also spontaneously generate examples of pretense themselves, and do not show any confusion as they switch back and forth between pretense (the mental world), and reality (the physical world). In contrast, most children with autism produce little pretense, and often appear confused about what pretense is for, and when someone is or is not pretending.

ix. Whilst they can understand simple causes of emotion (such as reactions to *physical* situations), the majority of children with autism have difficulty understanding more *mentalistic* causes of emotion (such as beliefs) [Baron-Cohen, 1991a; Baron-Cohen, Spitz, and Cross, 1993]. For example, they can understand that if Jane *actually* falls over and cuts her knee, she will feel sad, and that if John *actually* gets a present, he will feel happy. But they are poor at understanding that if John *thinks* he's getting a present (even if in reality he is

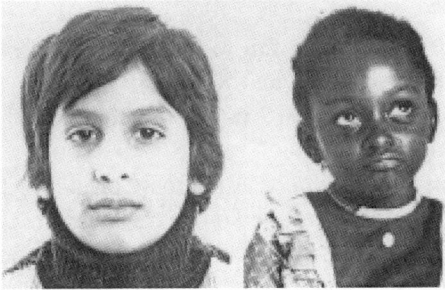

Figure 3.2. The "Which one is thinking?" test. Reproduced from Baron-Cohen and Cross (1992), with permission.

not), he will still feel happy. In contrast, normal 4 year old children comprehend such belief-based emotions.

x.   Most children with autism also fail to recognize *the eye-region of the face* as indicating when a person is *thinking* and what a person might *want* (Baron-Cohen and Cross, 1992; Baron-Cohen, Campbell, Karmiloff-Smith, Grant, and Walker, 1995). Children and adults without autism use gaze to infer both of these mental states.

For example, when presented with pairs of photos like those in Figure 3.2, normal 3-4 year olds easily identify the person looking upwards and away as the one who is thinking. Children with autism are less sure of this. And when shown a display like the one in Figure 3.3, normal 4 year olds identify the candy that Charlie is looking at as the one he wants. Children with autism mostly fail to pick up that gaze can be an indicator of what a person might want.

Figure 3.3. The "Which sweet does Charlie want?" test. Reproduced from Baron-Cohen et al. (1995), with permission.

In addition:

xi. Many children with autism fail to make the *accidental-intentional dis-tinction* (Phillips, Baron-Cohen, and Rutter, 1998). That is, they are poor at distinguishing if someone "meant" to do something, or if something simply happened accidentally.

xii. They also seem unable to *deceive* (Baron-Cohen, 1992; Sodian and Frith, 1992), a result that would be expected if one was unaware that people's beliefs can differ and therefore can be manipulated. In contrast, normal children of 4 begin to be quite adept at lying, thus revealing their awareness of the mental lives of others.

xiii. Most children with autism also have disproportionate difficulty on tests of understanding metaphor, sarcasm, and irony — these all being statements which cannot be decoded literally, but which are only meaningful by reference to the speaker's *intention* (Happé, 1994). An example would be understanding the phrase "the drinks are on the house," which one adult with autism (of above average IQ) could only interpret literally. This suggests that children with autism are aware of the physical (the actual words uttered), but are relatively unaware of the mental states (the intentions) behind them.

xiv. Indeed, most children with autism fail to produce most aspects of *prag-matics* in their speech (reviewed in Baron-Cohen, 1988; and Tager-Flusberg, 1993), and fail to recognize violations of pragmatic rules, such as the Gricean Maxims of conversational cooperation (Surian, Baron-Cohen, and Van der Lely, 1996). For example, one Gricean Maxim of conversation is "Be rel-evant." If someone replies to a question with an *irrelevant* answer, normal young children are very sensitive to this pragmatic failure, but most children with autism are not. Since many pragmatic rules involve tailoring one's speech to what the listener *expects*, or needs to *know*, or might be *interested* in, this can be seen as intrinsically linked to a sensitivity to another person's mental states.

xv. Crucially, most children with autism are unimpaired at understanding how *physical* representations (such as drawings, photos, maps, and models) work, even while they have difficulty understanding *mental* representations (such as beliefs) [Charman and Baron-Cohen, 1992, 1995; Leekam and Perner, 1991; Leslie and Thaiss, 1992].

xvi. They are also unimpaired on logical reasoning (about the physical world) even though they have difficulty in psychological reasoning (about the mental world) [Scott and Baron-Cohen, 1996).

This long list of experiments provides strong evidence for children with autism lacking the normal consciousness of the mental. For this reason, autism can be conceptualized as involving degrees of *mindblindness* (Baron-Cohen, 1990, 1995).

It is important to mention that a minority of children or adults with autism pass first-order false belief tests. First-order tests involve inferring what one person thinks. However, these individuals often fail second-order false belief tests (Baron-Cohen, 1989b), that is, tests of understanding what one character thinks another character thinks. Such second-order reasoning is usually understood by normal children of 5-6 years of age, and yet these tests are failed by individuals with autism with a mental age above this level.

We can therefore interpret these results in terms of there being a *specific developmental delay* in mind-reading at a number of different points (Baron-Cohen, 1991c). Some individuals with autism who are very high functioning (in terms of IQ and language level), and who are usually adults, may pass even second-order tests (Bowler, 1992; Ozonoff, Pennington, and Rogers, 1991; Happe, 1993). Those who can pass second-order tests correspondingly also pass the appropriate tests of understanding figurative language (Happé, 1993). However, their deficit shows up on tests of adult mind-reading (Baron-Cohen, Jolliffe, Mortimore, and Robertson, 1997). Thus, being able to pass a test designed for a 6 year old when you are an adult may mask persisting mind-reading deficits by ceiling effects.

In summary, there appears to be a relative lack of the normal consciousness of the mental in the majority of cases with autism. This finding has the potential to explain the social, communicative, and imaginative abnormalities that are diagnostic of the condition, since being able to reflect on one's own mental states (and those of others) would appear to be essential in all of these domains. This deficit has been found to correlate with real-life social skills, as measured by a modified version of the Vineland Adaptive Behaviour Scale (Frith, Happé, and Siddons, 1994). In the next section, we turn to consider the origins of this cognitive deficit.

3.2 *The brain basis of our consciousness of the mental*

One possibility arising from these studies is that there may be a particular part of the brain which in the normal case is responsible for our consciousness of mental states, and which is specifically impaired in autism. If this view is correct, the assumption is that this may be for genetic reasons, since autism appears to be strongly heritable (see Bolton and Rutter, 1990). The idea that the development of our consciousness of the mental is under genetic/biological control in the normal case is consistent with evidence from cross-cultural studies: Normally developing children from markedly different cultures seem to pass tests of 'mind-reading' at roughly the same ages (Avis and Harris, 1991).

Quite which parts of the brain might be involved in this is not yet clear, though candidate regions include right orbito-frontal cortex, which is active when subjects are thinking about mental state terms during functional imaging using SPECT (Baron-Cohen, Ring, et al., 1994); and left medial frontal cortex, which is active when subjects are drawing inferences about thoughts whilst being PET scanned (Fletcher et al., 1995; Goel et al., 1995). Other candidate regions include the superior temporal sulcus and the amygdala (for reasons explained below). These regions may form parts of a neural *circuit* supporting theory of mind processing (Baron-Cohen and Ring, 1994).

3.3 *Developmental origins of our consciousness of the mental*

In an influential article, Alan Leslie (1987) proposed that in the normal case, the developmental origins of mind-reading (or 'theory of mind') lie in the capacity for pretense; and that in the case of children with autism, the developmental origins of their mindblindness lies in their inability to pretend. In his model, pretense was the 'crucible' for theory of mind, as both involved the same computational complexity. Thus (according to Leslie), in order to understand that someone else might *think* "This banana is real," or *pretend* "This banana is real," the child would need to be able to represent the agent's *mental attitude* towards the proposition — since the only difference between these two states of affairs *is* the person's mental attitude. One idea, then, is that consciousness of the mental is first evident from about 18-24 months of age, in the normal toddler's emerging pretend play.

However, there is some evidence that this aspect of consciousness might have even earlier developmental origins. Soon after the first demonstrations of mindblindness in autism, Marian Sigman and her colleagues at UCLA also reported severe deficits in *joint attention* in children with autism (Sigman, Mundy, Ungerer, and Sherman, 1986). Joint attention refers to those behaviors produced by the child which involve monitoring or directing the target of attention of another person, so as to coordinate the child's own attention with that of somebody else (Bruner, 1983). Such behaviors include the pointing gesture, gaze-monitoring, and showing gestures, most of which are absent in most children with autism.

This was an important discovery because joint attention behaviors are normally well-developed by 14 months of age (Scaife and Bruner, 1975; Butterworth, 1991), so their absence in autism signifies a very early-occurring deficit. This was also important because the traditional mind-reading skills referred to above are mostly those one would expect to see in a 3-4 year old normal child. Deficits in these areas cannot therefore be the developmentally earliest signs of autism, since we know that autism is present from at least the second year of life (Rutter, 1978), if not earlier.

Implicit in the idea of joint attention deficits in autism was the notion that these might relate to a failure to appreciate other people's point of view (Sigman et al, 1986). Bretherton, McNew, and Beeghly-Smith (1981) had also suggested joint attention should be understood as an "implicit theory of mind" — or an implicit awareness of the mental. Baron-Cohen (1989c, d, 1991b) explicitly argued that the joint attention and mind-reading deficits in autism were no coincidence, and proposed that joint attention was a *precursor* to the development of mind-reading. In one study (Baron-Cohen, 1989c), young children with autism (under 5 years old) were shown to produce one form of the pointing gesture (imperative pointing, or pointing to request) whilst failing to produce another form of pointing (declarative pointing, or pointing to share interest).

This dissociation was interpreted in terms of the declarative form of pointing alone being an indicator of the child monitoring another person's mental state — in this case, the mental state of "interest" or "attention." More recent laboratory studies have confirmed the lack of spontaneous gaze-monitoring (Leekam, Baron-Cohen, Brown, Perrett, and Milders, 1997; Phillips, Baron-Cohen, and Rutter, 1992; Phillips, Gomez, Baron-Cohen, Riviere, and Laa, 1995). Early diagnosis studies have also borne this out (Baron-Cohen, Allen, and Gillberg, 1992; Baron-Cohen, Cox, Baird, Swettenham, Drew, Nightingale, and Charman, 1996). The demonstration of a joint attention deficit in autism, and the role that the superior temporal sulcus in the monkey brain plays in the monitoring of gaze-direction (Perrett et al., 1985) has led to the idea that the superior temporal sulcus may be involved in the development of our consciousness of the mental (Baron-Cohen, 1994, 1995; Baron-Cohen and Ring, 1994). Brothers (1990) also reviews evidence suggesting the amygdala contains cells sensitive to gaze and facial expressions of mental states. A recent neuroimaging study using fMRI confirms the role of the amygdala in normal mind-reading, and its under-activity in autism (Baron-Cohen, Ring, Williams, Wheelwright, Bullmore, and Simmons, 1999).

4. Conclusions

Autism may give us an important clue that the brain in fact allows at least *two distinct kinds of conscious experience*: consciousness of the physical (e.g., seeing an object) on the one hand, and consciousness of the mental (e.g., thinking about seeing an object) on the other. The latter is likely to be parasitic on the former, and whilst the former involves direct stimulation of perceptual systems, the mechanisms underlying the latter are still relatively unknown. As our understanding of the neurobiology of autism unfolds, so also our understanding of this second-order level of consciousness should too.

Notes

I was supported by grants from the Medical Research Council (UK), the Wellcome Trust, and the Gatsby Foundation during the period of this work. Parts of this chapter appeared in Cohen and Volkmar (Eds., 1996). Peter Grossenbacher gave valuable feedback on the first draft of this chapter.

1.   In the following list of studies, all of the tests mentioned are at the level of a normal 4 year old child.

References

Avis, J. & Harris, P. 1991. Belief-desire reasoning among Baka children: evidence for a universal conception of mind. *Child Development*, 62, 460-467.
Baron-Cohen, S. 1987. Autism and symbolic play. *British Journal of Developmental Psychology*, 5, 139-148.
Baron-Cohen, S. 1988. Social and pragmatic deficits in autism: cognitive or affective? *Journal of Autism and Developmental Disorders*, 18, 379-402.
Baron-Cohen, S. 1989a. Are autistic children behaviourists? An examination of their mental-physical and appearance-reality distinctions. *Journal of Autism and Developmental Disorders*, 19, 579-600.

Baron-Cohen, S. 1989b. The autistic child's theory of mind: a case of specific developmental delay. *Journal of Child Psychology and Psychiatry*, 30, 285-298.

Baron-Cohen, S. 1989c. Perceptual role-taking and protodeclarative pointing in autism. *British Journal of Developmental Psychology*, 7, 113-127.

Baron-Cohen, S. 1989d. Joint attention deficits in autism: towards a cognitive analysis. *Development and Psychopathology*, 1, 185-189.

Baron-Cohen, S. 1990. Autism: a specific cognitive disorder of "mind-blindness." *International Review of Psychiatry*, 2, 79-88.

Baron-Cohen, S. 1991a. Do people with autism understand what causes emotion? *Child Development*, 62, 385-395.

Baron-Cohen, S. 1991b. Precursors to a theory of mind: Understanding attention in others. In A. Whiten (Ed) *Natural theories of mind*. Oxford: Basil Blackwell.

Baron-Cohen, S. 1991c. The development of a theory of mind in autism: deviance and delay? *Psychiatric Clinics of North America*, 14, 33-51.

Baron-Cohen, S. 1992. Out of sight or out of mind: another look at deception in autism. *Journal of Child Psychology and Psychiatry*, 33, 1141-1155.

Baron-Cohen, S. 1994. How to build a baby that can read minds: Cognitive mechanisms in mindreading. *Cahiers de Psychologie Cognitive/ Current Psychology of Cognition*, 13(5), 513-552.

Baron-Cohen, S. 1995. *Mindblindness: An essay on autism and theory of mind*. MIT Press/ Bradford Books.

Baron-Cohen, S., Allen, J. & Gillberg, C. 1992. Can autism be detected at 18 months? The needle, the haystack, and the CHAT. *British Journal of Psychiatry*, 161, 839-843.

Baron-Cohen, S., Campbell, R., Karmiloff-Smith, A., Grant, J. & Walker, J. 1995. Are children with autism blind to the mentalistic significance of the eyes? *British Journal of Developmental Psychology*, 13, 379-398.

Baron-Cohen, S., Cox, A., Baird, G., Swettenham, J., Drew, A., Nightingale, N. & Charman, T. 1996. Psychological markers of autism at 18 months of age in a large population. *British Journal of Psychiatry*, 168, 158-163.

Baron-Cohen, S. & Cross, P. 1992. Reading the eyes: evidence for the role of perception in the development of a theory of mind. *Mind and Language*, 6, 173-186.

Baron-Cohen, S. & Goodhart, F. 1994. The "seeing leads to knowing" deficit in autism: the Pratt and Bryant probe. *British Journal of Developmental Psychology*, 12, 397-402.

Baron-Cohen, S., Jollife, T., Mortimire, C. & Robertson, M. 1997. Another advanced test of theory of mind: evidence from very high functioning adults with autism or Asperger Syndrome. *Journal of Child Psychology and Psychiatry*, 38, 813-822.

Baron-Cohen, S., Leslie, A.M. & Frith, U. 1985. Does the autistic child have a 'theory of mind'? *Cognition*, 21, 37-46.

Baron-Cohen, S., Pickett, A., Gilbert, A., Rohrer, J. & Jolliffe, T. 1996. Can adults with autism or Asperger Syndrome read the mind in the face? Unpublished ms, University of Cambridge.

Baron-Cohen, S. & Ring, H. 1994. A model of the mindreading system: neuropsychological and neurobiological perspectives. In Mitchell, P. & Lewis, C. (Eds) *Origins of an Understanding of Mind*. Lawrence Erlbaum Associates.

Baron-Cohen, S. Ring, H., Moriarty, J., Shmitz, P., Costa, D. & Ell, P. 1994. Recognition of mental state terms: a clinical study of autism, and a functional neuroimaging study of normal adults. *British Journal of Psychiatry*, 165, 640-649.

Baron-Cohen, S., Ring, H., Williams, S., Wheelwright, S., Bullmore, E., Brammer, M. & Simmons, A. 1999. Social intelligence in the normal and autistic brain: an fMRI study. *European Journal of Neuroscience*, 11, 1891-1898.

Baron-Cohen, S., Spitz, A. & Cross, P. 1993. Can children with autism recognize surprise? *Cognition and Emotion*, 7, 507-516.

Baron-Cohen, S., Tager-Flusberg, H., and Cohen, D. J. Eds. 1993. *Understanding other minds: perspectives from autism*. Oxford University Press.

Baron-Cohen, S., Tager-Flusberg, H., and Cohen, D.J. , Eds. 2000. *Understanding other Minds: Perspectives from developmental cognitive neuroscience*. 2nd Edition. Oxford University Press.

Bloch, N. 1995. On a confusion about the function of consciousness. *Behavioural and Brain Sciences*, 18, 227-287.

Bolton, P. & Rutter, M. 1990. Genetic influences in autism. *International Review of Psychiatry*, 2, 67-80.

Bowler, D. M. 1992. Theory of Mind in Asperger Syndrome. *Journal of Child Psychology and Psychiatry*, 33, 877-893.

Brentano, F. von. 1874/1970. *Psychology from an empirical standpoint*. ed. O. Kraus trans. L. L. MacAllister. London: Routledge, and Kegan Paul.

Bretherton, I., McNew, S., & Beeghly-Smith, M. 1981. Early person knowledge as expressed in gestural and verbal communication: when do infants acquire a "theory of mind"? in M. Lamb & L. Sharrod (Eds) *Infant social cognition*. Hillsdale, New Jersey, Lawrence Erlbaum Associates. pp. 333-374.

Bruner, J. 1983. *Child's talk: learning to use language*. Oxford: Oxford University Press.

Butterworth, G. 1991. The ontogeny and phylogeny of joint visual attention. In A. Whiten (Ed) *Natural theories of mind*. Oxford: Basil Blackwell.

Charman, T. and Baron-Cohen, S. 1992. Understanding beliefs and drawings: a further test of the metarepresentation theory of autism. *Journal of Child Psychology and Psychiatry*, 33, 1105-1112.

Charman, T. & Baron-Cohen, S. 1995. Understanding models, photos, and beliefs: a test of the modularity thesis of metarepresentation. *Cognitive Development*, 10, 287-298.

Cosmides, L. 1989. The logic of social exchange: has natural selection shaped how humans reason? Studies with the Wason selection task. *Cognition*, 31, 187-276.

Crick, F. 1994. *The astonishing hypothesis*. London: Simon and Schuster.

Dennett, D. 1978. *Brainstorms: philosophical essays on mind and psychology*. Harvester Press, USA.

Dennett, D. 1993. *Consciousness explained*. Penguin Press.

Fletcher, P., Happé, F., Frith, U., Baker, S., Dolan, R., Frackowiak, R. & Frith, C. 1995. Other minds in the brain: a functional imaging study of "theory of mind" in story comprehension. *Cognition*, 57, 109-128.

Frith, U. 1989. *Autism: explaining the enigma*. Oxford: Basil Blackwell.

Frith, U. & Baron-Cohen, S. 1987. Perception in autistic children. In Cohen, D., Donnellan, A. & Paul, R. (Eds) *Handbook of autism and pervasive developmental disorders*, New York: Wiley and Sons.

THE PHYSICAL AND THE MENTAL 75

Frith, U. and Happé, F. 1994. Autism: beyond "theory of mind". *Cognition*, 50, 115-132.
Frith, U., Happé, F. & Siddons, F. 1994. Autism and theory of mind in everyday life. *Social Development*, 3, 108-124.
Goel, V., Grafman, J., Sadato, N. & Hallett, M. 1995. Modeling other minds. *Neuroreport*, 6, 1741-1746.
Gray, J. 1995. The contents of consciousness: a neuropsychological conjecture. *Behaviour and Brain Sciences*, 18, 659-722.
Happé, F. 1993. Communicative competence and theory of mind in autism: A test of Relevance Theory. *Cognition*, 48, 101-119.
Happé, F. 1994. An advanced test of theory of mind: Understanding of story characters' thoughts and feelings by able autistic, mentally handicapped, and normal children and adults. *Journal of Autism and Developmental Disorders*, 24, 129-154.
Hobson, R. P. 1984. Early childhood autism and the question of egocentrism. *Journal of Autism and Developmental Disorders*, 14, 85-104.
Jolliffe, T. & Baron-Cohen, S. 1997. Are people with autism and Asperger Syndrome faster than normal on the Embedded Figures Task? *Journal of Child Psychology and Psychiatry*, 38, 527-534.
Leekam, S., Baron-Cohen, S., Brown, S., Perrett, D. & Milders, M. In press. Eye-Direction Detection: a dissociation between geometric and joint-attention skills in autism. *British Journal of Developmental Psychology*, 15, 77-95.
Leekam, S. & Perner, J. 1991. Does the autistic child have a metarepresentational deficit? *Cognition*, 40, 203-218.
Leslie, A. M. 1987. Pretence and representation: the origins of "theory of mind". *Psychological Review*, 94, 412-426.
Leslie, A.M. & Frith, U. 1988. Autistic children's understanding of seeing, knowing, and believing. *British Journal of Developmental Psychology*, 6, 315-324.
Leslie, A.M. & Thaiss, L. 1992. Domain specificity in conceptual development: evidence from autism. *Cognition*, 43, 225-251.
Lewis, V. & Boucher, J. 1988. Spontaneous, instructed and elicited play in relatively able autistic children. *British Journal of Developmental Psychology*, 6, 325-339.
Ozonoff, S., Pennington, B. & Rogers, S. 1991. Executive function deficits in high-functioning autistic children: relationship to theory of mind. *Journal of Child Psychology and Psychiatry*, 32, 1081-1106.
Penrose, R. 1989. *The emperor's new mind*. Oxford: Oxford University Press.
Perner, J., Frith, U., Leslie, A. M. & Leekam, S. 1989. Exploration of the autistic child's theory of mind: knowledge, belief, and communication. *Child Development*, 60, 689-700.
Perrett, D., Smith, P., Potter, D., Mistlin, A., Head, A., Milner, A. & Jeeves, M. 1985. Visual cells in the temporal cortex sensitive to face view and gaze direction. *Proceedings of the Royal Society of London*, B223, 293-317.
Phillips, W., Baron-Cohen, S. & Rutter, M. 1992. The role of eye-contact in the detection of goals: evidence from normal toddlers, and children with autism or mental handicap. *Development and Psychopathology*, 4, 375-383.
Phillips, W., Baron-Cohen, S., & Rutter, M. 1998. Understanding intention in normal development and in autism. *British Journal of Developmental Psychology*, 16, 337-348.

Phillips, W., Gomez, J-C., Baron-Cohen, S., Riviere, A. & Laa, V. 1995. Treating people as objects, agents, or subjects: How young children with and without autism make requests. *Journal of Child Psychology and Psychiatry*, 36, 1383-1398.
Reed, T. & Peterson, C. 1990. A comparative study of autistic subjects' performance at two levels of visual and cognitive perspective taking. *Journal of Autism and Developmental Disorders*, 20, 555-568.
Rutter, M. 1978. Diagnosis and definition. In Rutter, M. & Schopler, E. (eds) *Autism: A reappraisal of concepts and treatment*. New York: Plenum Press.
Scaife, M. & Bruner, J. 1975. The capacity for joint visual attention in the infant. *Nature*, 253, 265-266.
Scott, F. & Baron-Cohen, S. 1996. Logical, analogical, and psychological reasoning in autism: a test of the Cosmides theory. *Development and Psychopathology*, 8, 235-246.
Shah, A. 1988. *Visuospatial Islets of Abilities and Intellectual Functioning in Autism*. Unpublished Ph.D. thesis, University of London.
Shah, A. & Frith, U. 1983. An islet of ability in autism: a research note. *Journal of Child Psychology and Psychiatry*, 24, 613-620.
Shah, A. & Frith, U. 1993. Why do autistic individuals show superior performance on the block design test? *Journal of Child Psychology and Psychiatry*, 34, 1351-1364.
Sigman, M., Ungerer, J., Mundy, P. & Sherman, T. 1987. Cognition in autistic children In Cohen, D., Donnellan, A. & Paul, R. (eds) *Handbook of Autism and Pervasive Developmental Disorders*, New York: Wiley and Sons.
Sigman, M., Mundy, P., Ungerer, J. & Sherman, T. 1986. Social interactions of autistic, mentally retarded, and normal children and their caregivers. *Journal of Child Psychology and Psychiatry*, 27, 647-656.
Sodian, B. & Frith U. 1992. Deception and sabotage in autistic, retarded, and normal children. *Journal of Child Psychology and Psychiatry*, 33, 591-606.
Surian, L., Baron-Cohen, S. & Van der Lely, H. 1996. Are children with autism deaf to Gricean Maxims? *Cognitive Neuropsychiatry*, 1, 55-72.
Swettenham, J. 1996. Can children be taught to understand false belief using computers? *Journal of Child Psychology and Psychiatry*, 37, 157-166.
Tager-Flusberg, H. 1992. Autistic children's talk about psychological states: deficits in the early acquisition of a theory of mind. *Child Development*, 63, 161-172.
Tager-Flusberg, H. 1993. What language reveals about the understanding of minds in children with autism. In Baron-Cohen, S., Tager-Flusberg, H. & Cohen, D. J. (eds) *Understanding other minds: perspectives from autism*. Oxford University Press.
Wellman, H. 1990. *Children's Theories of Mind*. Harvard: MIT Press.
Wimmer, H. & Perner, J. 1983. Beliefs about beliefs: Representation and constraining function of wrong beliefs in young children's understanding of deception. *Cognition*, 13, 103-128.
Wing, L. 1976. *Early Childhood Autism*. Pergamon Press.
Wing, L., Gould, J., Yeates, S. R. & Brierley, L. M. 1977. Symbolic play in severely mentally retarded and in autistic children. *Journal of Child Psychology and Psychiatry*, 18, 167-178.

Section II

Mental Content and Action

Introduction

The content of awareness fluctuates according to what is going on in the mind. At one point in time, it may be a perceptual experience which fills awareness, while a moment later this could change to the guidance of an intricate body movement, or day dreaming, the creation of mental images not tethered to currently available sensations. The ongoing stream of conscious content can be influenced by personal intent, or volition, but often proceeds into unintended topics. The phrase "attending a class meeting" means getting your body into the classroom, a behavior that requires physical presence. But sitting in a room does not guarantee that *you* are fully present in the class, devoting undivided attention to the class discussion. As instructed, your mind *might* be focused on the discussion topic. Or, you might instead be focusing on some other current event, for example, the appearance of a nearby classmate. Or you could be "miles away," thinking of something from your past, or a possible future. These examples show how the content of consciousness is largely determined by *attention*, the ongoing direction of mind.

Each of us might admit that we are often less than 100% successful in keeping attention focused on the task at hand. It is not uncommon for attention be drawn this way and that against our will. As an exercise in phenomenology, it can be instructive to *become aware of how your attention is currently directed*. Try this now by mentally noting what passes through as current content.

||••||

What determines the contents of conscious awareness and the direction of attention? In finding an answer to this question, we must consider some of the complex ways that mental representations interact with systems that control attention. Whereas the previous section of this book dealt with boundaries which restrict the scope of conscious experience, this section concerns those brain functions which are required for consciousness.

In Chapter 4, Stephen Kosslyn explores the neural basis for conscious phenomena in visual imagery. Based on findings from functional brain activation studies, he boldly suggests a new principle for predicting which neurally-represented information can become conscious.

The conscious content in human experience is at times directed by events in the world around us. In Chapter 5, Barry Stein and Mark Wallace explain those neural mechanisms which enable the capture of attention by stimuli such as the sound of a barking dog or the visual flicker of a flying bird.

Subjectively, it sometimes seems possible to voluntarily control the content of consciousness. Conscious awareness can certainly be influenced by any ongoing actions being performed by the subject. In Chapter 6, Yves Rossetti reports on conscious and nonconscious sensory processing in healthy people, and patients with brain damage, while they make controlled bodily movements.

When awake, the living human brain produces conscious experience which ranges over the domains of imagery, perception, and action. Each chapter in this section reveals how one of these domains of conscious experience depends on brain activity.

CHAPTER 4

Visual Consciousness

Stephen M. Kosslyn
Harvard University

Mental imagery has had a long and checkered history, perhaps in part because it is so tightly tied to consciousness. Indeed, part of the definition of mental imagery is an experience of "seeing," "hearing," or otherwise being aware of something that is not in fact present. The fact that imagery is accessible only via introspection is, of course, fraught with problems. For example, if two people disagree, there is no easy way to resolve the disagreements; the contents of consciousness are not open to public observation. This problem began to be solved when researchers devised ways to measure subtle behavioral consequences of imagery, which allowed them to "externalize" properties of imagery. For example, researchers recorded the amount of time subjects required to rotate (e.g., Shepard & Cooper, 1982), scan (Kosslyn, 1973), and otherwise manipulate images (for reviews, see Kosslyn, 1980, 1994). Even so, my sense is that the study of mental imagery did not become entirely respectable (if it has indeed attained this status!) until brain-scanning studies allowed the neural footprints of imagery to be captured, providing the kind of "public" data that has long been preferred in science (for a review, see Thompson & Kosslyn, in press).

However, the fact that imagery is accessible only via introspection is actually a virtue in some respects. The fact that we are only aware of mental images when we have the experience of "seeing with the mind's eye" or the like provides an opportunity to study the relation between the qualities of visual experience, the underlying information processing, and the neural bases of that information processing — which in turn reflect back on the nature of visual consciousness itself. The sow's ear can be made into a silk purse. Because much more is known about visual mental imagery, and the neural bases of visual processing, than other modalities, I will focus here purely on visual imagery; I expect that parallel cases could be made for other modalities.

This paper has three parts. First, I review evidence that qualities of our conscious experience of visual imagery do in fact reflect properties of information processing. Second, I show how these properties of information processing can be explained in a straightforward way by reference to properties of the brain. And third, I then speculate about why we are conscious of some aspects of neural processing but not others.

1.   Introspection reveals properties of information processing

Visual mental imagery is a complex phenomenon that has many distinct facets (e.g., for reviews, see Kosslyn, 1980, 1994). I cannot hope to address all of the properties or uses of imagery in this chapter, and hence will restrict myself to a few findings that reflect the nature of imagery itself. Specifically, I will summarize briefly some results that indicate that objects in visual mental images have spatial extent, that there are limits to their spatial extent, and that they have a grain.

1.1 Spatial extent

First, let me try to illustrate the kinds of experiences we will consider. Can you recall how many windows there are in your home? Most people report that in order to answer this question, they visualize each room and mentally scan the walls, counting each window they "see." (I put quotes around the word *see* because their eyes are closed, and they are not actually seeing the windows.) One seems to be scanning across walls and along surfaces. Such introspections suggest that the representations underlying the experience of imagery (which are used in information processing) embody spatial extent.

Here is another illustration: Try to visualize a horse, standing so that it is facing to the right. Now imagine that you are "mentally staring" at the place where its tail meets its body. Having fixated your mental gaze in this location, now decide whether the animal's ears protrude above the top of its skull. Once you have tried this, repeat the exercise, but with one change: Now begin by "mentally staring" at the center of its body, and then consider the ears. And finally, try this again, but now begin by "mentally staring" at the head before shifting your attention to the ears.

Most people report that in the first case, they have the experience of scanning along the horse's back, up its neck, and to its head, at which point they "see" the ears. In the second case, people report having to scan a shorter distance along the flank and up the neck to get to the ears. And in the third case, they report that almost no scanning at all is necessary.

This second example is interesting because it lends itself to an experimental technique initially developed by Kosslyn (1973). In this technique, the time to make the decision in the various conditions is measured. Time, in this case, is being treated as a kind of "mental tape measure." The logic is that if people really do have to scan farther distances in one situation than another, this additional scanning should be reflected by additional amounts of time. And in fact, if we had measured the time you required to answer questions in the examples above, you would have required more time when you had to scan greater distances over the object (for reviews, see Kosslyn, 1980, 1994).

In the actual experiments, subjects memorized drawings and later closed their eyes and mentally fixated on a given location. They then were asked to "find" another location on the image, and to press one button as soon as they could find it and another if they could not. The time to scan to a second location increased linearly with the distance scanned.

This early research came under attack from two quarters. First, one group of researchers (e.g., Intons-Peterson, 1983; Goldston, Hinrichs, & Richman, 1985) claimed that the data were produced because subjects complied with demand characteristics. That is, they claimed that the experimenters conveyed their expectations to subjects, who — good soldiers that they are — were only too happy to try to please the experimenters by producing the desired results. Intons-Peterson (1983), for example, conducted an experiment in which subjects scanned visualized maps or perceived maps, and told the experimenters different predictions about the expected outcomes. In one case, experimenters were told that subjects should scan the perceived map faster than the imaged one, and in another case the experimenters were given the reverse "predictions." And in fact, depending on what the experimenters believed, the subjects required different amounts of time to respond. Intons-Peterson recorded the way the experimenters delivered the instructions, and reports that the experimenters were more animated when delivering instructions for the condition in which they expected the subjects to scan faster, which led the subjects to respond more vigorously. The instructions only altered the overall times, not the slopes (the time to scan each increment of distance), and it is the slopes that I take to reflect the spatial nature of the underlying representation. Nevertheless, Intons-Peterson argued that the increases in time with distance (which is the important aspect of the results) could have been a consequence of such demand characteristics.

Jolicoeur and Kosslyn (1985) examined this possibility in detail. In one experiment, subjects visualized and scanned maps that were in black-and-white or in color. We adopted the technique used by Intons-Peterson, and varied the "predictions" given to the experimenters. We told the experimenters either that the subjects should scan the black-and-white map faster than the colored one (because it was less complex), or that they should scan the colored map faster (because it would be more vivid). The experimenters were told that these manipulations should affect scanning per se, and hence alter the slope of the function relating response time to distance. In no case, however, did such expectations affect the slopes. In fact, we failed to replicate Intons-Peterson, possibly because we always use written instructions and thereby eliminated a major avenue whereby expectations are communicated to subjects.

In addition, Pylyshyn (1981) claimed that the scanning results were due not to experimenter expectancy effects, but rather to implicit task demands. That is, if subjects believe that the task is to scan an image, and believe that scanning time should increase with distance (based, presumably, on perceptual experience), then they may simply take longer to respond when they believe they should have taken more time to scan. This increase in time would reflect an implicit belief, which could be stored verbally (or as a "proposition," which captures the gist of the content of the verbal expression). Such beliefs need not be conscious, and thus simply interviewing the subjects would not allow one to determine whether this theory is correct. Instead, Jolicoeur and Kosslyn (1985) asked subjects to visualize objects and focus on one side, and then asked them to determine whether the object had a given property. For example, they might be asked to visualize a honeybee and mentally focus on its stinger, and then be asked whether the insect has a dark head. We stressed that the subjects did not need to use the image to perform the task, and selected the items so that half involved subtle visual properties (which we expected to require imagery) and half involved more obvious properties (which we expected not to require imagery; see Chapter 9 of Kosslyn, 1980). Half the time the subjects focused on the end of the object where the probed property was located, and half the time they focused on the opposite side. As expected, the subjects required more time when they had to scan to the other side to verify a subtle property, but not for an obvious property. Not only did we not require the subjects to use imagery in this task, but we never mentioned the word "scanning" nor did we discuss any related concepts.

Probably the best counter to Pylyshyn's claims, however, is a series of experiments reported by Finke and Pinker and their colleagues (Finke & Pinker, 1982, 1983; Pinker, Choate, & Finke, 1984). In these experiments, subjects first saw a set of random dots. The dots were removed, and an arrow appeared at a random location. The subjects were asked to decide whether the arrow would have pointed to one of the dots, if the dots had remained on the screen. Not only did the response times increase with the distance from the arrow to the target dot, but the magnitude of increase was about the same as that observed in the earlier scanning experiments. In these experiments, however, no mention was made of imagery at all, let alone of scanning.[1]

In short, there is good evidence that our introspection about this aspect of imagery inform us about properties of the underlying functional representation. Namely, image representations embody spatial extent.

1.2 *Limits on spatial extent*

Now try this: Visualize a rabbit, as if it was seen from the side and far off in the distance. Now imagine that you are walking toward it, so that it seems to loom larger in your image. As you seem to approach the object in your image, do you eventually get so close that all of the edges of the rabbit cannot all be "seen" all at once? Now try this same exercise with an elephant. Visualize it off in the distance, and imagine that you are walking close to it, and "stop" when the edges of the animal start to overflow the image. Did one of the animals seem closer to you at the point when it seemed to begin to overflow? I have asked this question to hundreds of people as part of a demonstration, and the vast majority of people reply that the rabbit seems closer than the elephant at the point it overflows. Introspectively, it appears as if the objects overflow the "space" available for images, almost as if a picture on a screen can only subtends a certain visual angle before it overflows a screen.

If there is in fact the equivalent of a "mental screen" with a fixed size, then we would expect objects to seem to overflow it at the same visual angle. To test this hypothesis, we performed the "mental walk" task carefully. We showed subjects pictures (for example, of black rectangles that had random heights and widths), and asked them to perform the "imagery walk" task with the rectangles and then indicate the distance at which the stimuli overflowed by positioning a tripod apparatus from a wall; we then measured the distance from the tripod to the wall. And in fact, we found that the larger the object, the farther away it seems when it seems to begin to overflow the image. In addition, for stimuli with well-defined edges (such as rectangles or drawings of animals), the stimuli overflowed at a constant angle. Finally, as will become pertinent shortly, when the stimuli were actually physically present (pasted to a wall), and subjects performed the tasks on what they saw, very similar results were obtained as were found in the corresponding imagery condition (for additional details on these results, see Kosslyn, 1978, and Kosslyn, 1994).

In sum, the introspection that objects in images can only seem to subtend a certain visual angle before "overflowing" does in fact predict behavioral data.

1.3 Resolution limitations in imagery

Again, consider what you experience when you try the following task. Visualize a butterfly as if it were standing on one of your fingers, held at arm's length. Now, what color is its head? Compare this experience to what happens when you visualize the butterfly as close as possible while not overflowing. Most people report that they had to "zoom in" to see the butterfly's head in the first situation, but not the second. And in fact, when subjects are asked to begin by visualizing an object so that it subtends a small visual angle, they take more time to evaluate its properties than when they are asked to begin by visualizing an object at a larger size (e.g., see Kosslyn, 1975).

In addition, when subjects were asked to visualize objects and evaluate properties, they found larger properties faster than smaller ones — even though the smaller ones were more strongly associated with the object (Kosslyn, 1976). For example, when they used imagery, subjects could affirm that a mouse has a back faster than that a mouse has whiskers. In contrast, when they did not use imagery, the results were in the opposite direction, with association strength determining the response times. It is clear, then, that the size of a part or object affects the time to evaluate objects in images.

In short, there is good evidence that our introspections about properties of objects in visual mental images reflect properties of the underlying representation of those objects. Just as is evident to us introspectively, the representations appear to have a spatial extent, a fixed spatial extent, and a grain.

2. Neural bases of image representations

The behavioral results summarized in the previous section were inspired in part by introspection and in part by a metaphor. The metaphor likened images to displays on a cathode ray tube, which were generated by a computer program (see Kosslyn, 1975). This metaphor has the virtue of suggesting that images are spatial representations that occur in a medium that has a limited spatial extent and a grain. However, it has a major drawback: as a theory, it is clearly incorrect. There is no actual screen in one's head. And if there were, who would look at it? The representation of images must correspond, in some way, to neural activity; the brain is the machine that performs human information processing, and hence properties of information processing must reflect (directly or indirectly) properties of the brain.

Imagery is a cognitive ability that can be easily related to brain function because it shares mechanisms with visual perception (e.g., for reviews see Farah, 1988; Kosslyn, 1994), and we know an enormous amount about the neural substrate of vision. In addition, imagery clearly relies on memory, and we also know a lot about the neural mechanisms underlying memory (e.g., Squire, 1987). One reason we know so much about vision and memory is that nonhuman primates have similar systems, and so animal models can be studied to understand these abilities. Animal models are not available for many other cognitive abilities, such as language. In the following two sections I outline some ways in which findings about the neural substrates of vision and memory can illuminate the nature of human visual mental imagery, which in turn leads to some conjectures about the neural bases of visual consciousness.

One of the fundamental facts about the visual system is that it is composed of many areas (32 cortical areas, at last count; see Felleman & Van Essen, 1991). Approximately 15 of these areas are retinotopically organized. That is, the pattern of stimulation on the retina is projected onto the cortical surface of these areas, which thus approximately preserve the spatial layout of the stimulus. Each of these areas has some interesting properties, including:

2.1 *Spatial extent*

The representations (patterns of neural activation) within these areas are spatial, functionally as well as physically. That is, the fact that the representations are physically spatial is irrelevant if the processes that access and use the representations do not extract information from variations in the spatial properties. For example, letters of the alphabet are spatial patterns, but the spatial aspects do not play a key role in how they convey information: a large letter has the same interpretation as a small one. In contrast, the pattern v could represent a picture of a flock of geese heading south for the winter. In this case, a larger pattern, such as V, would indicate more geese, or geese spread over a larger area; the spatial patterns of the representation do in fact convey information. Similarly, the spatial properties in retinotopic areas convey information; these representations are spatial both physically and functionally.

2.2 *Limited spatial extent*

Similarly, the retinotopic areas are capable of representing only a limited spatial extent. The brain evolved to process input from the eyes, which subtend only a limited visual angle. Moreover, greater amounts of tissue are devoted to the input from the high-resolution, central part of the eye, the fovea. In addition, the receptive fields become larger for neurons that register input farther from the fovea, more towards the periphery. Thus, resolution decreases gradually from the high-resolution central region. All of these properties dovetail neatly with the results reported by Kosslyn (1978), as well as the phenomenology reported by many people when they observe objects gradually overflowing an image.

2.3 *Limited resolution*

The neurons in the retinotopically organized regions respond when stimuli are presented in specific places in the visual field (this is part and parcel of their being organized retinotopically). If two stimuli are close enough together, however, they fall into the same receptive field, and may not be discriminated. This property of "spatial summation" produces a resolution limitation.

     In short, the retinotopically organized regions of the brain have three properties that might underlie the aspects of imagery reviewed earlier. These aspects of imagery reflect properties of the representations used in information processing, and properties of the brain must underlie (directly or indirectly) such functional properties. This observation led my colleagues and me to hypothesize that visual mental images are patterns of activation in at least some of the retinotopically organized visual areas.

     The hypothesis that imagery arises when retinotopically organized parts of the visual system are activated has now received much support. I will focus on studies that rely on positron emission tomography (PET). Kosslyn et al. (1993) asked subjects to visualize letters of the alphabet. They found that when subjects visualized upper case letters in grids, more blood flow occurred in primary visual cortex (Area 17, which is known to be retinotopically organized in humans) than when they merely viewed these grids without visualizing. Primary visual cortex is the largest retinotopically organized area in the human brain, and thus activation in it is easiest to detect.

In another experiment, Kosslyn et al. (1993) asked subjects to close their eyes and listen to cues of the following sort: "B.....curved." A letter was named, followed four seconds later by a cue. Prior to the experiment, the subjects learned the meanings of the cue words. In this example, "curved" required them to determine whether the upper case letter had any curved lines. The subjects participated in two conditions: In one, they visualized the letters as small as possible while still remaining visible; in the other, they visualized them as large as possible without overflowing. The subjects were to form the image of the letter as soon as they heard its name, and then were to maintain the image at the appropriate size until they heard the cue word.

The logic of this experiment was as follows: When subjects visualized the small image, the parts of retinotopically mapped areas that represent central, foveal stimulation should be more strongly activated than when subjects visualized the large images; greater spatial variation would need to be preserved in that region with the small image (the entire pattern is there, as opposed to empty space or a single segment with the large image), and hence more neural activation should occur. In contrast, when subjects visualized letters at the large size, the parts of retinotopically mapped areas that represent peripheral stimulation should be more activated than when subjects visualized letters at the small size; parts of the large images extend into these regions, whereas they are not stimulated by the small images.

And in fact, when blood flow induced by the large images was subtracted from that induced by the small images, the part of Area 17 that represents small, very central stimuli proved to be more activated; in contrast, when blood flow induced by the small images was subtracted from that induced by the large images, the part of Area 17 that represents peripheral stimuli was more activated.

Some researchers were not convinced by the results of the experiment just described. For example, some argued (informally, so far), that two different areas were activated — one by the large images and one by the small images. Moreover, the fact that the two conditions were compared to each other only indicated relative amounts of activation, it did not allow one to localize activation in the individual conditions. But there appeared to be a more serious problem with all of our imagery experiments: Roland and Gulyas (1994) reported that when they performed similar experiments, they found no evidence of activation of retinotopically organized areas.

We were greatly puzzled over the fact that Roland and his colleagues could not replicate our results, and thus looked very carefully at what they had actually done. Kosslyn and Ochsner (1994) suggested a number of reasons why they had failed to obtain the results we did, and Kosslyn, Thompson, Kim and Alpert (1995) tested one we found particularly plausible. We had noticed that Roland and his colleagues always compared blood flow in imagery conditions to blood flow in a resting baseline condition. Indeed, they asked subjects to close their eyes in darkness and to "visualize blackness" and then subtracted cerebral blood flow in this condition from that in the imagery condition. It seemed to us that Area 17 might be activated in this baseline condition, and thus subtracting activation in it from that in the imagery condition would remove all evidence of activation during imagery.

To test this idea, we had subjects participate in five conditions. They participated in two baseline conditions. One was the same resting baseline used by Roland and colleagues. The other required them to listen to stimuli of the following form, "anchor.... right-higher" and to press a response pedal as soon as they heard the spatial terms. The subjects participated in this "listening baseline" before having any idea about the nature of the experimental conditions. After participating in these conditions, they memorized a set of line drawings of common objects. The imagery conditions had trials of the following form: They first heard the name of one of the drawings and visualized it (with their eyes closed), and then heard a spatial descriptor, at which point they decided whether it characterized the drawing (e.g., was the right side of the drawing higher than the left?). Before each of the three imagery conditions the subjects studied a square, which subtended .25, 4, or 16 degrees of visual angle. They were told to form images, and hold them, at the size of the square during all the trials in that condition. Counterbalancing ensured that the same auditory stimuli occurred equally often in the listening baseline condition and in each of the three imagery conditions, and also ensured that images at each size were formed equally often in each of the three serial orders.

The most important results were as follows: First, when we simply compared the two baseline conditions, we found that Area 17 was more activated during the resting baseline than during the listening baseline. Thus, our original conjecture was correct. Second, when we compared the imagery trials to the resting baseline we found just what Roland and colleagues found: No activation in Area 17 (but instead evidence for parieto-occipital activation, in an area very close to the one reported by Roland & Friberg, 1985). Third, in contrast, when we compared the imagery trials to the listening baseline, we now found activation in primary visual cortex. But more than that, the small images activated the portion of this area that registers central stimuli that subtend small visual angles, the medium images activated cortex that registers objects that subtend medium sized visual angles, and the large images activated cortex that registers objects that subtend large visual angles.

These findings, then, appeared to resolve a lingering controversy about the PET findings (but, given the history of this work, it won't be surprising that contrary results were soon reported (see Mellet et al., 1998; Thompson & Kosslyn, 2000). Fortunately, many other studies have reported activation in Area 17 during visual mental imagery (e.g., Damasio et al., 1993; Le Bihan et al., 1993; Menon et al., 1993; Kosslyn et al., 1999), and it is unlikely such activation is merely a fluke. It is clear that in some tasks, imagery activates Area 17).

But what evidence do we have that imagery in fact relies on activation retinotopically organized brain areas? Three sorts: First, Farah, Soso and Dasheiff (1992) used the "mental walk" task described earlier to study the visual angle subtended by images in a patient before and after she had one occipital lobe removed (for medical reasons). The visual angle subtended by images shrank by about a half following the operation. This was only true along the horizontal extent, which was as expected because each retinotopically organized area typically includes only the opposite half-field. In contrast, the vertical extent was preserved after surgery, as expected because areas in each hemisphere include the entire vertical extent. We (unpublished data) have extended these results with a series of patients, and provided converging evidence that damage to primary visual cortex per se disrupts the ability to form sharp images in the affected visual field.

Second, Kosslyn, Thompson, Kim and Alpert (1996) reasoned that if activation of primary visual cortex gives rise to the representations of imagery used in information processing, then characteristics of the neural activity ought to be reflected in performance. That is, by varying size we had shown that characteristics of the image are reflected in patterns of neural activation, but now we wanted to go the other direction and show that properties of the brain predict properties of information processing. Thus, we analyzed the data from the 16 subjects tested by Kosslyn et al. (1993) in the original size-variation experiment. We now pooled the data over size, and normalized each person's brain to the same mean blood flow. We then examined the relative amount of flow in Area 17 (compared to the global mean), and correlated this measure with the mean time the person required to evaluate the images (which was recorded while the subject was being scanned). We found that the slowest subjects were in fact those who had the least amount of blood flow in Area 17 (r = .65). Third, Kosslyn et al. (1999) used transcranial magnetic stimulation (TMS) to impair processing in Area 17 (and probably at least part of Area 18) before participants engaged in an imagery task that had previously been shown to activate Area 17. This prior stimulation impaired performance on the imagery task, and did so to the same degree that it impaired performance on the analogous perceptual task.

In short, we have good evidence that imagery does arise, in at least some types of imagery tasks, when retinotopically organized areas of the visual system are activated (but, as reviewed by Mellet et al., 1998, not all types of imagery tasks engender such imagery; see also Thompson & Kosslyn, 2000). This suggests that some of the properties of these areas underlie properties of imagery that are evident to introspection. However, the fact that brain areas that have spatial extent, limited extent, and limited resolution give rise to imagery does not imply that these areas are directly responsible for our experiences during imagery. In the following section I consider how images are formed and then speculate about the origins of consciousness.

3.   The neural bases of consciousness: Some speculations

The argument so far is as follows: First, introspectively, objects in visual mental images seem to have spatial extent, but they can only be so large before overflowing, and if they are too small, parts are obscured. Second, these properties evident to introspection reflect functional properties of the internal representations that underlie imagery, as revealed by their effects on response times and judgments. Third, these properties of the representations used in information processing directly reflect properties of the neural structures in which such spatial representations occur. Having said all this, it is still possible that consciousness does not arise directly from the existence of representations in Area 17 for other topographically organized cortices, but rather is a consequence of how these representations are processed later in the processing stream.

Indeed, Crick and Koch (1995) have argued that consciousness does not arise from activity in primary visual cortex. The most compelling evidence they cite is that Area 17 can register properties of stimuli that we do not sense consciously. Very fine stripes are registered in Area 17 and can affect later processing even though all we are aware of seeing is homogenous gray; for example, viewing such sub-conscious gratings can cause an orientation-specific decrement in sensitivity to visible gratings (MacLeod & He, as summarized in Crick & Koch, 1995). I want to argue that consciousness is not simply a byproduct of activation in any one area, or even a set of areas, but rather occurs when processing going downstream must mesh with processing going upstream. To explain this hypothesis fully I must first outline more about the system in which visual mental images are formed and interpreted.

3.1 Six major components of imagery processing

In this section I briefly describe the major component processes used in the later phases of visual perception and visual mental imagery. PET results from our laboratory have implicated these components in both types of tasks (for a review, see Kosslyn, 1994; Kossly, Thompson & Alpert, 1997). In each case, I will describe the role of a subsystem in vision before turning to imagery. The architecture of the system is illustrated in Figure 4.1.

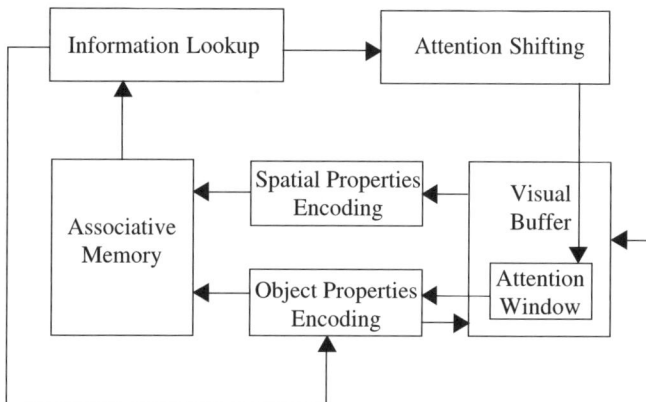

Figure 4.1. The major subsystems postulated by Kosslyn (1994) to underlie higher-level visual processing. See text for explanation.

3.1.1 *Visual buffer*

The visual buffer is a set of topographically organized areas of cortex. There are at least 15 such maps in the primate brain (e.g., for reviews, see Felleman & Van Essen, 1991, and Van Essen, Felleman, DeYoe, Olavarria, & Knierim, 1990). I conceive of these structures as forming a single functional structure that I call the visual buffer. The areas subsumed by this structure are localized in the occipital lobe.

The visual buffer corresponds to the array in the theory of Kosslyn (1980). The topographically mapped areas of cortex receive connections not only from the lower visual areas, but also from the higher ones. This feature of the neuroanatomy is consistent with the claim that a visual mental image is a pattern of activation in the visual buffer that is induced by stored information, as opposed to input from the eyes (which induces a pattern of activation during perception).

Kosslyn (1980) treated the visual buffer as a static structure, exactly analogous to an array in a computer. This clearly is overly simplistic. My present view is that the visual buffer itself performs much computation. I suspect that we do not store very complete information in long-term memory, and that when an image is generated the buffer itself must fill in many gaps in patterns. This filling-in process may rely on bottom-up processes that complete fragments that are collinear, fill in regions of the same color or texture, and so forth. This sort of processing would allow stored fragments to engender a more complete pattern.

3.1.2 *Attention window*

The visual buffer typically contains more information than can be processed during perception (there are more cells in these areas than there are projections to other visual areas; cf. Van Essen, 1985). Hence, some information must be given a high priority for further processing whereas other information must be placed in the background. The attention window selects a region within the visual buffer for detailed further processing. The size of the window in the visual buffer can be altered (cf. Larsen & Bundesen, 1978; Treisman & Gelade, 1980). Indeed, Larsen and Bundesen (1978) and Cave and Kosslyn (1989) showed that the time necessary to adjust the size of the attention window increases linearly with the amount of adjustment necessary.

In addition, the location of the attention window in the visual buffer can be shifted, independently of any overt attention shift. As noted earlier, people can scan visual mental images, even when their eyes are closed, and the farther they scan across the imaged object, the more time is required. However, we do not "bump into the edge" of the visual buffer when we scan; rather, we can scan to portions of objects that initially were "off screen" (see Kosslyn, 1980, for evidence). This can be accomplished if new portions of an image are introduced on one side of the visual buffer and the pattern is slid towards the opposite side (rather like an image on a TV screen as the camera scans over a scene). Similarly, when we "zoom in" on an imaged object, further details of the object become apparent. Thus, there must be a means of fixing a portion of a pattern in the attention window, and adding more details to the pattern as the window is expanded.

3.1.3 Object-properties encoding

A major anatomical pathway runs from the occipital lobes down to the inferior temporal lobes, which has been shown to be involved in the representation of object properties such as shape and color (e.g., Maunsell & Newsome, 1987; Mishkin, Ungerleider, & Macko, 1983; and Ungerleider & Mishkin, 1982). Indeed, the anterior portions of this system appear to be the site where visual memories are stored (for a review, see Kosslyn, 1994); during visual perception, input is compared to these stored representations, and if a match is found the object is recognized. This system receives the information that is selected by the attention window. Kosslyn (1994) divides this system into various subsystems, but this is not necessary for present purposes.

According to the theory, visual mental images of individual shapes are formed by activating visual memories top-down, and this process in turn induces a pattern of activity in the visual buffer. The areas that presumably are involved in storing visual memories are not topographically organized (e.g., Fujita, Tanaka, Ito & Cheng, 1992); thus, in order to make local geometric relations explicit, it is necessary to use such stored information to produce a representation in a spatial format.

The activation of a visual memory is but one component of visual image generation. We can create composite images, such as an image of the current American President shaking hands with George Washington; such images require combining stored memories in novel ways. In order to understand this ability, we need to consider additional components of the system.

3.1.4 *Spatial properties encoding*

A second major cortical pathway projects dorsally from the occipital lobes, up to the parietal lobes. This "dorsal system" is concerned with spatial properties, such as location, size, and orientation (see Maunsell & Newsome, 1987). Indeed, Ungerleider and Mishkin (1982) identify the ventral and dorsal systems as being concerned with "what" and "where," respectively. The dorsal system apparently receives information from the visual buffer at the same time as the ventral system.

The distinction between an object-properties-encoding and a spatial-properties-encoding system, embodied in the inferior temporal and posterior parietal lobes, respectively, receives support from experiments by Pohl (1973) and Ungerleider and Mishkin (1982). In these experiments, monkeys discriminated between patterns on food lids or between the locations of a small "landmark." When the animals' parietal lobes were removed, their performance on the landmark task was devastated, but they performed the pattern task well; this result is consistent with the idea that the parietal lobes are critically involved in encoding location. In contrast, when animal's inferior temporal lobes were removed, their performance on the pattern discrimination task was devastated, but they performed the location task well; this result has been taken to show that the inferior temporal lobes encode shape.

3.1.5 *Associative memory*

The simple fact that people can report from memory where furniture is placed in their living rooms indicates that the outputs from the dorsal and ventral systems are conjoined downstream. I infer an associative memory in which such associations are stored. During perception, the outputs from the ventral and dorsal systems are matched in parallel in associative memory to parts and relations of stored objects. If an object is seen close up (so that it is examined over the course of multiple eye fixations) or is seen from an unusual point of view or in impoverished circumstances (so that more than one glance is necessary to identify it), then associative memory will be used to build up a composite representation of the object and to identify it. The system converges on the identity of the object being viewed by finding the stored representation that is most consistent with the encoded parts and their spatial relations. When such evidence exceeds a threshold (which presumably can be varied, depending on context), identification occurs.

Associative memory plays a critical role in imagery for at least two reasons. First, this is where information is associated with an object's name. We often form images upon hearing the name of an object. Second, because associative memory integrates the outputs from the dorsal and ventral systems, it must contain representations of the structure of scenes and objects. To form an image that is composed of more than one part, we must access information about the structure of the object and use this information to activate the appropriate visual memories and the appropriate spatial relations representations. This process involves additional subsystems, as noted below.

3.1.6 *Information lookup*
We see only about 2° of visual angle with high resolution. Thus, we often must move our eyes over an object or scene. Logically, there are only three ways in which we can guide eye movements: randomly, on the basis of bottom-up information (e.g., motion), or using stored information. Yarbus (1967) provides ample evidence that knowledge is often used to guide one's sequence of attention fixations. If, for example, we see an object in an unusual position (e.g., a cat sleeping in a contorted position), we may recognize a part or two but not recognize the object with confidence. At this point, distinctive properties (which distinguish the best-matching object from other similar objects) are looked up in associative memory, and this information is used in two ways: (1) The location of a distinctive part or characteristic is sent to mechanisms that shift attention; and, (2) the representation of the expected part or characteristic is primed in the ventral system, making the part or characteristic easier to encode. This priming mechanism lies at the heart of image generation: According to this theory, visual mental images are formed when the ventral system is primed so strongly that an image is projected back into the visual buffer.

3.1.7 *Attention shifting*
The attention shifting subsystem guides the movement of the body, head and eyes, and also adjusts the attention window in the visual buffer (both in perception and visual mental imagery). These mechanisms are important for several reasons. Most important for present purposes, they play a critical role when parts or characteristics are added to an image. In this case, one shifts attention to the location where the image must be further fleshed out. For example, when asked to form an image of a cat, and then asked whether it has curved front claws, most people report that the image did not contain the claws

initially. Rather, the claws were added only when they were needed. In this case, the resolution limits of the visual buffer may have prevented one from finding the location where the claws belong, and thus the image first was expanded (one "zoomed in" on it) until the front of the paw was "visible," and then image of the part was formed at that location (for a detailed discussion of how such processing may occur, see Chapters 4 and 9 of Kosslyn, 1994).

3.2 *Summary and critical distinctions*

The logic used to develop the theory of imagery hinges on the idea that perceptual mechanisms are used in imagery. Thus, I will summarize the way the system operates during perception proper before returning to imagery.

3.2.1 *Identifying objects*
An object is identified by first positioning the attention window in the appropriate part of the visual buffer. Once the image of the object is enveloped by the attention window, it is sent simultaneously to the dorsal and ventral systems for further processing. The ventral system, which encodes object properties, attempts to organize perceptual units and match them to those of stored shapes. The dorsal system, which encodes spatial properties, converts retinal location to spatiotopic coordinates and encodes spatial relations. An object can be recognized at first glance if the match to a stored shape in the ventral system is very good. However, if the match does not definitively implicate a single object, then the identity of the closest matching object is treated as an hypothesis to be tested.

Hypothesis testing requires first accessing associative memory and looking up distinctive parts or characteristics of the candidate object and their locations on the object. This information is then used to position the attention window to the location of the most distinguished property, and to prime the ventral system to encode that representation. Once the attention window (and perhaps the eyes, head, and body) is properly shifted, the portion of the image at that location is then encoded via the ventral and dorsal systems. The subsequent output of these systems, which is sent to associative memory, may provide evidence in favor of the hypothesis or may lead to the formulation of a new hypothesis. The top-down hypothesis-testing cycle is repeated as many times as necessary until the stimulus has been identified.

3.2.2 *Visualizing objects*

The visual buffer functions to make explicit the local geometry of surfaces of objects. An image is a pattern of activation in topographically organized areas. Images of individual remembered shapes are formed by activating stored visual memories in the ventral system; the process used to prime this system during perception can activate the stored representations so strongly that a pattern of activation is produced in the visual buffer, which is an image representation. If the image is composed of multiple parts, a representation of a part is looked up in associative memory and attention is shifted to the location on the present image where the part belongs. The image is adjusted (scaled up or down, as necessary) until the location of the part is clearly evident, at which point an image of the part is formed. This process is repeated until all of the appropriate parts are included in the image.

3.3 *The neural bases of visual consciousness?*

I earlier (Kosslyn, 1992) outlined a theory of consciousness called "parity theory." The key idea was that consciousness arises when disparate neural events must be coordinated; it is an emergent property of such events in the same way that a chord is an emergent property of the simultaneous playing of sets of notes. I now want to extend this idea: It seems particularly important to coordinate different representations when different types (formats) of internal codes must be related to each other. For example, the vast majority of visual areas that send fibers downstream to another area also receive feedback fibers from that area (and these feedback connections are of comparable size to those flowing downstream). The areas early in the visual processing sequence are retinotopically organized, whereas those later in the sequence are not; the later areas — which store visual memories and match input to those memories — appear to use a "population code," where properties are specified by activation of sets of neurons that represent components. And here is the new idea: It is possible that *consciousness arises at junctures where different types of representations meet and feedback and feedforward flows must be coordinated.*

If this view is correct, we should be conscious of information that crosses from the visual buffer to the object-properties encoding subsystem and from the visual buffer to the spatial-properties encoding subsystem. In both cases, the higher-level representations are not retinotopic. At least in my case, I do seem to have awareness of a world of objects with shape, color, and texture (which presumably arises from the interaction of the visual buffer and object-properties-encoding subsystem) and I do seem to have awareness of a world of objects with location, size, and orientation (which presumably arises from the interaction of the visual buffer and spatial-properties-encoding subsystem).

In addition, there is reason to believe, at least as of this writing, that there is a qualitative distinction between the type of code used in the ventral and dorsal systems and in associative memory. The modality-specific systems appear to use representations that are tailor-made for the kinds of contents that must be represented. In contrast, associative memory must use representations that can be addressed via input from multiple modalities. If so, then we would expect to be conscious of the "meanings" of inputs, and I — for one — do have such experience.

In contrast, the format of the code used in associative memory may be the same as that used to direct the attention shifting subsystem, and thus I would not expect us to be conscious of that type of representation. Similarly, the kind of representations used in associative memory may be the same as those used in many sorts of planning, in which case I would not be aware of them when used in this way. I would be aware of the process of planning when the representations used are different from those used in the processing that leads up to planning — as occurs when I employ visual images to help me plan the best route to the airport or the like. In addition, if a problem is very difficult, I may be aware of trying to solve it because the representations have been converted into a form that can be used in "working memory"; in many circumstances, problem solving (and planning) may occur "automatically" via constraint satisfaction processes (e.g., see Kosslyn & Koenig, 1992), in which case I would not be aware of such processing.

On this view, then, we are not directly conscious of the contents of primary visual cortex. Rather, our consciousness arises when back-projections from higher areas meet projections from retinotopically organized areas, and we become conscious of the representations in the lower-level retinotopic areas.

However, this theory does not immediately explain why people generally are not aware of the fact that the time to form an image increases linearly with the number of parts that must be added to an image. Each part corresponds to a representation in the visual buffer, and I have just suggested that we are aware of such representations when they produce input to later areas that must mesh properly with feedback from those areas. One possible account of this apparent inconsistency is that the top-down generation of parts is faster than the process whereby the image information meshes with the top-down information. If the generation process is sufficiently fast, it could occur before consciousness of the experience can arise. Indeed, Libet and his colleagues (for a review, see Libet, 1987) have shown that conscious experience of a touch may lag several hundred milliseconds (as much as 500) behind the brain's initial registration of a stimulus. Kosslyn (1980) estimates that between 50 and 150 milliseconds are required to generate each part. Thus, it is plausible that much of the process of fleshing out an image can occur prior to one's being consciously aware of it.

4. Conclusions

Introspectively, visual mental images seem to portray spatial extent, have a limited spatial extent, and have a limited resolution. Behavioral results suggest that the internal representations of visual mental images do in fact incorporate these properties. Moreover, brain scanning results indicate that regions of the brain that have these characteristics are usually activated during imagery, and if these areas are damaged imagery is impaired. In addition, the relative amount of blood flow in these areas predicts how quickly subjects can form and use visual mental images.

These results suggest that conscious experience reflects properties of specific neural areas; other parts of the visual system do not have the characteristics that appear evident in images. However, it is not necessarily the case that consciousness simply reflects the activation of individual areas. Indeed, representations in primary visual cortex, which is usually activated in imagery, include properties that are not evident to introspection. The alternative view suggested here is that visual consciousness arises when the outputs from retinotopically organized areas mesh with top-down feedback from areas that are not so organized. If so, then we should be conscious of properties of representations that are at the points where a code must be transformed into a different type. This hypothesis appears consistent with my personal experience, for what that's worth, if we assume that the process of meshing representation must occur for a few hundred milliseconds before one is aware of it.

Notes

Preparation of this chapter was supported by NIA Grant AG 126750-01 and grant F49620-98-1-0334 from the Air Force Office of Sponsored Research. The third section of this chapter was adapted in part from Kosslyn (1991), with the permission of both the publisher of that chapter and the present publisher. I thank Irene Kim, Carolyn Rabin and Kayt Sekul for careful reading of an earlier draft of this chapter.

1. This brief review just touches on the high points of the debate over the best interpretation of the scanning findings; for more thorough reviews, see Chapters 4 and 8 of Kosslyn (1980), Chapters 1 and 10 of Kosslyn (1994), and Denis & Kosslyn (1999).

References

Andersen, R. A. 1987. The role of the inferior parietal lobule in spatial perception and visual-motor integration. In F. Plum, V. B. Mountcastle, and S. R. Geiger (Eds.), *Handbook of Physiology: The Nervous System V, Higher Functions of the Brain Part 2.* Bethesda, MD: American Physiological Society.

Cave, K. R., and Kosslyn, S. M. 1989. Varieties of size-specific visual selection. *Journal of Experimental Psychology: General, 118,* 148-164.

Cooper, L. A., and Shepard, R. N. 1973. Chronometric studies of the rotation of mental images. In W. G. Chase (Ed.), *Visual Information Processing.* New York: Academic Press.

Corbetta, M., Miezin, F. M., Dobmeyer, S., Shulman, G. L., and Petersen, S. E. 1990. Attentional modulation of neural processing of shape, color, and velocity in humans. *Science, 248,* 1556-1559.

Crick, F., & Koch, C. 1994. Are we aware of neural activity in primary visual cortex? *Nature.* 375, 121-123.

Damasio, H., Grabowski, T. J., Damasio, A., Tranel, D., Boles-Ponto, L., Watkins, G. L. and Hichwa, R. D. 1993. Visual recall with eyes closed and covered activates early visual cortices. *Society for Neuroscience Abstract, 19, 1603.*

Denis, M., and Kosslyn, S.M. 1999. Scanning visual mental images: A window on the mind. (Target article). *Cahiers de Psychologie Cognitive/Current Psychology of Cognition, 18,* 409-465.

Farah, M. J. 1988. Is visual imagery really visual? Overlooked evidence from neuropsychology. *Psychological Review, 95,* 307 - 317.

Farah, M. J., Soso, M. J., & Dasheiff, R. M. 1992. Visual angle of the mind's eye before and after unilateral occipital lobectomy. *Journal of Experimental Psychology: Human Perception and Performance, 18,* 241-246.

Felleman, D. J., & Van Essen, D. C. 1991. Distributed hierarchical processing in the primate cerebral cortex. *Cerebral Cortex, 1,* 1-47.

Finke, R. A., & Pinker, S. 1982. Spontaneous imagery scanning in mental extrapolation. *Journal of Experimental Psychology: Learning, Memory, and Cognition, 8,* 142-147.

Finke, R. A., & Pinker, S. 1983. Directional scanning of remembered visual patterns.

Journal of Experimental Psychology: Learning, Memory, and Cognition, 9, 398-410.

Fujita, I., Tanaka, K., Ito, M., & Cheng, K. 1992. Columns for visual features of objects in monkey inferotemporal cortex. *Nature, 360*, 343-346.

Goldston, D. B., Hinrichs, J. V., & Richman, C. L. 1985. Subjects' expectations, individual variability, and the scanning of mental images. *Memory and Cognition, 13*, 365-370.

Gross, C. G., Desimone, R., Albright, T. D., and Schwartz, E. L. 1984. Inferior temporal cortex as a visual integration area. In F. Reinoso-Suarez & Ajmone-Marsan (Eds.), *Cortical Integration*. New York: Raven Press, pp. 291-315.

Hyvarinen, J. 1982. Posterior parietal lobe of the primate brain. *Physiological Review, 62*, 1060-1129.

Intons-Peterson, M. J. 1983. Imagery paradigms: How vulnerable are they to experimenters' expectations? *Journal of Experimental Psychology: Human Perception and Performance, 9*, 394-412.

Jolicoeur, P., and Kosslyn, S. M. 1985. Is time to scan visual images due to demand characteristics? *Memory and Cognition, 13*, 320 - 332.

Kosslyn, S. M. 1973. Scanning visual images: some structural implications. *Perception and Psychophysics, 14*, 90-94.

Kosslyn, S. M. 1975. Information representation in visual images. *Cognitive Psychology, 7*, 341-370.

Kosslyn, S. M. 1976. Can imagery be distinguished from other forms of internal representation? Evidence from studies of information retrieval times. *Memory and Cognition, 4*, 291-297.

Kosslyn, S. M. 1978. Measuring the visual angle of the mind's eye. *Cognitive Psychology, 10*, 356-389.

Kosslyn, S. M. 1980. *Image and Mind*. Cambridge, MA: Harvard University Press.

Kosslyn, S. M. 1991. A cognitive neuroscience of visual cognition: further developments. In R. Logie and M. Denis (Eds.), *Mental Images in Human Cognition*. New York: North-Holland.

Kosslyn, S. M. 1992. Cognitive neuroscience and the human self. In A. Harrington (Ed.), *So Human a Brain*. New York: Pergamon.

Kosslyn, S. M. 1994. *Image and Brain: The Resolution of the Imagery Debate*. Cambridge, MA: MIT Press.

Kosslyn, S. M., & Ochsner, K. 1994. In search of occipital activation during visual mental imagery. *Trends in Neurosciences, 17*, 290-292.

Kosslyn, S. M., Alpert, N. M., Thompson, W. L., Maljkovic, V., Weise, S. B., Chabris, C. F., Hamilton, S. E., and Buonano, F. S. 1993. Visual mental imagery activates topographically organized visual cortex: PET investigations. *Journal of Cognitive Neuroscience, 5*, 263-287.

Kosslyn, S. M., Cave, C. B., Provost, D., and Von Gierke, S. 1988. Sequential processes in image generation. *Cognitive Psychology, 20*, 319-343.

Kosslyn, S.M., Pascual-Leone, A., Felician, O., Camposano, S., Keenan, J.P., Thompson, W.L., Ganis, G., Sukel, K.E. and Alpert, N.M. 1999. The role of Area 17 in visual imagery: Convergent evidence from PET and rTMS. *Science* 284, 167-170.

Kosslyn, S.M., Thompson, W.L., and Alpert, N.M. 1997. Neural systems shared by visual imagery and visual perception: A positron emission tomography study. *NeuroImage, 6*, 320-334.

Kosslyn, S.M., Thompson, W.L., Kim, I.J., and Alpert, N.M. 1995. Topographical representations of mental images in primary visual cortex. *Nature, 378,* 496-498.

Kosslyn, S.M., Thompson, W.L., Kim, I.J., Rauch, S.L. and Alpert, N.M. 1996. Individual differences in cerebral blood flow in area 17 predict the time to evaluate visualized letters. *Journal of Cognitive Neuroscience, 8,* 78-82.

Larsen, A., and Bundesen, C. 1978. Size scaling in visual pattern recognition. *Journal of Experimental Psychology: Human Perception and Performance, 4,* 1 - 20.

Le Bihan, D., Turner, R., Zeffiro, T. A., Cuénod, C. A., Jezzard, P., & Bonnerot, V. 1993. Activation of human primary visual cortex during mental imagery. In D. Le Bhihan, R. Turner, M. Mosley, & J. Hyde (Vol. Ed.) (Sec. Ed.), *Functional MRI of the Brain: A Workshop presented by the Society of Magnetic Resonance in Medicine and the Society for Magnetic Resonance Imaging.* Arlington, VA: Society of Magnetic Resonance in Medicine, Inc.

Libet, B. 1987. Consciousness: Conscious, subjective experience. In G. Adelman (Ed.), *Encyclopedia of Neuroscience.* Boston: Birkhauser.

Marr, D. 1982. *Vision.* San Francisco, CA: W. H. Freeman.

Maunsell, J. H. R., and Newsome, W. T. 1987. Visual processing in monkey extrastriate cortex. *Annual Review of Neuroscience, 10,* 363-401.

Mellet, E., Petit, L., Mazoyer, B., Denis, M., & Tzourio, N. 1998. Reopening the mental imagery debate: Lessons from functional neuroanatomy. *NeuroImage, 8,* 129-139.

Menon, R., Ogawa, S., Tank, D. W., Ellermann, J., Merkele, H., & Ugurbil, K. 1993. Visual mental imagery by functional brain MRI. In D. Le Bihan, R. Turner, M. Mosley, & J. Hyde (Vol. Ed.) (Sec. Ed.), *Functional MRI of the Brain: A Workshop presented by the Society of Magnetic Resonance in Medicine and the Society for Magnetic Resonance Imaging.* Arlington, VA: Society of Magnetic Resonance in Medicine, Inc.

Mishkin, M., Ungerleider, L. G., and Macko, K. A. 1983. Object vision and spatial vision: Two cortical pathways. *Trends in Neurosciences, 6,* 414-417.

Paivio, A. 1971. *Imagery and Verbal Processes.* New York: Holt, Rinehart & Winston.

Perrett, D. I., Smith, P. A. J., Potter, D. D., Mistlin, A. J., Head, A. S., Milner, A. D., and Jeeves, M. A. 1985. Visual cells in the temporal cortex sensitive to face view and gaze direction. *Proceedings of the Royal Society London, B 223,* 293-317.

Pinker, S., Choate, P., & Finke, R. A. 1984. Mental extrapolation in patterns constructed from memory. *Memory and Cognition, 12,* 207-218.

Podgorny, P., and Shepard, R. N. 1978. Functional representations common to visual perception and imagination. *Journal of Experimental Psychology: Human Perception and Performance, 4,* 21-35.

Pohl, W. 1973. Dissociation of spatial discrimination deficits following frontal and parietal lesions in monkeys. *Journal of Comparative and Physiological Psychology, 82,* 227-239.

Posner, M. I., Inhoff, A. W., Friedrich, F. J., and Cohen, A. 1987. Isolating attentional systems: A cognitive-anatomical analysis. *Psychobiology, 15,* 107 - 121.

Pylyshyn, Z. W. 1981. The imagery debate: Analogue media versus tacit knowledge. *Psychological Review, 87,* 16-45.

Roland, P. E., & Friberg, L. 1985. Localization of cortical areas activated by thinking. *Journal of Neurophysiology, 53,* 1219-1243.

Roland, P. E., and Gulyas, B. 1994. Visual imagery and visual representation. *Trends in Neurosciences, 17,* 281-287.

Shepard, R. N., & Cooper, L. A. 1982. *Mental Images and their Transformations.* Cambridge, MA: MIT Press.

Sperling, G. 1960. The information available in brief visual presentations. *Psychological Monographs, 74 (11, Whole no. 498).*

Squire, L. R. 1987. *Memory and Brain.* New York: Oxford University Press.

Thompson, W.L., and Kosslyn, S.M. In press. Neural systems activated during visual mental imagery: A review and meta-analyses. In J. Mazziotta and A. Toga (Eds.), *Brain Mapping II: The Applications.* New York: Academic Press.

Treisman, A. M., and Gelade, G. 1980. A feature integration theory of attention. *Cognitive Psychology, 12,* 97-136.

Ungerleider, L. G., and Mishkin, M. 1982. Two cortical visual systems. In D. J. Ingle, M. A. Goodale, and R. J. W. Mansfield (Eds.), *Analysis of Visual Behavior.* Cambridge, MA: MIT Press.

Van Essen, D. 1985. Functional organization of primate visual cortex. In A. Peters and E. G. Jones (Eds.), *Cerebral Cortex (Vol. 3).* New York: Plenum Press.

Van Essen, D. C., Felleman, D. J., DeYoe, E. A., Olavarria, J., & Knierim, J. 1990. Modular and hierarchical organization of extrastriate visual cortex in the macaque monkey. *Cold Spring Harbor Symposia on Quantitative Biology, 55,* 679-696.

Yarbus 1967. *Eye Movements and Vision.* New York: Plenum Press.

CHAPTER 5

Intersensory Integration
Underlying neural mechanisms

Barry E. Stein and Mark T. Wallace
Wake Forest University School of Medicine

1. Perceptual consequences of integrating multiple sensory cues

We tend to have great faith in the accuracy with which our senses reflect the world. So it is either disconcerting or delightful, depending on one's age and circumstance, to discover how easily we can be fooled by them. The clever use of background, perspective, and shading in cinematography makes it almost believable that a common lizard can tower over Tokyo, terrorize its citizens, and do battle with an interloping triceratops (actually a small chameleon with its casque elevated).

The use of perspective in art and photography to create the illusion of dimensionality is now so commonplace in our experience that we readily accept the idea that visual cues are relative. There is also little problem in accepting that other sensory modalities function in similar fashion, so that the relative nature of within-modality cues is a potent factor in how any given stimulus is evaluated. Less pervasive, however, is acceptance of the fact that sensory judgments are also very much dependent on cues from other sensory modalities. Although cross-modality influences are as common in daily experience as within-modality influences, the substantial effect of a brief, low-intensity auditory stimulus on one's perception of the brightness of a light (Stein et al., 1996) is generally met with greater surprise than the observation that one looks slimmer wearing vertical stripes than horizontal stripes.

That influences across modalities are less apparent than the relative nature of within-modality cues is due, in part, to the peculiar subjective impressions associated with each sensory modality. There is simply no equivalent for tickle and itch in vision or audition, nor are there somatosensory counterparts to hue and pitch. These modality-specific subjective qualities encourage the impression that sensory systems function in independent realms. Yet, this is only an illusion. The potent effects of one sense on another are quite evident when discordant cues from different sensory modalities are synthesized. Under such circumstances, perception suffers.

The effect of nonvisual cues on visual perception is of considerable interest to pilots, who experience major shifts in the gravitational-inertial field during take-off, landing and other accelerations. This results in conflicting visual and interoceptive (e.g., proprioceptive, vestibular) cues, creating what is known as the "oculogravic illusion" (Graybiel, 1952; Clark and Graybiel, 1966). For example, a pilot taking off from an aircraft carrier is accelerating very rapidly and his/her body seems to be tilted too far back in the seat and the instrument panel seems to be rising too rapidly. The impression is that the nose of the aircraft is rising too fast and immediate corrective action must be taken. However, the instrument panel will indicate the illusory nature of this experience. Pilots must learn to trust their instruments under these conditions to avoid making life-threatening decisions. The sensitivity of visual perception to proprioceptive cues is also evident when the muscles on one side of the neck are artificially vibrated. Now a visual target in a darkened room appears to be displaced contralaterally or to be moving in the direction of the vibration (Biguer et al., 1988).

For most of us the examples given above are hardly daily occurrences. However, we all engage in daily conversation, and the effect of discordant visual and auditory cues creates impressive auditory-visual illusions in these circumstances as well. An example of such an illusion is the "McGurk effect" (McGurk and MacDonald, 1976). In this study, naive subjects who were presented with the sounds "ba-ba," but who saw the mouth form "ga-ga," reported the perception of neither the former nor the latter, but "da-da," a synthesis of the two.

There are quite a number of these cross-modality illusions (e.g., see Welch and Warren, 1986; Stein and Meredith, 1993), but the few described here are sufficient to make the point: disrupting the normal coherence among sensory cues alters perception. This is because the different sensory modalities have evolved to work in concert, and normally, different sensory cues that originate from the same event are concordant and interact synergistically to enhance the salience of the event. Seeing a speaker's face makes it far easier to understand the spoken message, especially in a noisy room (Sumby and Pollack, 1954). The visual and auditory cues produce interactions among input channels in the central nervous system and visualizing lip movements actually alters evoked activity in auditory cortex (Sams et al., 1991). Cross-modality synergy is also expressed as a decrease in reaction time when concordant inputs are received from different modalities. Thus, visual and auditory cues derived from the same location in space markedly decrease the reaction time of eye movements to that location (Hughes et al., 1994; Frens et al., 1995).

Despite the fact that we are generally unaware of the process of cross-modality or "multisensory" integration, it is continuously engaged and determines perception and behavior on a moment-to-moment basis. Although ultimately we are concerned with how the combination of different sensory stimuli affects the human CNS and our own behaviors and perceptions, it is important to note that the ability to integrate cues from different sensory modalities did not arise with the appearance of humans. Rather, it is an ancient scheme that antedates mammals and even the evolution of the nervous system. The simplest neural plan for such integration is when inputs from different receptors converge on the same cell. This multisensory convergence scheme is found in unicellular organisms and has been retained in some form throughout multicellular speciation and the evolution of complex animals, including humans (Stein and Meredith, 1993). In fact, it would be unprecedented to find an animal in which there exists a complete segregation of sensory processing. Presumably, during the process of developing and differentiating sensory systems, mechanisms were preserved and/or elaborated for using their combined action to provide information that would be unavailable from their individual operation.

Yet, we are only beginning to understand how multisensory integration enhances information processing in the nervous system. When compared to our appreciation of the processing of unimodal sensory cues, our knowledge of how information from the different sensory modalities is combined and synthesized is rudimentary. We do know, however, that whenever inputs from different sensory modalities converge in the central nervous system there is the opportunity for multisensory integration, and that this is not as rare as might be supposed. As one ascends the neuraxis from brainstem to cortex in

higher animals, there exists an interesting duality — modality-specific structures and pathways coexisting with those in which multiple sensory inputs
converge. Thus, there are many sites in the brainstem, thalamus, and cortex in
which individual neurons are dedicated to preserving modality-specific information, while others serve to synthesize this information (see Stein and
Meredith, 1993 for a review). It is the various populations of neurons that
integrate multiple sensory modalities that are of primary interest here, and it is
curious to note that despite the fact that different populations of multisensory
neurons may participate in very different functions, there are many common-
alities in the way in which they integrate their inputs. This is a point to which
we shall return later.

2.   Multisensory convergence

2.1 *Converging sensory inputs and multisensory integration in the cat superior colliculus*

One of the best-studied multisensory structures is the cat superior colliculus
(SC). This laminated midbrain structure plays an important role in attentive and
orientation behaviors directed toward visual, auditory and somatosensory
stimuli (Fig. 5.1; also see Sprague and Meikle, 1965; Schneider, 1969;
Casagrande et al., 1972; Goodale and Murison, 1975; Stein, 1984a; Sparks,
1986; Stein and Meredith, 1991, Wallace, Meredith, and Stein, 1993). Neurons
responsive to two or more sensory modalities (i.e., multisensory neurons) as
well as neurons responsive to a single sensory modality (i.e., unimodal neurons)
are found in this structure.

By virtue of receiving two or more different sensory inputs, each multisensory neuron has two or more receptive fields (e.g., a somatosensory-visual
multisensory neuron has both a somatosensory and a visual receptive field).
These receptive fields are well-defined and show a striking correspondence to
one another, so that if the somatosensory-visual neuron described above had a
somatosensory receptive field on the forelimb, it would have a visual receptive field in a corresponding location in inferior nasal visual space (Fig. 5.2).

By determining such receptive field correspondence in each multisensory
neuron over a large population in the SC, each sensory representation can be
"mapped" in another modality's spatial coordinate frame. Thus, the body has

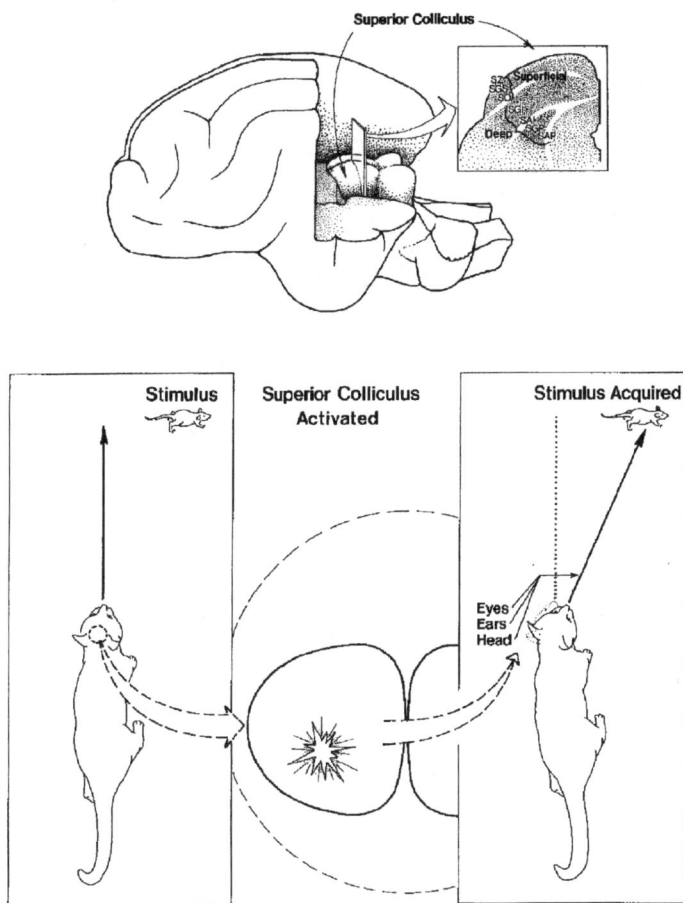

Figure 5.1. The cat superior colliculus plays an integral role in orientation behavior. Top figure shows the location and lamination pattern of the cat superior colliculus. In this schematic, the posterior portion of the cerebral cortex has been removed to show the underlying midbrain. Lamination pattern of the superior colliculus is shown in a representative coronal section. The superficial layers are comprised of: SZ—stratum zonale, SGS—stratum griseum superficiale and SO—stratum opticum. The deep layers are comprised of: SGI—stratum griseum intermediale, SAI-stratum album intermediale, SGP—stratum griseum profundum and SAP—stratum album profundum. Bottom figure shows the behavioral role of the superior colliculus. A sensory stimulus activates a localized region of the superior colliculus which, in turn, directs the eyes, ears and head toward the stimulus. Adapted from Stein and Meredith (1993).

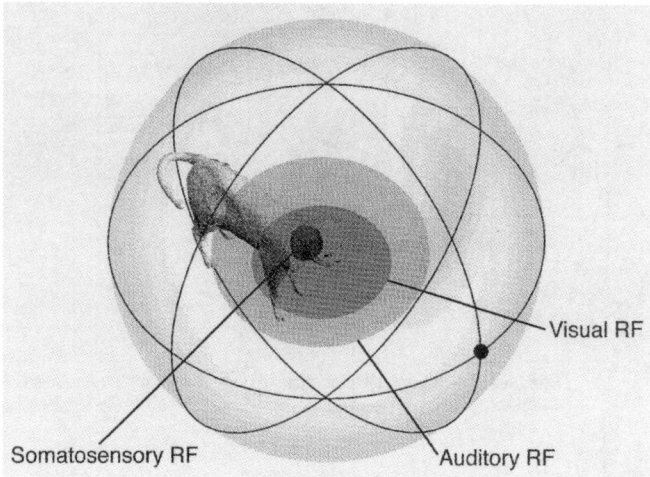

Figure 5.2. Receptive field correspondence characterizes multisensory neurons in the cat SC. In this representational scheme, the cat is positioned with its head in the center of a hypothetical sphere, the surface of which represents a plane in visual and auditory space (somatosensory space is represented on the animal's body). In this example of a trimodal neuron, a somatosensory receptive field on the neck, shoulder and upper forelimb is linked to visual and auditory receptive fields in corresponding locations of superior space. The filled circle represents the intersection of the horizontal and vertical meridians.

been mapped in visual coordinates in cat (Fig. 5.3; see Stein et al., 1976; Meredith, Clemo and Stein, 1991; Meredith and Stein, 1993) and, using several variations of this method, an excellent topographic correspondence among the maps has been reported in rodent (Drager and Hubel, 1975; Chalupa and Rhoades, 1977; McHaffie et al., 1989), as well as in the nonmammalian homologue of the SC, the optic tectum, of both bird (Knudsen and Brainard, 1995) and reptile (Stein and Gaither, 1981). Our recent studies also show a similar organizational scheme in monkey (Wallace, Wilkinson and Stein, 1996). The utility of such a correspondence is that a sensory cue, regardless of modality, activates neurons in the same general SC location — the location that corresponds to the position of the stimulus in sensory space. Cues positioned forward in space (or on the front of the body) activate neurons in the rostral SC, whereas those in caudal (i.e., temporal) space activate neurons in the caudal SC. Similarly, cues in superior space (i.e., above the animal or on its upper body) activate neurons in the medial SC, whereas those in lower sensory space activate neurons in the lateral SC (Stein, Magalhaes-Castro and Kruger, 1976).

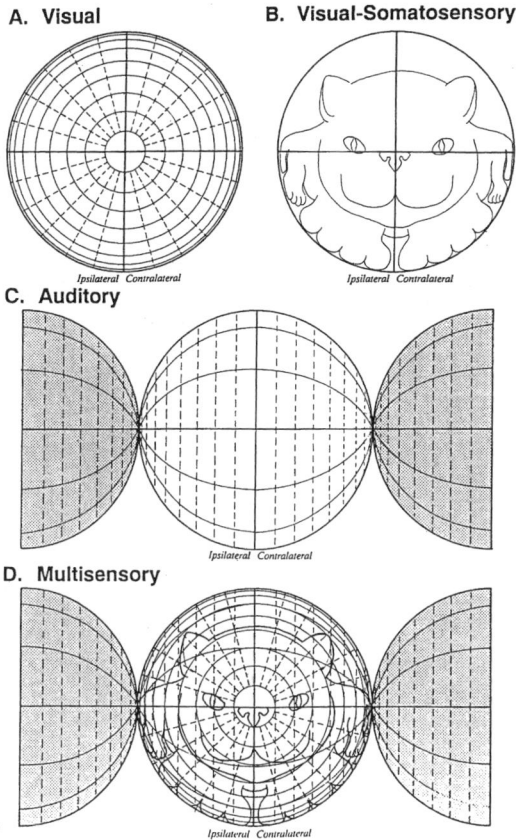

A. Visual

B. Visual-Somatosensory

Ipsilateral Contralateral

C. Auditory

Ipsilateral Contralateral

D. Multisensory

Ipsilateral Contralateral

Figure 5.3. A multisensory spatial coordinate system. **A:** Visual space is depicted by a double-pole coordinate system with the intersection of the vertical and horizontal meridians at the area centralis. **B:** The representation of the body in the superior colliculus as it would appear if mapped in visual coordinates. This map was constructed by plotting the body part at the center of the visual receptive field in each visual-somatosensory neuron sampled. For example, bimodal neurons with somatosensory receptive fields on the nose had visual receptive fields near the area centralis, and those with somatosensory receptive fields on the forepaw had inferior and nasal visual receptive fields. **C:** Auditory space shown in double-pole coordinates. Positions anterior to a line through the ears (the interaural axis) are depicted within the middle circle, while those posterior to it are represented by the shaded crescents on each side. These crescents were detached from one another at the posterior midline (180°) and folded forward in order to flatten auditory space into a two-dimensional representation. Curved lines represent elevations 30° and 60° above and below the horizontal meridian. Dashed lines represent azimuth in 10° increments. **D:** Auditory space is aligned with the overlapping visual and somatosensory representations to produce a schematic of multisensory space. Adapted from Stein and Meredith, 1993.

This scheme is not only an economical way to represent sensory space but is also an efficient way to match incoming sensory information with outgoing motor-related information, for these sensory maps are also in register with the premotor maps found in the SC (Harris, 1980; McIlwain, 1986, 1990; Roucoux and Crommelinck, 1976; Stein and Clamann, 1981; Stein, Clamann and Goldberg, 1980). In fact, many SC neurons have both sensory and premotor properties (Guitton and Munoz, 1991; Munoz and Guitton, 1985; Peck, 1987; Peck et al., 1995; Hartline et al., 1995; also see Sparks, 1986; Wurtz and Albano, 1980). Thus, neurons in the rostral SC not only respond to cues in frontal sensory space, they also serve to direct the eyes, ears, and head

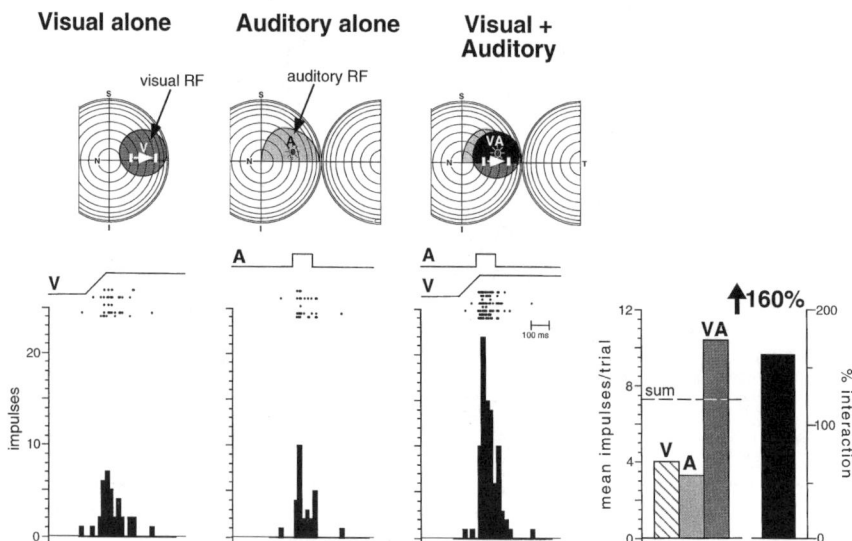

Figure 5.4. Spatially-coincident stimuli give rise to response enhancement. The top panels show the individual receptive fields (gray shading) of this visual-auditory SC neuron, as well as the region of overlap of these receptive fields (black shading). In addition, the position of the stimuli used in unimodal and multisensory trials are shown. The visual stimulus (V) was a moving bar of light (direction of movement shown by the arrow). The auditory stimulus (A), a broad band noise burst, was delivered from a stationary speaker indicated by the icon. Bottom panels contain rasters and histograms representing the neuron's response to the unimodal and multisensory trials, as well as bar graphs summarizing the mean response and multisensory interaction. Similar conventions are used in subsequent figures. Note that the spatially-coincident pairing resulted in a 160% response enhancement, well above the sum of the unimodal responses.

forward. Neurons in the medial SC respond to superior sensory cues and direct movements upward, etc. Such a sensory-motor register is in keeping with the role of the SC in attentive and orientation behaviors.

The effects of combinations of different sensory stimuli on neuronal responses in the SC are quite striking, and it is these profound changes in the activity of multisensory neurons that are of primary interest to us here. On the one hand, combinations of stimuli from different modalities have been shown to result in a dramatic increase in the number of impulses evoked above that elicited by the same stimuli presented alone; an effect defined as *response enhancement* (Fig. 5.4). On the other hand, the same pair of unimodal stimuli may result in far fewer impulses compared to their individual presentation. Their combination may even eliminate responses altogether (Fig. 5.5). This effect is defined as *response depression* and can occur even if a unimodal stimulus is highly effective when presented alone. Both response enhancement and response depression can be exhibited by the same neuron.

Whether response enhancement or response depression will be evoked by stimulus pairing depends on a variety of factors, most notably the temporal and spatial relationships between the stimuli and their respective receptive fields, as well as the effectiveness of the individual stimuli. Thus, although maximal interactions often occur when stimuli are presented at the same time, the temporal "window" during which multisensory interactions can take place has been found to be surprisingly long (Meredith et al., 1986). This allows stimuli from two sensory modalities to interact despite the fact that they may have very different input latencies. In general, response enhancements are most dramatic when the individual unimodal stimuli are weak and difficult to perceive and/or ambiguous, a principle referred to as *inverse effectiveness* (Fig. 5.6) (Stein and Meredith, 1993). As the individual stimuli become more effective, the magnitude of the multisensory interaction declines.

Because the different modality-specific receptive fields of a multisensory neuron are in topographic register (Meredith and Stein, 1996), stimuli from the same position in sensory space fall within each of the neuron's excitatory receptive fields, resulting in an enhanced response (Fig. 5.4). If one stimulus is moved so that it is spatially disparate from the other and falls within its receptive field's suppressive surround, it will inhibit responses to the second stimulus (Fig. 5.5) (Kadunce et al., 1997). Thus, not only is stimulus position important, but the spatial overlap among the receptive fields of a multisensory neuron is critical for normal integration. Such a finding makes intuitive sense.

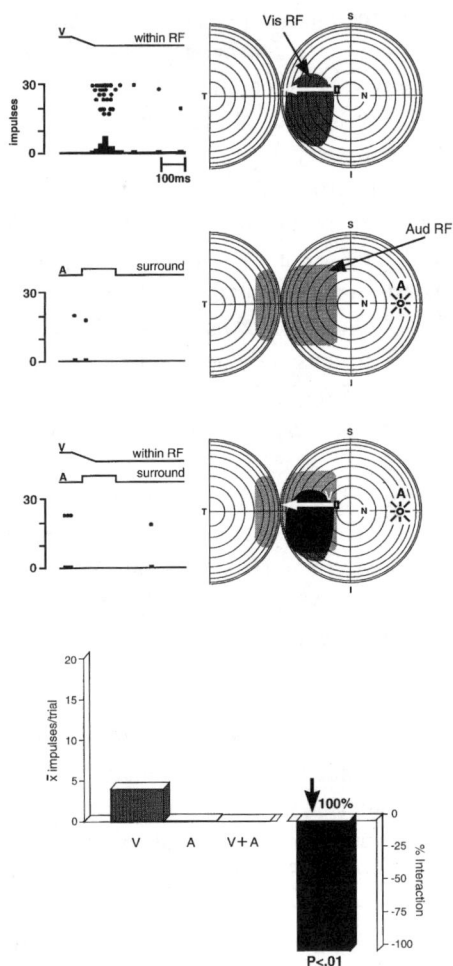

Figure 5.5. Spatially-disparate stimuli can result in response depression. Top series of panels show this visual-auditory neuron's response to: (top) a within field visual stimulus; (center) an auditory stimulus outside of its receptive field borders (i.e., in the "surround"); and (bottom) the combination. Bottom bar graphs summarize the mean responses and interactive effect. Note that the combination of an effective visual stimulus with an auditory stimulus in the surround results in an abolition of this neuron's response. Adapted from Kadunce et al., 1997.

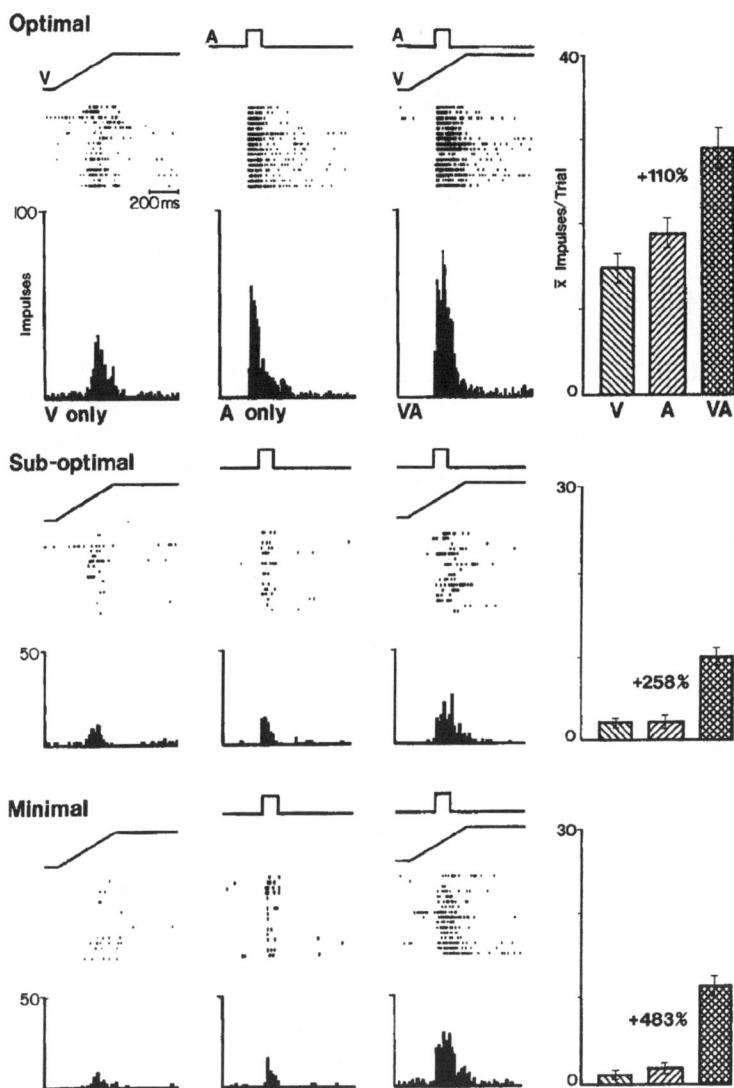

Figure 5.6. Multisensory enhancement increases as unimodal stimulus effectiveness decreases. As the physical properties (e.g., size, intensity, etc.) of the single-modality stimuli are systematically changed so that progressively fewer discharges are evoked, the percentage response enhancement evoked by their combination increases. From Meredith and Stein, 1986.

Stimuli that are related because they arise from the same event originate from the same location and enhance one another's effect. Unrelated stimuli are likely to originate from disparate spatial locations and will either fail to enhance, or actually inhibit, one another.

A major component of the behavior-effecting output of the SC comes from multisensory neurons. Antidromic activation of the tectoreticulospinal tract, the principal crossed tectofugal projection pathway to the brainstem and spinal cord, demonstrated that 94% of activated neurons responded to sensory stimuli, and that 84% of these neurons were multisensory and had integrative properties like those described above (Meredith et al., 1992, Wallace et al., 1993). Such a scheme represents a parsimonious way to give different sensory modalities access to the same targets of the SC and, when possible, to synthesize information from multiple sources before it influences the motor output circuitry of the brainstem and spinal cord.

2.2 *Attention and orienting in space*

Given the behavioral role of the SC and the integrative characteristics of its constituent neurons, stimulus configurations that enhance stimulus salience (as defined by increased neural activity) should enhance SC-mediated attentive and orientation behaviors, and those that decrease stimulus salience should have the opposite effect. This is, indeed, the result that has been obtained in experiments examining the attentive and orientation responses of cats to unimodal and multisensory stimuli (Stein et al., 1989).

Animals were trained in a perimetry apparatus containing pairs of light-emitting diodes (LEDs) and audio speakers, as shown in Figure 5.7. The animal was trained to fixate directly ahead and then orient toward, and directly approach, a briefly illuminated LED. Some animals were trained to ignore an auditory stimulus; others were never exposed to this stimulus. During training, the intensities of the LED's were reduced in order to make the task difficult. During the test sessions unimodal and multisensory stimuli were presented in an interleaved manner. In some trials only the LED was presented, in others the auditory stimulus was linked to the illuminated LED, either at the same location (spatial coincidence) or at a different location (spatial disparity). All animals were far better able to detect and orient to the LED when the auditory stimulus was presented simultaneously and in spatial coincidence; all animals performed significantly worse when the auditory cue was presented simultaneously but spatially disparate than when the LED was presented alone (Fig. 5.7).

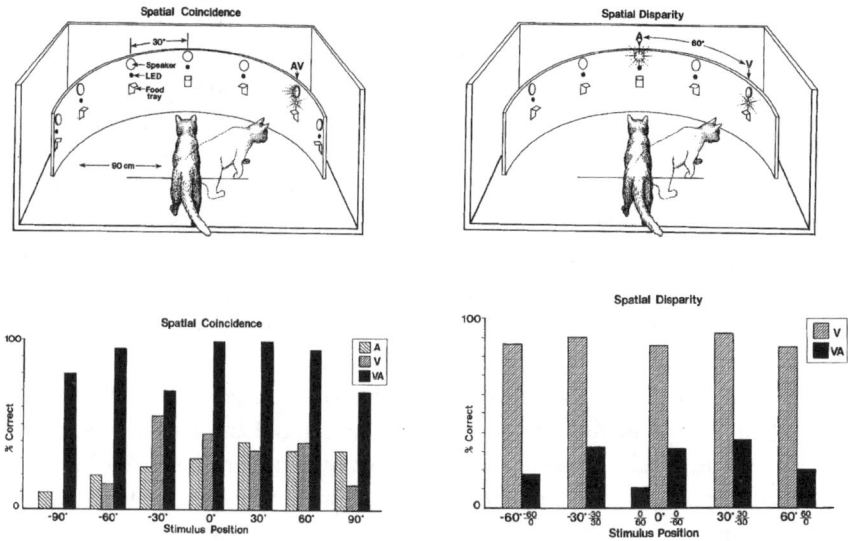

Figure 5.7. Spatially-coincident stimuli result in enhanced multisensory orientation, whereas spatially-disparate stimuli result in depressed multisensory orientation. An array of speakers (larger circles) and LEDs (smaller circles) are located in vertical pairs above a food tray at each of seven regularly spaced (30°) intervals. In the spatially-coincident paradigm (left), during training an animal was required to orient to and move directly toward a visual or auditory stimulus to receive a food reward. During testing, low-intensity stimuli were presented individually and then in combination (AV) at the same location at each of the seven eccentricities. The animal's ability to detect and approach the correct position was enhanced by combined-modality stimuli at every location (bottom left). In the spatially-disparate paradigm (right), animals were trained to approach a visual stimulus (V) but to ignore an auditory stimulus (A). During testing, the visual stimulus was presented alone or in combination with an auditory stimulus that was 60° out of register with it (e.g., A at 0°, V at 60°). The intensity of the visual stimulus was such that a high percentage of correct responses were elicited to it alone. When a visual stimulus was combined with a spatially-disparate auditory stimulus, orientation to the visual stimulus was depressed. Adapted from Stein et al., 1989.

Thus, the animal's behavior closely paralleled the behavior of SC multi-sensory neurons. As it turned out, the key to the multisensory integration seen both physiologically and behaviorally was to be found in cortex.

2.3 *Cortex is critical for multisensory integration in SC*

Neurons in the SC become multisensory by receiving converging inputs from both ascending and descending (i.e., cortical) sources (Stein and Meredith, 1993). In a number of neurons, converging ascending inputs are sufficient to render the neuron multisensory (e.g., convergence of visual and somatosensory inputs onto a single neuron makes that neuron multisensory). However, descending cortical inputs, particularly those from a region of association cortex, the anterior ectosylvian sulcus (AES), turned out to be vitally important for multisensory integration in cats. Inputs from somatosensory, auditory, and visual neurons in AES converge onto individual SC neurons (Wallace et al., 1993). Although unimodal and multisensory neurons abound in AES, only the unimodal neurons are corticotectal (Wallace et al., 1992; 1993). Most importantly, these unimodal AES inputs were found to be essential for multisensory integration in most SC neurons (Wallace and Stein, 1994). Reversible deactivation of AES eliminated multisensory integration in SC neurons, yet unimodal responses were retained (Fig. 5.8). Apparently, some of the associative functions of AES are accomplished via its target neurons in the SC, perhaps providing a means for this cortex to control SC-mediated multisensory behaviors.

Based on the earlier discussion, one would predict that overt multisensory behaviors would depend on the integrity of AES. Confirmation of this prediction was obtained when animals, trained as described earlier, were examined with AES temporarily deactivated (i.e., by means of lidocaine injected into indwelling cannulae). Whereas deactivating AES had no effect on orientation to visual or auditory cues presented individually, animals lost the ability to integrate these cues (Fig. 5.9). Thus, the significant enhancement in correct responses seen to spatially coincident stimuli, and the significant depression seen to spatially disparate stimuli, were lost (Wilkinson et al., 1996). These effects were not seen when other cortical regions were deactivated, or when only saline was injected into the AES.

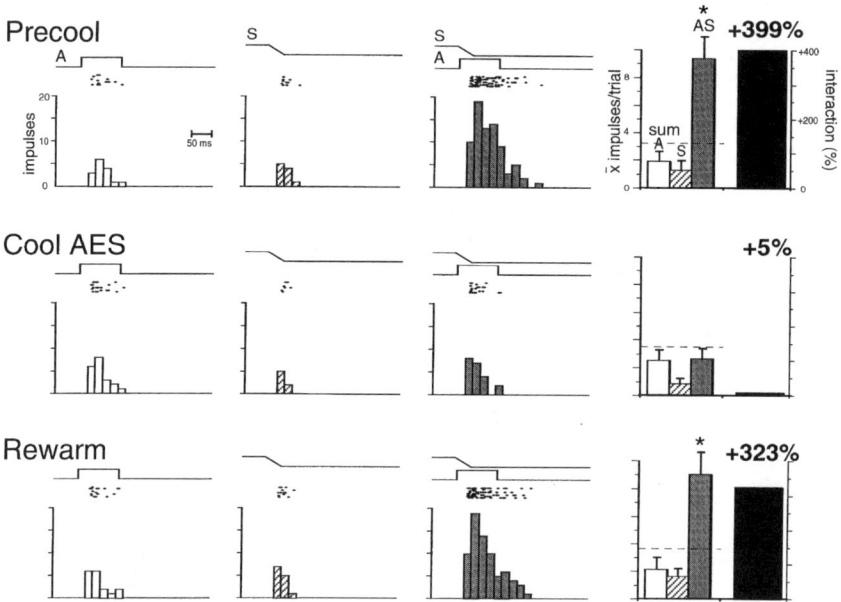

Figure 5.8. AES deactivation eliminates multisensory enhancement in the SC. Far left: the presentation (n=8) of an auditory stimulus ('A', square wave) evoked a modest response in this neuron. The neuron's responses are shown before (precool), during (cool AES) and after (rewarm) cryogenic deactivation of AES cortex. Left center panels show the neuron's response to a somatosensory stimulus ('S', ramp), and right center panels show the response to the auditory-somatosensory stimulus combination. At the far right is shown the averaged data for both the unimodal (A and S) and multisensory (AS) responses, as well as the magnitude of the multisensory interaction (black bars). Dashed lines depict the expected result to multisensory combinations if the unimodal responses were simply summed. Asterisks show significant ($p < 0.01$) multisensory enhancements. Note that cooling of AES eliminated multisensory integration, but did not preclude unimodal responses to either somatosensory or auditory stimuli.

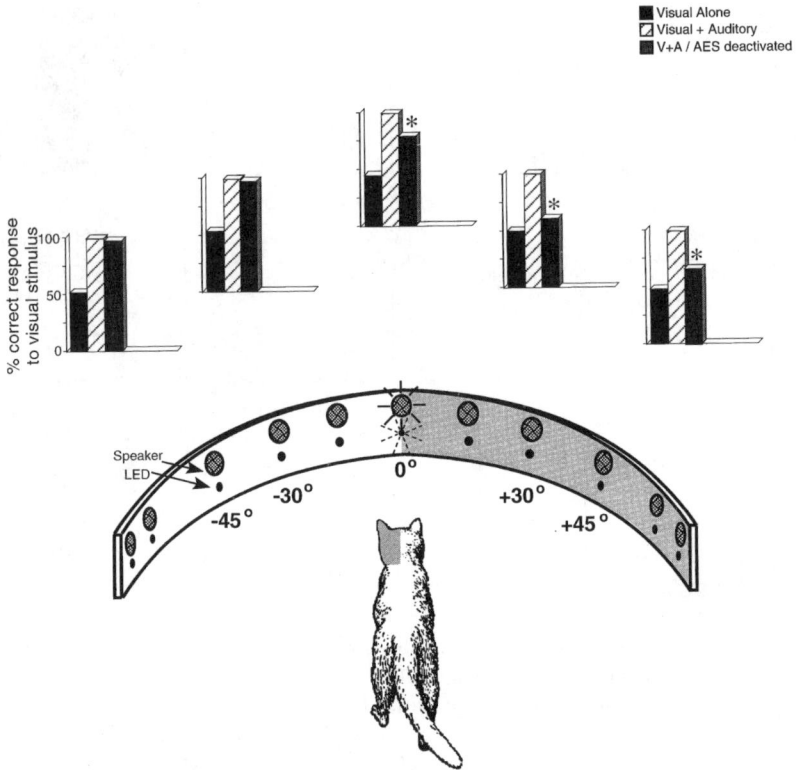

Figure 5.9. Deactivation of AES interferes with multisensory orientation behavior in contralateral sensory space. Bottom panel shows the cat in the perimetry device (see figure 5.7). Shading on the cat depicts the side of cortical deactivation, and shading on the perimetry device depicts the affected region of sensory space. Bar graphs in the top panel illustrate the results of cortical deactivation (via lidocaine injection through indwelling cannulae on multisensory orientation to spatially-coincident visual and auditory stimuli. Five locations were examined: 45° on the side ipsilateral to the deactivated cortex (-45°), -30°, 0°, +30° (contralateral) and +45°. Each bar set shows the results to three conditions: visual alone, visual + auditory, and visual + auditory during AES deactivation. Note the enhanced orientation to the stimulus combination under normal conditions, and the significant impact AES deactivation has on this enhancement to stimuli positioned in contralateral space. Asterisks denote significant differences (p < .05) between stimulus pairings in nondeactivated and deactivated conditions.

2.4 *Multisensory convergence in monkey SC*

Despite a paucity of comparable data, there was good reason to suspect that the multisensory organization of the primate SC would parallel that seen in the cat. Indeed, recent theories of primate SC function have relied heavily on the presence of a topographic arrangement among the different sensory representations (Groh and Sparks, 1992; 1996; Jay and Sparks, 1984; Schiller and Stryker, 1972; Frens et al., 1995; Hughes et al., 1994). Our recent observations have confirmed this organization. Topographically aligned visual, auditory and somatosensory maps were found in the deep layers of the monkey SC (Wallace et al., 1996). Multisensory neurons were common in the deep layers and showed a convergence of somatosensory, visual, and auditory inputs onto individual neurons (Fig. 5.10). The characteristic topographic relationship among the different sensory receptive fields of multisensory neurons is illustrated by the examples in Figure 5.11.

Just as in cat, an integration of information from the different modalities typified neurons in monkey SC. Thus, dramatic response enhancements were observed as long as both stimuli were presented within their respective receptive fields. When one of the stimuli was outside its receptive field, either response depression or no integration was observed (Fig. 5.12).

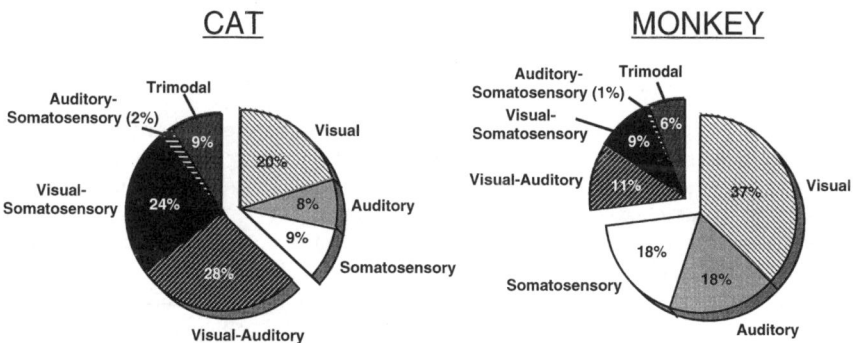

Figure 5.10. Modality distributions in the deep layers of the cat (left) and monkey (right) superior colliculus. Multisensory neurons comprise 63% of the cat population, and 27% of the monkey population. Adapted from Wallace and Stein, 1996.

Visual-Somatosensory   **Visual-Somatosensory**   **Visual-Auditory**

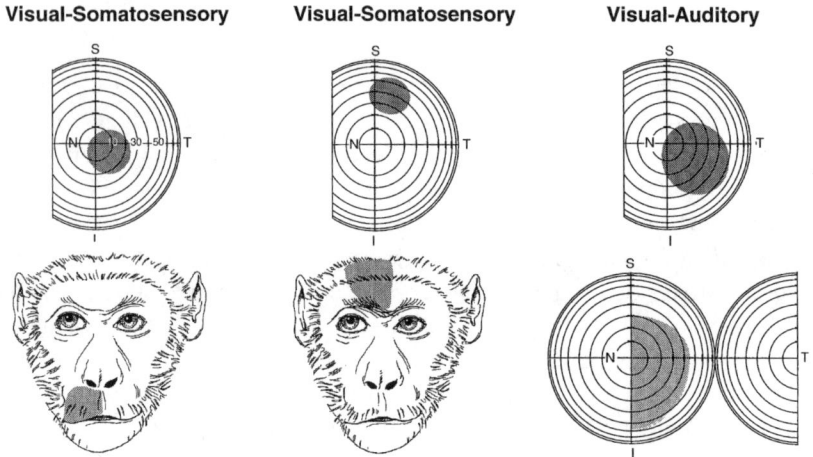

Figure 5.11. Receptive field correspondence in three multisensory neurons in the primate SC. The visual-somatosensory neuron shown on the left had a nasal and inferior visual receptive field and a somatosensory receptive field above the mouth. The visual-somatosensory neuron shown in the center had a nasal and superior visual receptive field and a somatosensory field near the top of the head. The visual-auditory neuron shown on the right had a superior and temporal visual field that corresponded to its auditory receptive field. In this figure, caudal auditory space on the right side is represented by the hemisphere attached to the central sphere. For purposes of representation this depiction of caudal auditory space has been folded forward. Adapted from Wallace et al., 1996.

$\rightarrow$

Figure 5.12. Multisensory neurons in primate SC integrate cues from the different modalities based on their spatial properties. In this visual-auditory neuron the visual and auditory receptive fields are illustrated on the left (visual and auditory space are plotted on a "bubble" which surrounds the monkey; 0° is directly in front of the animal, and the 90° cone shows the interaural plane). The visual and auditory receptive fields are depicted by shading and two locations of the auditory (A_1 and A_2) and visual (a bar whose movement is indicated by the arrow) stimuli are also shown. The middle panels of rasters, peristimulus histograms and bar graphs show the neuronal responses when the stimuli were positioned at different locations. Receptive fields are depicted separately in space for clarity in the right panels. Spatially coincident visual and auditory stimuli (top and middle) resulted in a characteristic response enhancement. However, when the auditory stimulus was positioned outside the excitatory receptive field (bottom), and thus out of spatial register with the visual stimulus, response depression resulted. From Wallace et al., 1996.

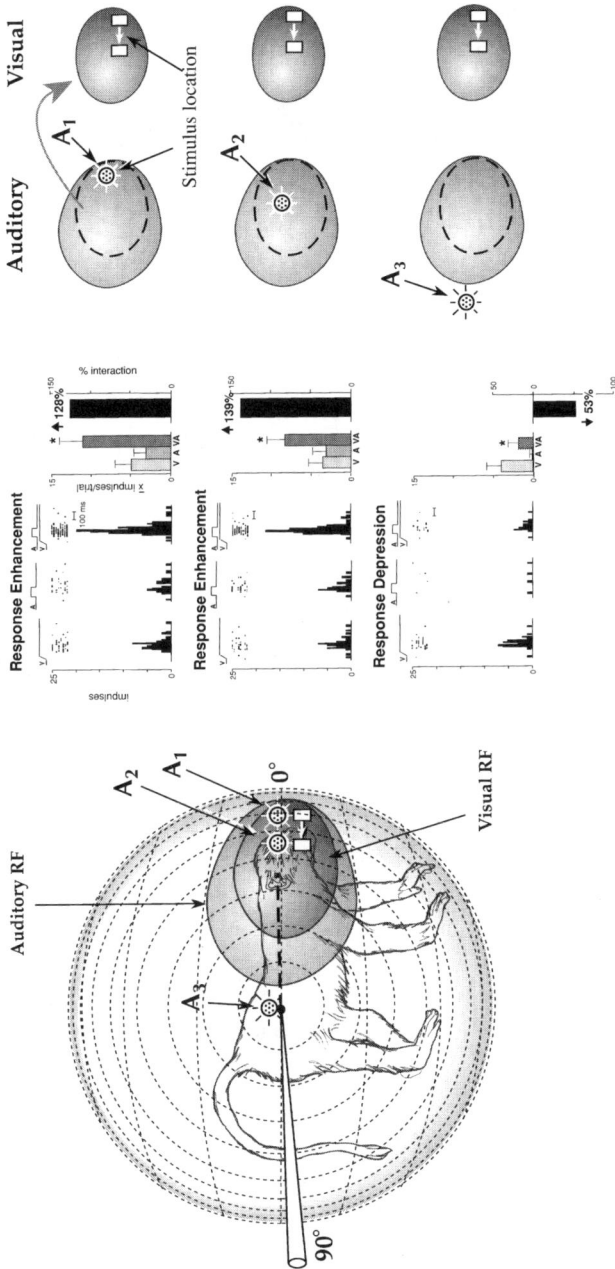

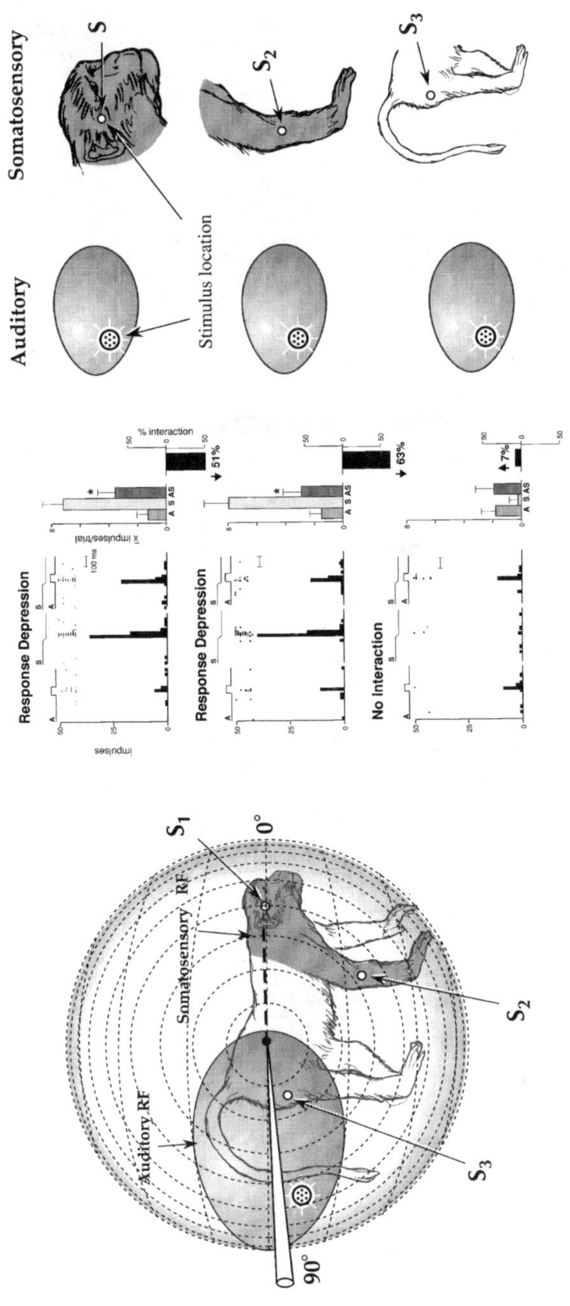
Somatosensory
Auditory
Stimulus location
Response Depression
Response Depression
No Interaction
% interaction
51%
63%
7%
Auditory RF
Somatosensory RF
90°
0°

←

Figure 5.13. Multisensory neurons lacking receptive field correspondence also lacked the normal capacity for multisensory integration. In this auditory-somatosensory neuron there was a gross mismatch between the locations of the different sensory receptive fields. The auditory receptive field was in far caudal (i.e., temporal) space, whereas the somatosensory receptive field was in far rostral sensory space (i.e., on the face and forelimb). When stimuli were presented within their receptive fields (top and middle panels) they could not be in spatial register with one another. The result was response depression rather than the characteristic response enhancement (see figure 5.12). When the somatosensory stimulus was placed outside of the somatosensory receptive field so that it was in spatial correspondence with the auditory stimulus, no interaction occurred. From Wallace et al. 1996.

Interestingly, two striking cases of receptive field mismatch were encountered in the monkey SC — "anomalous" examples which had never been encountered previously. Such neurons offered an unexpected opportunity to evaluate the spatial principle of multisensory integration. For example, in the somatosensory-auditory neuron shown in Figure 5.13, it was not possible for somatosensory and auditory stimuli originating from the same location in space to fall simultaneously within their respective receptive fields. Regardless of how the pair of stimuli were positioned, if one was in its receptive field, the other was outside its receptive field. As a result, coupling these stimuli in space produced no interaction. In order for the stimuli to fall within their respective excitatory receptive fields they had to be disparate in space. The result of this pairing was surprising and revealing: the two stimuli inhibited one another.

This example underscores the adaptive nature of constructing overlapping sensory representations in order to enhance the salience of stimuli derived from the same event. The convergence of inputs on the same neurons not only allows each of the modalities to have access to the same premotor circuitry in order to evoke appropriate eye, head, and limb movements using the same spatial coordinate system (as noted above), but is also adapted to combining weak and/or subthreshold inputs from different input channels in order to increase the probability of appropriate attentive and orientation behaviors. However, because it depends on topographic register among sensory representations, and this register is not always maintained in the behaving animal, the utility of such an organizational scheme is not as straightforward as it first seems.

2.5 *Eye movements displace the map of visual space in the SC and disrupt*
    multisensory alignment

When the head remains stationary but the eyes move, the alignment of the
visual and nonvisual coordinate frames of sensory space is disturbed. A visual
target which initially fell on one region of the retina will now fall onto a
different portion of the retina. Consequently, spatially-coincident visual and
nonvisual stimuli that previously fell within the respective excitatory receptive fields of a multisensory neuron and produced response enhancement may
no longer do so. Indeed, passively displacing the ears or eyes in anesthetized
cats can dissociate a multisensory neuron's receptive fields. Under these
circumstances, spatially coincident visual and auditory stimuli produce response depression rather than response enhancement (Meredith and Stein,
1996). Yet, it would hardly be adaptive if the nature of multisensory integration changed in this way every time the animal decided to move only its eyes
(or another peripheral sensory organ).

There are two obvious solutions to this problem: either a self-generated
(as opposed to the passive movements above) eye movement causes the
nonvisual maps to shift to maintain alignment with the visual map (perhaps via
an "efference copy" mechanism), or the animal keeps the major sensory
organs in alignment whenever multisensory integration is of greatest value.
Examples of the latter are evident in the hunting dog's stance when focused on
a quail's hiding place, and a cat's intense and aligned posture directed toward
something it detects in the grass. Evidence for the first solution was first
reported in an elegant series of studies conducted by Jay and Sparks (1984) in
the awake monkey. They described a population of SC neurons in which the
location of an optimal auditory stimulus would shift to "compensate" for small
changes in eye position. Large eye movements provoked only partial shifts.
The inference was that auditory receptive fields are dynamic and shift to
realign the visual and auditory maps. The presence of such an effect in cat SC
was initially denied (Harris et al., 1980), but such neurons have recently been
noted by others (Peck et al., 1995; Hartline et al., 1995). Similarly, Groh
(1996) has shown that the magnitude of a somatosensory response is altered
by changes in eye position.

It is clear from these studies that not all SC neurons shift their receptive
fields (or responses) with shifts in eye position. Furthermore, the two solutions
proposed above are not mutually exclusive. Nonetheless, both indicate that

maintaining alignment among the different sensory representations is of substantial importance. It is our contention that multisensory integration depends on the correspondence of receptive fields in multisensory neurons and that these attempts to maintain the alignment of maps reflect adaptations that operate to ensure that multisensory integration is not disrupted. This contention is consistent with the general finding of aligned multisensory receptive fields even in areas with poorly organized sensory topographies (see below).

2.6 *Similarities in multisensory integration across species and structures*

Receptive field alignment in multisensory neurons is a characteristic of these neurons in very different structures and species. As is evident from the foregoing discussion, this alignment is seen in both cat and monkey SC neurons. However, it has also been noted in SC neurons in hamster (Chalupa and Rhoades, 1977; Finlay et al., 1978; Stein and Dixon, 1979; Tiao and Blakemore, 1976), mouse (Drager and Hubel, 1975), rat (McHaffie et al., 1989) and guinea pig (King and Palmer, 1985). Multisensory receptive field alignment has also been noted in the optic tectum of nonmammalian species, including birds (Knudsen, 1982), reptiles (Stein and Gaither, 1981, Hartline et al., 1978) and fish (see Bullock, 1984), suggesting that it is a very old scheme of midbrain organization that antedates the evolution and radiation of mammals.

The registry of receptive fields in individual multisensory neurons is also characteristic of cortex, where such neurons are likely to be involved in very different functional roles. This organization is somewhat surprising in at least one area of the cat cortex, the anterior ectosylvian sulcus, where little topography is seen in the visual and auditory representations (Wallace et al., 1992). Nonetheless, multisensory neurons here have a very good spatial correspondence in their individual receptive fields. In addition, multisensory cortical neurons have integrative properties very much like those seen in SC neurons (Wallace et al., 1992, Ramachandran et al., 1993, Stein et al., 1993). These features appear to be common to multisensory neurons regardless of where they are found.

3. Concluding remarks

These observations are consistent with the idea that the fundamental principles of multisensory integration are general and supersede structure, function, and species.

This should not be interpreted to mean that there are no differences associated with different structures. Undoubtedly there are specialized multisensory properties in different brain regions which underlie their specialized functions (e.g., multisensory depression appears to be less frequent and less pronounced in cortex than in SC). Rather, it is likely that these specialized properties are overlaid on a fundamental core of common properties, such as those found here to be shared by the midbrain and cortex. Presumably, a constancy in the integrative principles at different levels of the neuraxis would provide coherence by ensuring that the salience of the same multisensory stimulus combinations will be roughly equivalent in areas of the brain involved in different response components of an integrated behavior.

Similarly, there are likely to be significant species differences in the impact that different stimuli will have on a neuron's responses (e.g., high frequency sounds are more effective in rodent than primate). Yet, based on current evidence, the most parsimonious hypothesis is that there are principles of multisensory integration that have been preserved in polysensory regions of the brain despite the rich diversity of mammalian species that radiated from a common ancestor more than 200 million years ago.

Notes

We thank Nancy London for her excellent assistance. Research described here was supported by NIH grants NS22543 and EY06562.

References

Biguer, B., Donaldson, I.M.L., Hein, A. and Jeannerod, M. 1988. Neck muscle vibration modifies the representation of visual motion and direction in man. *Brain* 111:1405-1424
Bullock, T.H. 1984. Physiology of the tectum mesencephali in elasmobranchs. In *Comparative Neurology of the Optic Tectum*, H. Vanegas (ed.), New York: Plenum, pp. 47-68.
Casagrande, V.A., Harting, J.K., Hall, W.C. and Diamond, I.T. 1972. Superior colliculus of the tree shrew: a structural and functional subdivision into superficial and deep layers.

Science 177:444-447.

Chalupa, L.M. and Rhoades, R.W. 1977. Responses of visual, somatosensory, and auditory neurones in the golden hamster's superior colliculus. *J. Physiol.* (Lond.) 207:595-626.

Clark, B. and Graybiel, A. 1966. Contributing factors in the perception of the oculogravic illusion. *Amer. J. Psychol.* 79:377-388.

Drager, U.C. and Hubel, D.H. 1975. Responses to visual stimulation and relationship between visual, auditory and somatosensory inputs in mouse superior colliculus. *J. Neurophysiol.* 38:690-713.

Finlay, B.L., Schneps, S.E., Wilson, K.G. and Schneider, G.E. 1978. Topography of visual and somatosensory projections to the superior colliculus of the golden hamster. *Brain Res.* 142:223-235.

Frens, M.A., Van Opstal, A.J. and Van der Willigen, R.F. 1995. Spatial and temporal factors determine audio-visual interactions in human saccadic eye movements. *Percept. Psychophysics* 57:802-816.

Goodale, M.A. and Murison, R.C.C. 1975. The effects of lesions of the superior colliculus on locomotor orientation and the orienting reflex in the rat. *Brain Res.* 88:243-261.

Graybiel, A. 1952. Oculogravic illusion. *AMA Arch. Ophthalmol.* 48:605-615.

Groh, J.M. and Sparks, D.L. 1992. Two models for transforming auditory signals from head-centered to eye-centered coordinates. *Biol. Cybernet.* 67:291-302.

Groh, J.M. and Sparks, D.L. 1996. Saccades to somatosensory targets. III. Eye-position-dependent somatosensory activity in primate superior colliculus. *J. Neurophysiol.* 75: 439-453.

Guitton, D. and Munoz, D.P. 1991. Control of orienting gaze shifts by the tectoreticulospinal system in the head-free cat. I. Identification, localization, and effects of behavior on sensory responses. *J. Neurophysiol.* 66:1605-1623.

Harris, L.R. 1980. The superior colliculus and movements of the head and eyes in cats. *J. Physiol.* 300:367-391.

Hartline, P.H., Kass, L. and Loop, M.S. 1978. Merging of modalities in the optic tectum: Infrared and visual integration in rattlesnakes. *Science* 199:1225-1229.

Hartline, P.H., Pandey Vimal, R.L., King, A.J., Kurylo, D.D. and Northmore, D.P.M. 1995. Effects of eye position on auditory localization and neural representation of space in superior colliculus of cats. *Exp. Brain Res.* 104:402-408.

Hughes, H.C., Reuter-Lorenz, P.A., Nozawa, G. and Fendrich, R. 1994. Visual-auditory interactions in sensorimotor processing: saccades versus manual responses. *J. Exp. Psychol. Human Percept. Perf.* 20:131-153.

Jay, M.F. and Sparks, D.L. 1984. Auditory receptive fields in primate superior colliculus shift with changes in eye position. *Nature* 309:345-347.

Kadunce, D.C., Vaughan, J.W., Wallace, M.T. Benedek, G. and Stein, B.E. 1997. Mechanisms of within- and cross-modality suppression in the superior colliculus. *J. Neurophysiol.* 78: 2834-2847.

King, A.J. and Palmer, A.R. 1985. Integration of visual and auditory information in bimodal neurones in the guinea-pig superior colliculus. *Exp. Brain Res.* 60:492-500.

Knudsen, E.I. 1982. Auditory and visual maps of space in the optic tectum of the owl. *J. Neurosci.* 2:1177-1194.

Knudsen, E.I. and Brainard, M.S. 1995. Creating a unified representation of visual and auditory space in the brain. *Ann. Rev. Neurosci.* 18:19-43.

McGurk, H. and MacDonald, J. 1976. Hearing lips and seeing voices. *Nature* 264:746-748.
McHaffie, J.G., Kao, C.-Q. and Stein, B.E. 1989. Nociceptive neurons in rat superior colliculus: Response properties, topography and functional implications. *J. Neurophysiol.* 62:510-525.
McIlwain, J.T. 1986 Effects of eye position on saccades evoked electrically from superior colliculus of alert cats. *J. Neurophysiol.* 55:97-112.
McIlwain, J.T. 1990. Topography of eye-position sensitivity of saccades evoked electrically from the cat's superior colliculus. *Visual Neurosci.* 4:289-298.
Meredith, M.A., Clemo, H.R. and Stein, B.E. 1991. Somatotopic component of the multisensory map in the deep laminae of the cat superior colliculus. *J. Comp. Neurol.* 312:353-370.
Meredith, M.A., Nemitz, J.W. and Stein, B.E. 1987. Determinants of multisensory integration in superior colliculus neurons. I. Temporal factors. *J. Neurosci.* 10:3215-3229.
Meredith, M.A. and Stein, B.E. 1986. Visual, auditory and somatosensory convergence on cells in superior colliculus results in multisensory integration. *J. Neurophysiol.* 56:640-662.
Meredith, M.A. and Stein, B.E. 1996. Spatial determinants of multisensory integration in cat superior colliculus. *J. Neurophysiol.* 75: 1843-1857.
Meredith, M.A., Wallace, M.T., and Stein, B.E. 1992. Visual, auditory and somatosensory convergence in output neurons of the cat superior colliculus: Multisensory properties of the tecto-reticulo-spinal projection. *Exp. Brain Res.* 88:181-186.
Munoz, D.P. and Guitton, D. 1985. Tectospinal neurons in the cat have discharges coding gaze position error. *Brain Res.* 341:184-188.
Peck, C.K. 1987. Visual-auditory interactions in cat superior colliculus: their role in the control of gaze. *Brain Res.* 420:162-166.
Peck, C.K., Baro, J.A. and Warder, S.M. 1995. Effects of eye position on saccadic eye movements and on the neural responses to auditory and visual stimuli in cat superior colliculus. *Exp. Brain Res.* 103:227-242.
Ramachandran, R., Wallace, M.T., Clemo, H.R. and Stein, B.E. 1993. Multisensory convergence and integration in rat cortex. *Soc. Neurosci. Abst.* 19:1447.
Roucoux, A. and Crommelinck, M. 1976. Eye movements evoked by superior colliculus stimulation in the alert cat. *Brain Res.* 106:349-363.
Sams, M. Aulanko, R., Hamalainen, M., Hari, R., Lounasmaa, O.V., Lu, S.-T. and Simola, J. 1991. Seeing speech: Visual information from lip movements modified activity in the human auditory cortex. *Neurosci. Lett.* 127:141-145.
Schiller, P.H. and Stryker, M. 1972. Single-unit recording and stimulation in superior colliculus of the alert rhesus monkey. *J. Neurophysiol.* 35:915-924.
Schneider, G.E. 1969. Two visual systems: brain mechanisms for localization and discrimination are dissociated by tectal and cortical lesions. *Science* 163:895-902.
Sparks, D.L. 1986. Translation of sensory signals into commands for control of saccadic eye movements: Role of primate superior colliculus. *Physiol. Rev.* 66:116-177.
Sprague, J.M. and Meikle, T.H. Jr. 1965. The role of the superior colliculus in visually guided behavior. *Exp. Neurol.* 11:115-146.
Stein, B.E. and Clamann, H.P. 1981. Control of pinna movements and sensorimotor register in cat superior colliculus. *Brain Behav. Evol.* 19:180-192.

Stein, B.E., Clamann, H.P. and Goldberg, S.J.1976. The control of eye movements by the superior colliculus in the alert cat. *Brain Res.* 118:469-474.

Stein, B.E. and Dixon, J.P. 1979. Properties of superior colliculus neurons in the golden hamster. *J. Comp. Neurol.* 183:269-284.

Stein, B.E. and Gaither, N. 1981. Sensory representation in reptilian optic tectum: Some comparisons with mammals. *J. Comp. Neurol.* 202:69-87.

Stein, B.E., London, N., Wilkinson, L.K. and Price, D.D. 1996. Enhancement of perceived visual intensity by auditory stimuli: a psychophysical analysis. *J. Cogn. Neurosci.* 8: 497-506.

Stein, B.E., Magalhaes-Castro, B. and Kruger, L. 1976b. Relationship between visual and tactile representation in cat superior colliculus. *J. Neurophysiol.* 39:401-419.

Stein, B.E. and Meredith, M.A. The Merging Of The Senses. MIT Press, Cambridge, Mass., 1993.

Stein, B.E., Meredith, M.A., Huneycutt, W.S., and McDade, L. 1989. Behavioral indices of multisensory integration: Orientation to visual cues is affected by auditory stimuli. *J. Cogn Neurosci.* 1:12-24.

Stein, B.E. and Meredith, M.A. 1991. Functional organization of the superior colliculus. In: *The Neural Basis of Visual Function*, A.G. Leventhal (ed.), Hampshire, U.K.: Macmillan, pp. 85-110.

Stein, B.E., Meredith, M.A. and Wallace, M.T. 1993. The visually responsive neuron and beyond: multisensory integration in cat and monkey. *Prog. Brain Res.*, 95: 79-90.

Sumby, W.H. and Pollack, I. 1954. Visual contribution to speech intelligibility in noise. *J. Acoust. Soc. Am.* 26:212-215.

Tiao, Y.-C. and Blakemore, C. 1976. Functional organization in the superior colliculus of the golden hamster. *J. Comp. Neurol.* 168: 483-504.

Wallace, M.T., Meredith, M.A. and Stein, B.E. 1992. Integration of multiple sensory modalities in cat cortex. *Exp. Brain Res.* 91:484-488.

Wallace, M.T., Meredith, M.A. and Stein, B.E. 1993. Converging influences from visual, auditory and somatosensory cortices onto output neurons of the superior colliculus. *J. Neurophysiol.* 69:1797-1809.

Wallace, M.T. and Stein, B.E. 1994. Cross-modal synthesis in the midbrain depends on input from association cortex. *J. Neurophysiol.* 71:429-432.

Wallace, M.T. and Stein, B.E. 1996. Sensory organization of the superior colliculus in cat and monkey. *Prog. Brain Res.* 112: 301-311.

Wallace, M.T., Wilkinson, L.K. and Stein, B.E. 1996. Representation and integration of multiple sensory inputs in primate superior colliculus. *J. Neurophysiol.* 76: 1246-1266.

Welch, R.B. and Warren, D.H. 1986. Intersensory interactions. In *Handbook of Perception and Human Performance, Volume I: Sensory Processes and Perception*, K.R. Boff, L. Kaufman and J.P. Thomas (Eds.), New York:Wiley, pp. 1-36.

Wilkinson, L.K., Meredith, M.A. and Stein, B.E. 1996. The role of anterior ectosylvian cortex in cross-modality orientation and approach behavior. *Exp. Brain Res.* 112: 1-10.

Wilkinson, L.K., Price, D.D., London, N. and Stein, B.E. 1993. Multisensory integration influences perception of visual intensity. *Soc. Neurosci. Abst.* 19:1802.

Wurtz, R.H. and Albano, J.E. 1980. Visual-motor function of the primate superior colliculus. *Ann. Rev. Neurosci.* 3:189-226.

CHAPTER 6

Implicit Perception in Action

Short-lived motor representations of space

Yves Rossetti
Vision et Motricité, I.N.S.E.R.M.

Perception is often conscious, which allows one to report about the object of perception, and to elaborate deliberate actions in the environment. But there is a considerable body of evidence that action does not always result primarily from such elaborated perceptual processes. These two statements are illustrated in Figure 6.1.

The experimental data reported in this chapter represent an attempt to summarize several instances of implicit (or non-conscious) use of sensory information during action. These data make it clear that the idea of a pure serial processing of sensory information from mental representation to action (see Figure 6.1: upper panel) is out of date. Examples of such implicit sensory representation will be obtained from various experimental fields ranging from psychology to neurophysiology and neuropsychology. These empirical data not only make the case for a dissociation between conscious awareness and motor representations of sensory targets, but also provide a basis for understanding how these two representations can interact (see Figure 6.1).

Current theoretical and experimental work about consciousness seems to make the assumption that implicit processing may be an intermediate level between brain mechanisms and consciousness (e.g. Rossetti, 1992; Bock and

Consciousness and Cognition, Vol. 7, No. 3, 520-558. Reprinted with modifications by permission of Academic Press © 1998.

Mind to Body Pathway

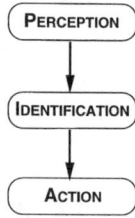

Descartes 1662

Pathway Dissociation

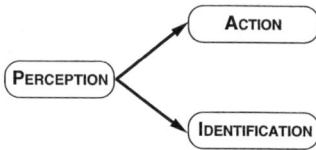

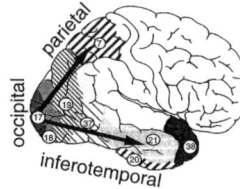

Pathway Interaction ?

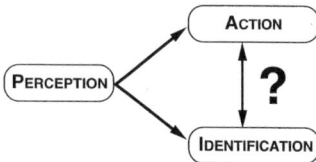

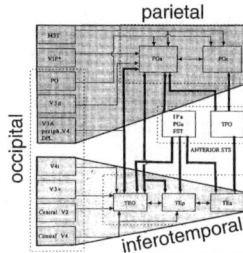

From Morel et Bullier 1990

Figure 6.1. Three conceptions of how light comes into muscles.
This figure displays three main conceptions of vision that are discussed in this chapter (as well as the concurrent evolution of scientific illustration). The Cartesian view of how light comes into muscles is clearly linear (higher panel). The pineal gland, seat of the mind, here considered as interface between sensory input and motor output, could be cognitively described as the potential locus of spatial representation. The modern conception of sensory processing now often focuses on dissociations between spatial and object vision, or vision for action and vision for perception (middle panel). These functionally segregated types of processing would fit two separable anatomical pathways leading visual information to the posterior parietal cortex (dorsal pathway) and to the inferior temporal cortex (ventral pathway). Given this functional dissociation, it is however worth noticing that several anatomical cross-connections between the two main streams have been described in monkey (Morel and Bullier, 1990) (lower panel).

Marsh, 1993). The qualifications of the 'cognitive unconscious' (Kihlstrom, 1987) may indeed be shared with conscious processing. It is therefore of prime interest to study how sensory inputs can be processed implicitly in the brain, and to investigate whether this processing can be distinguished from conscious operations. Dissociations between implicit and explicit information processing have been described in psychological functions such as memory, perception, motor behavior, aphasia, and prosopagnosia (cf. Weiskrantz, 1991). Other kinds of dissociations reported in perception or memory may also be tentatively listed together here. The terminology used to describe these dissociations can be clustered in two main groups.

On the one hand, the attention was drawn to the perceptual side, e.g. conscious vs. unconscious aspect of processing (see Bridgeman, 1992); localization vs. identification of the stimulus (e.g. Schneider, 1969; Paillard et al., 1983); spatial vs. object vision (Ungerleider & Mishkin, 1982); direct parameter specification vs. conscious representation (Neumann and Klotz, 1994); procedural vs. declarative (Cohen & Squire, 1980) and implicit vs. explicit (Shacter, 1987) memory systems.

On the other hand, the dichotomy was based on the possible responses provided by the subject: experiential vs. action (Goodale, 1983); cognitive vs. motor (Bridgeman, 1991); cognitive or representational vs. sensorimotor (Paillard 1987, 1991); reaching vs. grasping visuo-motor channels (Jeannerod, 1981); sensorimotor vs. conceptual components in memory (Perrig & Hofer, 1989); direct parameter specification vs. conscious perception ("what") vs. action ("how") visual processing (Goodale and Milner, 1992; Milner and Goodale, 1993) and pragmatic vs. semantic representations (Jeannerod and Rossetti, 1993; Jeannerod, 1994a).

This non-exhaustive list of dissociations described in the literature illustrates the great variety of approaches to implicit brain processing and shows the type of confusion that may result from attempts to reconcile their various concepts. In particular, a partial agreement can be found between several theories of automatic vs. controlled processes, or between several conceptions of dissociations within vision, but it appears difficult to unify these different views within one single line of thought.

The aim of this chapter will be twofold: first, I will review some evidence for implicit processing of sensory information during action from the two main lines of research outlined above. As sketched on Figure 6.1, I will summarize data indicating that explicit sensory processing and implicit processing for action can be dissociated, but also that they can interfere. Second, I will focus our analysis on a restricted number of parameters, that will allow proposing a common feature for most of the data reviewed. Indeed, this review will highlight the crucial role played by time-factors in many of the distinctions quoted above (see also Pisella and Rossetti, 2000; Rossetti, Pisella and Rode, 2000). Further attempts to integrate the concepts developed in these experimental fields within a unified framework may benefit from this observation.

1. Movement fundamentals

1.1 Eye movement

Following Bridgeman (1992: 76), it can be stated about exploratory movements that "The vast majority of behavioral acts are saccadic jumps of the eye, unaccompanied by any other behaviors". It is particularly striking that most eye movements are not consciously elicited. This is particularly true for the low amplitude saccades like the microsaccades occurring during fixation. But as is shown below, even larger saccades performed in response to a target jump exhibit a similar automatic component.

The orientation of gaze toward an eccentric target (presented as a step from an initial fixation point) is composed of a main saccade that usually undershoots the target, and a corrective saccade. Corrective saccades have an amplitude of about 10 % of the target eccentricity, and their latency is about half that of primary saccades (Becker and Fuchs, 1969). This reduced latency does not mean that corrective saccades are fully pre-programmed. Indeed, when the target jumps again during the main saccade (so called double-step stimulus), an appropriate corrective saccade is elicited with a short latency (Prablanc and Jeannerod, 1975). When the second jump is larger than about 4 degrees, then a new decision has to be made, resulting in an increase in latency of the secondary saccade, in the same range as the latency of the initial saccade. The authors thus suggested that the planning of a corrective saccade at the end of the main saccade can by-pass the normal decision time, i.e. that this fast eye movement can be unconsciously elicited.

Another interesting phenomenon related to eye movements is called **saccadic suppression**. As early as in 1900, Dodge noted that seeing his own eye motion in a mirror was impossible. Indeed it is easily demonstrated that one's own eyes can be seen in successive positions, but never in motion. Psychophysical studies revealed that human subjects are unaware of displacements occurring within the visual world if these displacements are tightly timed during the saccade (ex: Bridgeman, Hendry, Stark, 1975). Saccadic suppression thus refers to the apparent loss of perception occurring during saccades (Campbell and Wurtz, 1978).

The experimental paradigm using double-step stimuli with the second step synchronized with the first saccade has proved to be a powerful tool for investigating both eye and hand motor control. Later investigation of visual perception during saccadic eye movements demonstrated that eye and hand movements do not become disoriented after saccades as could be expected from the perceptual effect. Indeed, the eyes can saccade accurately to a target that is flashed during a previous saccade (Hallett and Lightstone, 1976, but see Honda, 1990). This maintenance of visually guided behavior may appear contradictory to the loss of perceptual information described as the saccadic suppression. One possible solution to this paradox comes from the comparison of the responses used. Saccadic suppression experiments required a symbolic response (verbal report or key-press), whereas maintenance of fairly accurate eye orientation requires a quantitative information (cf. Bridgeman, 1992: 79). This crucial distinction will be followed throughout the present chapter.

1.2 Arm movement

1.2.1 Reaching
As for eye saccades, two phases are classically described in arm movements. Reaching movements are initiated in a ballistic way, but are then subjected to several sensory feedback loops. Given this parallelism, the effect of abrupt stimulus change on arm movements has been investigated in conditions allowing or not the conscious detection of this change by the subject (see review in Rossetti et al., 1998).

1.2.1.1 *Pointing.* The saccadic suppression paradigm described above has also been applied to arm responses. In one early experiment, subjects were asked to point to the position of a target that had been displaced during the saccade (by a stroboscopic induced motion) and then extinguished (Bridgeman et al., 1979). These authors made similar conflicting observations for eye movements: the saccadic suppression effect was not followed by a motor disorientation. Moreover, it was found that a pointing movement following a target jump remained accurate irrespective of whether this displacement could be verbally reported or not. These experiments therefore suggested that two psychophysically separable visual systems can be distinguished, one system for a "cognitive" response, and a second one for motor behavior.

Experiments aimed at exploring the type of sensory information involved in motor control further explored this interesting phenomenon. Goodale, Pélisson and Prablanc (Goodale et al., 1986; Pélisson et al., 1986) asked subjects to point as fast and as accurately as possible to visual targets presented in the dark. In half of the trials, the target simply jumped from a central position to randomly selected positions in the peripheral visual field. In the other trials, the target made a further jump time-locked to the saccade, so that the second target was either closer or farther than the first, but always in the same direction. Subjects were never aware of the second target jump, and could not even guess its direction. Nevertheless, it was clearly shown that not only the eye (after a corrective saccade) but also the hand reached the target in all cases, although they were both initially directed toward the first target. It was concluded that vision of the moving hand was not necessary to control the movement, and that movement trajectory could be updated without the subject knowing it. This study thus demonstrated that perception of target position could be dissociated from visuomotor response directed to that target, i.e. that different types of visual computation are made for visual perception and visuomotor control. Similar results were obtained when the second target jump altered movement direction instead of its amplitude (Prablanc and Martin, 1992). Again, neither the target change in location nor a kinesthetic sensation of correction were consciously detected. Since there was no visual information available apart from the target, the encoding of target in an external frame of reference by the conscious perceptual system could in both experimental situations be mislead to assume that the position of the target, because it was stable before the saccade, remained unchanged. The motor coding of target location was nevertheless correctly performed with respect to an egocentric reference allowing accurate movements.

One interesting feature of this automatic sensory processing in action is the particularly short latency that is measured between the target jump elicited in the environment and the motor reaction to it (see Rossetti et al., 1999). In the previously described experiments, usual visuomotor delays were about

110 ms. Strikingly, this value is very similar to that obtained when the target jump was not synchronized with the saccade and therefore could be detected by the subjects (Soechting and Lacquaniti, 1983; Komilis et al., 1994), indicating that conscious awareness may be dissociated from the automatic sensorimotor reaction.

1.2.1.2 *Grasping versus vocalizing.* In addition to these pointing experiments, automatic corrections were explored for more complex grasping movements. Paulignan et al. (1991a) reported a similar delay of motor response (about 100 ms) to a perturbation of the location of the object to be grasped. When the perturbation instead affected the object orientation or size, the motor reaction time (concerning the transport component of the action) increased to respectively about 110 ms (Desmurget et al., 1996) and 300 ms (Paulignan et al., 1991b). Related studies stressed the high speed of motor correction and investigated the delay of subjective awareness of the perturbations (Castiello et al., 1991; Castiello and Jeannerod, 1991). In these experiments, a simple vocal utterance (Tah!) was used to signal subject's awareness of the object perturbation. Comparison of the hand motor reaction time and the vocal reaction time showed that the vocal response consistently occurred after the motor corrections. As in the Paulignan et al.'s experiments (1991a, b), the onset of motor correction was about 110 ms after the object displacement, and about 280 ms after the change in object size. However, the vocal responses occurred in both cases about 420 ms after the object's perturbation. It was concluded that conscious awareness of the object perturbation lagged behind the motor reaction to this perturbation.

1.2.1.3 *Altering or prohibiting vision.* There is additional evidence for the implicit use of sensory information in motor control *per se.* In contrast to the above manipulations of the target, another possible approach is to manipulate sensory information about either the target or the acting arm. Jakobson and Goodale (1989) showed that exposure to an about three degree shift of vision through wedge prisms could not be detected by uninformed subjects. Nevertheless, subjects having to point at visual targets during such exposure demonstrated an on-line correction of the prism-induced bias in movement direction, resulting in a modified hand-path curvature of the first few movements realized during prism exposure. Once again, these results suggest that the sensorimotor system can be responsive to consciously undetected sensory events. Moreover, different arm trajectory types were observed between pre-exposure and post-exposure phases, which suggests that visuo-motor adaptation took place in uninformed subjects.

Another experiment investigated pointing movements made under prism exposure (Rossetti et al., 1993; Rossetti and Koga, 1996). In this experiment, subjects were asked to point as fast as possible toward visual targets. Finger trajectory and eye movement analysis showed that subjects took into account terminal errors (knowledge of result) in the processing of the next movement in the sequence. Subjects could initiate their pointing in the appropriate direction within a few trials of prism exposure. However, the terminal part of their movements exhibited an 'attraction' toward the virtual location of the target (displaced through the prisms), which resulted in an increased path curvature. A dissociation was thus observed between the initial movement direction, which subjects could easily update between trials, and the terminal movement direction which escaped this updating. Terminal movement direction was indeed subjected to an automatic sensorimotor processing driving the hand off the physical target (i.e. toward the seen virtual target). Trajectory analysis revealed that this automatic processing may be based on an on-line comparison of the proprioceptively defined hand position with the visually defined target location.

Still another example of non-conscious integration of sensory information used for action is provided by studies on the encoding of hand position prior to movement onset (Prablanc et al., 1979; Elliott et al., 1991; Rossetti et al., 1994a; Desmurget et al., 1995). Although subjects are not aware of using visual information about their hand prior to movement, they perform with better accuracy when this information is available. This implicit use of visual information was best demonstrated when view of the hand was displaced by wedge prisms whereas the target, located outside the prism field, was seen normally (Rossetti, Desmurget, Prablanc, 1995). Subjects performing pointing movements without sight of their moving hand exhibited a pointing bias reflecting their implicit use of visual information about hand position prior to movement. Interestingly, another study demonstrated that pointing accuracy was degraded when the view of the hand was removed two seconds prior to movement onset (Elliott et al., 1991), suggesting that such information has to be used immediately. Similar experiments can be performed by altering proprioceptive input about hand location, by vibrating an arm muscle tendon just prior to movement onset (Velay, Roll and Rossetti, unpublished). It is known that tendon vibration induces illusory motion of the adjacent joint. An interesting effect of time was again reported, since this illusory effect of the vibration needs more time to develop than the effect observed on pointing.

Gentilucci et al. (1995a) designed an experiment in which subjects had to grasp a cylinder without seeing their hand. Visual information about the object was provided in a mirror, so that the apparent visual size of the object could be dissociated from its actual size encoded through repeated grasping movements. Although the experimental manipulation of the cylinder to be grasped did not reach subject's awareness, measurement of the movement grasp parameters were affected by object size, in such a way that subjects adapted their grip size to the actual size of the grasped object. This experiment suggested that a motor representation of the object could be implicitly constructed from somesthetic inputs.

1.2.1.4 *Conclusions.* All experiments summarized in this section suggest that an implicit use of various sources of sensory information can be made before and during goal directed movements. Extension of the double-step paradigm to arm movements led to the hypothesis that the saccadic suppression effect assessed only a cognitive component of the visual system. The paradigm of fast motor corrections, applied to both reportable and non-reportable target or sensory perturbations, further suggests that the neural pathways leading to visual awareness are distinct from those involved in pure visuomotor processing.

1.2.2 *Visual illusions*
Another means to distinguish between perceptual and motor responses to visual stimuli is to take advantage of visual illusions. The main idea is that visual perception would be more sensitive to illusion than the visuomotor behavior. Substantial experimental support for this hypothesis can be found (Rossetti, 1999).

When a large structured background is displaced during visual fixation of a small target, the latter appears to move in the opposite direction. This phenomenon can be observed for both smooth (induced motion) and step (induced displacement) background shifts. Bridgeman et al. (1981) replicated a finding made on eye movements (Wong and Mack, 1981), and compared the amount of perceptual illusory effect with the pointing response to the extinguished target. They showed that the motor system was much less affected by the apparent motion than the "cognitive" system. It was concluded that apparent target displacement affected only perception whereas real target displacement affected only motor behavior, which provides a case for a double dissociation between "cognitive" and motor function. Interestingly, a detailed subject-by-subject analysis of a similar experiment showed that only half of the subjects exhibited a motor effect of the visual illusion (Bridgeman, 1991). This observation became all the more interesting when it was observed that interposing an 8 s delay before the response forced all of the subjects to use spatial information that is biased by the perceptual illusion, again replicating the finding made on eye movements (Wong and Mack, 1981). This result suggested that subjects may switch from motor to cognitive modes of sensory processing at differing delays after stimulus offset.

Agloti et al. (1995) made use of size-contrast illusions (Titchener's circle illusion) to explore the effect of visual illusion on a grasping action. In Titchener's circle illusion, two circles in the center of two circular array each composed of circles of either smaller or larger size appear to be different in size even though they are physically identical. The circle surrounded by larger circles appears smaller than the one surrounded by smaller circles. Using this principle, one can build configurations with central circles of physically different size that will appear perceptually equivalent in size. Using this smart version of the illusion adapted in pseudo-3D, Agloti et al. required subjects to grasp the central circle between thumb and index finger, and measured their maximal grip aperture during the reaching phase of the movement. Strikingly, they observed that grip size was largely determined by the true size of the circle to be grasped and not its illusory size. In a later study, Haffenden and Goodale (1998) compared the scaling of the grasp to a matching condition, in which subjects had to indicate the central circle size with thumb and index finger without reaching it. The effect of the illusion on the matching task was very similar to the mean difference in actual size required to produce perceptually identical circles, whereas it was significantly smaller in the grasp condition. This result suggests that matching object size with the fingers relies on an object representation similar to the perceptual representation. By contrast, the motor representation remained less affected by the illusion.

An elegant experiment was performed by Gentilucci et al. (1996) to explore the effect of static visual illusion on pointing behavior (see Figure 6.2). The Müller-Lyer illusion induces the perception of longer or shorter length of a line ended by arrows. When both arrows are directed to the center of the line, it appears shorter. When they are oriented away from the line, it

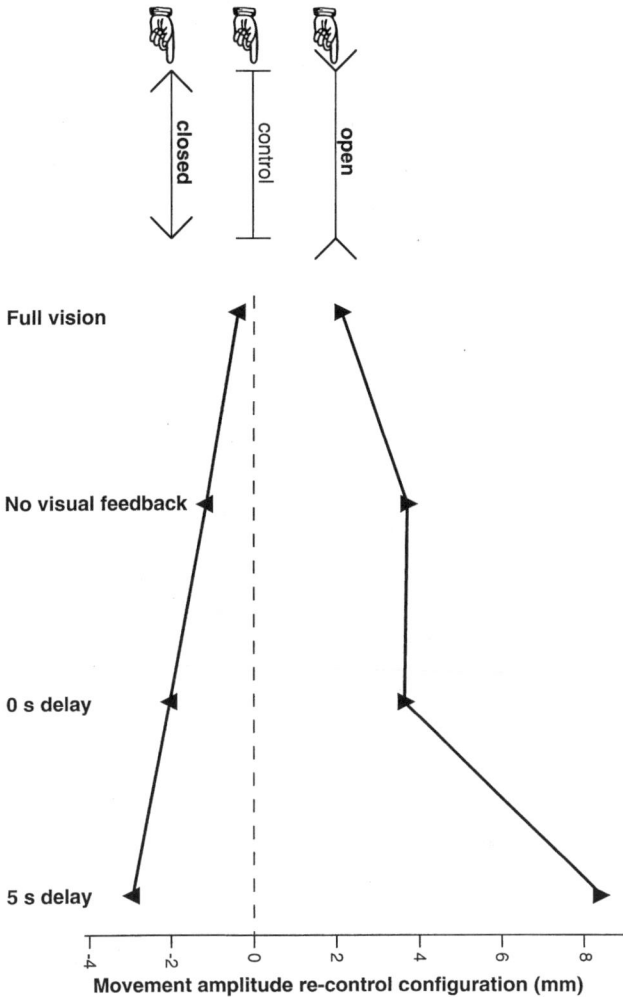

Figure 6.2. Visual illusion and action.
Pointing biases induced by the two configurations of the Müller-Lyer illusion. Movement amplitude tended to increase in the open configuration and to decrease in the closed configuration, i.e. in the same direction as the perceptual illusion. Values plotted on this figure were normalized by subtracting the value obtained for the control configuration. The effect of the illusion on pointing was very week in the full vision condition. It is noticeable that the effect of the illusion on movement amplitude increased when less information was available to the subject and when a delay was introduced between the stimulus presentation and the response. (Adapted from Gentilucci et al., 1996)

appears longer. Gentilucci et al. compared pointings made from one to the other end of lines ended by the two types of arrows used in the Müller-Lyer illusion, the subject having to look at the figure for two seconds prior to initiate his movement. Mean endpoints were significantly, influenced by the visual illusion, so that movement distance was increased or shortened according to the type of illusion produced (see Figure 6.2). The size of this effect is much smaller than the cognitive effect measured in a simple matching task to be 20% of line length (Rossetti, unpublished). Interestingly, early movement kinematics were altered, which suggests that the illusion affected the programming of the movement, not only its final execution. In addition, Gentilucci et al. showed that introducing delays between line observation and onset of movement proportionally increased the illusion effect on the pointing. These findings are very reminiscent of the idea that perceptual representation can influence the sensory processing devoted to action (Bridgeman et al., 1981). This influence becomes particularly noticeable as the delay between stimulus presentation and movement onset increases.

All experiments reported here provide evidence that visual illusions affect perception more intensely than motor behavior. They also raise an interesting point about the effect of the delay in responding to the stimulus. There is a clear convergence of several experimental paradigm to demonstrate that the effect of the illusion on motor behavior is strongly increased when the response delay increases.

1.2.3 *Visual masking*

Visual masking has been used as a probe to study conscious experience and cognition (see Price, chapter 2 of this book), and may explain some of the effects observed during saccadic suppression (Matin et al., 1972). Let us consider here some more specific implications of masking for action control. Taylor and McCloskey (1990) investigated the triggering of pre-programmed motor responses to masked stimuli. Three stimuli were tested: one *small* central LED with a 5 ms pulse, a *large* stimulus composed of the central LED plus four surrounding LEDs, and a *sequential* stimulus, where the central LED was shortly lit 50 ms prior to the onset of the surrounding LEDs. This last stimulus could evoke both metacontrast (masking by a surrounding shape) and backward masking (masking with a subsequent light of greater intensity than the small test light). Three motor responses of various complexity (from a single muscle group contraction to a predetermined movement sequence) were used. Reaction times (RT), as measured by EMG, were not affected by the masking of the small stimulus in the sequential condition. Comparison of RTs obtained for the large and for the sequential stimulus showed that motor response registered in the sequential condition was triggered by the short small stimulus preceding the masking surround. Although the simple response evoked a shorter RT, a similar effect of the masked stimulus was observed for the three types of movements tested. This experiment thus confirmed that motor reaction to a visual stimulus can be dissociated from the verbal report about detection of this stimulus (see also Fehrer and Biederman, 1962). As stated by Taylor and McCloskey (1990: 445), "the ability to react to such stimulus with a voluntary movement implies that sensory processing during reaction time does not have to be completed before motor processing can commence". Indeed motor RTs are usually shorter than the 500 ms delay that may be required before a conscious sensation can be elicited. Although these results confirmed that unconscious operations proceed faster than conscious ones, they cannot tell whether conscious perception and motor reaction are processed on parallel pathways with different thresholds, or whether these two responses can be elicited at different stages of serial sensory processing.

It appears that masking and metacontrast affect conscious perception of the stimulus although the ability to trigger a motor response remains largely intact. Neumann and Klotz (see 1994) have specifically explored several aspects of this phenomenon. They showed that similar effects can be observed on RT (measured by key pressing) even in a two-choice situation that required integrating form information with position information. In addition, this priming effect influenced error rate as well as speed of the motor response, and could appear despite of the use of variable stimulus-response couplings, showing that it is not restricted to pre-programmed responses.

The above results clearly questioned the classical sequential conception of sensory information processing for action. The psychophysical approach to the problem of sensorimotor coordination also suggests that early communication takes place between sensory and motor systems (Nandrino and El Massioui, 1995). This hypothesis was tested on auditory evoked potentials by requiring subjects to press a key with the left or the right hand in response to high and low tones presented dichotically. The main stages of information processing, target feature extraction, response choice and motor adjustments can be respectively affected by stimulus degradation, stimulus-response compatibility and presence or absence of a preparatory period. Manipulating these three variables may therefore specifically affect early or late components of event related potentials. However, an early interaction between stimulus degradation and preparatory period was observed, which suggests an overlapping of feature extraction and motor adjustments phases. This study provided more support for the possibility to prepare motor responses prior to stimulus-processing completion.

1.2.4 Conclusion
Three main conclusions can be drawn from the above experimental data. First, experiments on oculomotor and arm movement control demonstrate that an unconscious integration of visual information can take place during a simple action, and that an unconscious use of proprioceptive information can be made despite of a contemporaneous conflicting visual information. Second, masking experiments and visual illusions show that object perception can be dissociated from visuomotor processing of the same object. Third, several results suggest that the time required to elicit a motor response may be shorter than the delay observed between a stimulus occurrence and conscious awareness of this stimulus.

2. **Dissociation between conscious perception and action in brain-damaged patients**

One of the most striking dissociations between conscious and non-conscious processing observed in Neuropsychology is blindsight (reviews in Weiskrantz, 1989; Adams et al., 1990; Matthews and Kennard, 1993; Farah, 1994; Milner and Goodale, 1995; Stoerig and Cowey, 1997). After a lesion of primary visual areas, patients report no visual experience in whole or part of their visual field. However, some of them can still indicate the location of a contrasted visual stimulus through an eye or an arm movement (e.g. Pöppel et al., 1973; Perenin and Jeannerod, 1975). This phenomenon has raised new conceptions of extra-geniculostriate vision in human, and provided a model for questioning the neural and the phenomenal bases of consciousness (see Weiskrantz, 1991; Dennett and Kinsbourne, 1992; Stoerig and Cowey, 1993; Block, 1995). The discovery of a tactile equivalent of blindsight is more recent, but it has also stimulated both theoretical and empirical work (e.g. Lahav, 1993).

2.1 *Blindsight*

The discovery of the so called blindsight phenomenon has been made in a specific context of the knowledge about visual processes, and in particular about residual visual capacities in monkey (Humphreys and Weiskrantz, 1967). The search for functional dissociations within the animal visual system resulted shortly prior to the discovery of blindsight. The idea of a cortical 'focal vision' and a subcortical 'ambient vision' proposed by Trevarthen (1968) to account for dissociations observed in split-brain monkeys, and that of cortical blindness (impairing the 'What is it ?' system) and tectal blindness (impairing the 'Where is it ?' system) proposed by Schneider (1969) to account for observations made on rodents with occipital or tectal lesions, emerged a few years prior to the first report of blindsight (Pöppel et al., 1973).

2.1.1 *Historical context*
The early publications about blindsight described patients with lesions of the primary visual cortex, who exhibited remarkable residual capacities to orient their gaze or to direct their hand toward targets presented within their blind hemifield (Weiskrantz et al., 1974; Perenin and Jeannerod 1975). Provided the distinction made earlier from animal experiments (Schneider, 1969), this residual function was attributed to subcortical vision. The lack of awareness implied that patients usually felt like they were guessing, and was compatible with the idea that subcortical vision is unconscious. Indeed, similar results were then replicated in hemidecorticated subjects (e.g. Perenin and Jeannerod, 1978).

However, the two visual systems model as heralded by Schneider (1969) was rapidly challenged by new experiments and proved to be unsatisfactory (for a review, see Jeannerod and Rossetti, 1993: 442; Milner and Goodale, 1993:317). Another conception of vision as a dissociable function appeared, that considered both modes of vision as mediated by corticocortical pathways: the '**where**' function would depend on a dorsal stream projecting from primary visual cortex to posterior parietal lobule, and the '**what**' function on a ventral stream from primary visual cortex to inferotemporal cortex (see Mishkin et al., 1983). This well-known distinction has, however, not received unconditional support from recent electrophysiological and neuropsychological data (see Jeannerod and Rossetti, 1993: 443; Milner and Goodale, 1993: 317). The most recent experimental evidence is now converging toward a new interpretation of the cortical systems which emphasizes the final products of vision (see Figure 6.1: middle panel). It is now argued that the inferior parietal lobule of primates rather provides a set of modules specialized for visually directed action, whereas the inferotemporal cortex is primarily concerned with object recognition. The dorsal pathway would thus be concerned with **pragmatic** motor representations about **how** to act toward an object, and the ventral pathway would be involved in building more semantic representations about **what** the object is, in which the object appears as an identifiable identity (see Goodale and Milner, 1992; Jeannerod and Rossetti, 1993; Milner and Goodale, 1993; Jeannerod, 1994a).

Let us review some of the recent neuropsychological evidence that has prompted a reappraisal of the respective functions of the two cortical pathways and will be crucial for interpreting the blindsight phenomenon. Patients with optic ataxia, following a lesion of a restricted area of the posterior parietal lobule, have difficulties in directing actions to objects seen in peripheral vision, although they are not impaired in the recognition of these objects (Jeannerod, 1986; Perenin and Vighetto, 1988; Jakobson et al., 1991). They exhibit deficits not only in their ability to reach to the object, but also in adjusting the hand orienting and shaping during reaching. These results strongly suggest that the posterior parietal cortex plays a crucial role in the organization of object-oriented actions, whether the visual processing required for a given action is concerned with spatial vision (location) or with object vision (size or shape) (see Jeannerod, 1988; Jeannerod and Rossetti, 1993). Interestingly, a reciprocal dissociation was reported by Goodale et al. (1991) in a patient (D.F.) who developed a profound visual-form agnosia following a bilateral lesion of the occipitotemporal cortex. Strikingly, despite her inability to perceive the size, shape and orientation of visual objects, D.F. performed quite accurately when instructed to perform movements toward these objects. This observation suggests that, during action, D.F. could still process visual information about the objects intrinsic properties she could not perceive. Optic ataxia and visual agnosia patients clearly make the case for a double dissociation between perceptual recognition of objects and object oriented action.

Although they cannot accurately perform a goal directed action, optic ataxia patients are able to identify objects properly. Reciprocally, patient D.F. studied by Goodale et al. (1991) could not describe objects she was able to grasp. It may be emphasized here that her primary visual area was spared. As a consequence, processing of visual information may have been disrupted only in the ventral pathway and spared in the dorsal pathway, which would explain why she could perform visually directed movements. The question therefore arises whether blindsight patients, with V1 lesions, would also exhibit a similar dissociation between perception and action.

2.1.2 Blindsight in action

Although various residual functions have been reported in cortically blind hemifields, the majority of them are related to extrinsic properties of objects, that is mainly location and motion (cf. Weiskrantz, 1989; Adams et al., 1990, but see Stoerig and Cowey, 1992). Research on this phenomenon may however be gaining interest if we consider the motor performance studied in optic ataxia. It can be predicted that, since patients with V1 lesions showed an ability to direct an eye or an arm movement toward a target presented within their blind field (and thus not consciously seen), they may also be able to process unconsciously orientation, size or shape of visual stimuli during action. Indeed the useful parameters of objects, whose processing is required for guiding an action include metric properties of object other than its direction and distance. The following series of experiments was designed to test whether this prediction about action in blindsight patients can be verified.

In these experiments, several patients were tested for their ability to process orientation or size of visual objects. They were presented with slots of variable orientation or with rectangular objects (of equal surface) but variable horizontal length. Their performance was assessed in three kinds of tasks (see Figure 6.3). In the **verbal** task, they were asked to produce forced-choice verbal guesses about stimulus orientation or size. In the **motor** task, they had to insert a card in the orientable slot, or to grasp the rectangle between thumb and index finger. In addition, they had to perform a **matching** task, in which they were asked to match the slot orientation by wrist pro-supination movements, or to match object horizontal size with their thumb-index grip. Performances were recorded on video tapes and analyzed frame by frame (spatial accuracy was 0.5 cm).

One of these patient (P.J.G.) was a 32 year-old man who presented with a complete right hemianopia due to a left medial occipital lesion (see Figure 6.3; see Perenin and Rossetti, 1993, 1995, 1996). He could discriminate motion direction in his hemianopic field (Perenin, 1991), but remained unable to discriminate between simple geometric forms (e.g. circles vs. triangles). When

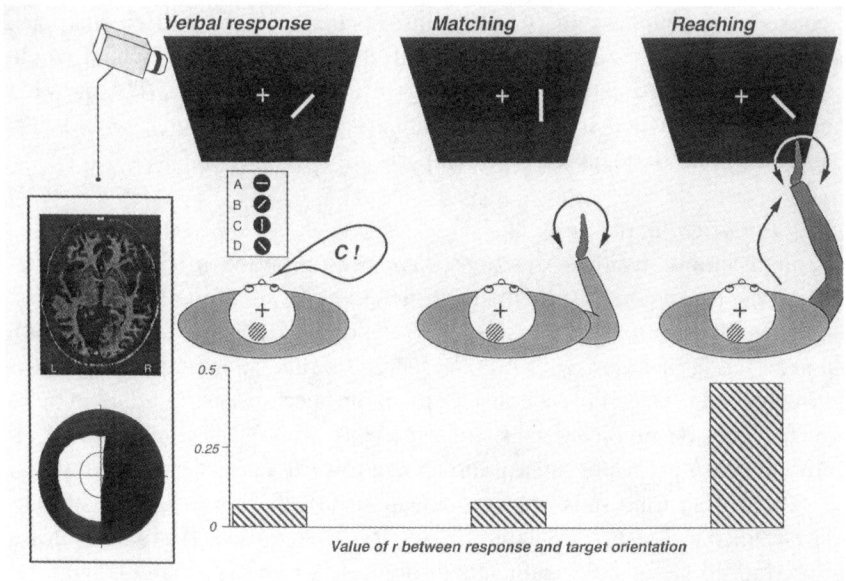

Figure 6.3. Dissociation between identification and action in blindsight.
Patient P.J.G. presented with a complete right hemianopia (see his visual field amputation).
He was tested for his ability to distinguish between four stimulus orientations by several
responses. The stimulus was a 18x3 cm slot presented 20 degrees left from fixation point in
the vertical black panel facing the subject. The slot was bordered by two bright white
stripes, producing a high contrast with the vertical panel. It was rotated between each trial
and presented in each of the four possible orientation (0, 45, 90, 135 degrees) in a random
order. Eye fixation was controlled during each trial. Verbal response was a forced choice
between the four possible orientations displayed on a sheet of paper. Matching response
consisted in showing the orientation of the target with wrist movements. Reaching response
was a natural aiming movement to the target, similar to posting a card in a mail box.
Performance was assessed by computing the correlation between the slot orientation and the
hand orientation, and a significant relationship was observed only in the motor task.
(adapted from Perenin and Rossetti, 1996).

instructed to perform each of the three tasks with his left hand in the normal
visual field, he performed as well as healthy subjects for either of the two types
of stimuli. When required to perform in the right side, he first explained that he
could not perform the task since he did not perceive the stimuli. After much
encouragement, he agreed to perform the task, performing verbal guesses and
making movements "by chance". Performance of P.J.G. in the orientation task

is displayed in Figure 6.3. The verbal guesses and the matching responses were at chance. However, a significant relationship between the slot orientation and hand orientation was obtained for the reaching responses (e.g. r = 0.463, p < 0.005). It should however be mentioned that, since he made consistent errors in reaching toward the panel, P.J.G. never succeeded in introducing the card in the slot even when it was well oriented (Perenin and Rossetti, 1996).

Similar results were obtained when P.J.G. had to grasp the horizontal objects. While he performed randomly in the verbal and in the matching task, his maximal finger grip aperture (measured during the transport phase) and his final grip aperture (measured at the time to contact with the horizontal panel) were both significantly correlated with the actual object size (rs > 0.414, ps < 0.01) (Perenin and Rossetti, 1996).

Another patient (N.S.) did not exhibit a constant off-target reaching, and thus was able to introduce the card in the slot to her own surprise (9 trials "in the slot" out of 40 trials, without being informed that only four orientations were used) (Rossetti et al., 1995). As P.J.G., she performed at chance level when asked to perform a verbal or matching task.

These data provide a further instance of dissociation between two modes of visual processing (knowing 'what' the object is vs. 'how' to grasp it). They indicate that the neural pathway responsible for space representation in action (or pragmatic representation) is much less dependent on V1 input than the pathway involved in visual discrimination, identification and perceptual awareness (see also Stoerig et al., 1985). This hypothesis is strongly supported by recent neurophysiological findings in the monkey. Indeed, selective brain cooling applied to V1 only partially affected the activity of visual areas (MT and V3A) that constitute the main input to the dorsal pathway, whereas it suppressed the visual activity of the ventral stream (see review in: Bullier et al., 1994; Girard, 1995). These results confirmed that neuronal activity of the dorsal pathway may arise from subcortical inputs such as colliculus and pulvinar (see Cowey and Stoerig, 1991; Bullier et al., 1994), and are likely to explain the ability of blindsight patients to process orientation and size to build a pragmatic representation of the goal to achieve.

2.2 Numb-sense: The sense of touch

About ten years after the discovery of blindsight, a case of implicit processing of somatic sensation following a lesion of central somatosensory areas was reported (Paillard et al., 1983). The patient was fully anaesthetized at the forearm level and could not report any tactile experience. However, she could point with her healthy hand to a location stimulated on the 'deafferented' forearm. At that time, this result was interpreted as a tactile analog of blindsight, i.e. as a dissociation between localization and identification. Other related observations have been described (Volpe et al., 1979; Weiskrantz and Zhang, 1987; Lahav, 1993; Brochier et al., 1994). These "numb-sense" observations have raised several questions relative to the type of representation involved in motor performance (cf. Rossetti, Rode, Boisson, 1995; 2000). Indeed several interpretations can account for it: dissociations between conscious-unconscious, motor vs. verbal, 'What' vs. 'Where' or 'What' vs. 'How' can be evoked (cf. Ettlinger, 1990).

A patient (J.A.) with a lesion of the thalamic relays of somesthetic afference allowed us to test these hypotheses. His lesion (left ventrolateral and ventroposterolateral nuclei) is shown on Figure 6.4. The tactile and proprioceptive deficit was complete on the right side of the body, and so stable that the patient could be tested over several years. To test J.A.'s tactile ability, stimuli were delivered to his right forearm and hand with the tip of a pencil, and left in place until the patient initiated his response. The investigator randomly stimulated locations that had been demonstrated on the left normal arm prior to the session. Since the patient did not feel the stimuli applied, he had to be instructed when to produce his response. No information was provided to the patient about his performance during the experiment. However, given the lack of explicit localization information, considerable encouragement was required.

The first experiment investigated J.A.'s ability to locate tactile stimuli applied to his right hand. The patient was blindfolded and motor and verbal performances were compared. Motor responses involved pointing movements using the left index finger. Verbal responses were obtained by a forced-choice paradigm, choosing among the possible stimulus locations demonstrated previously. When guesses were made by a pointing movement of the left hand to the stimulated right arm (pointing-on-arm condition), he consistently performed well above chance. Several sessions demonstrated that his performance improved when more distal areas were tested on the arm (from shoulder to hand), and when the number of possible stimuli was higher (from 3 to 8). Figure 6.4 provides an example of results obtained with 8 stimuli. The verbal forced-choice paradigm demonstrated that the deficit exhibited by J.A.

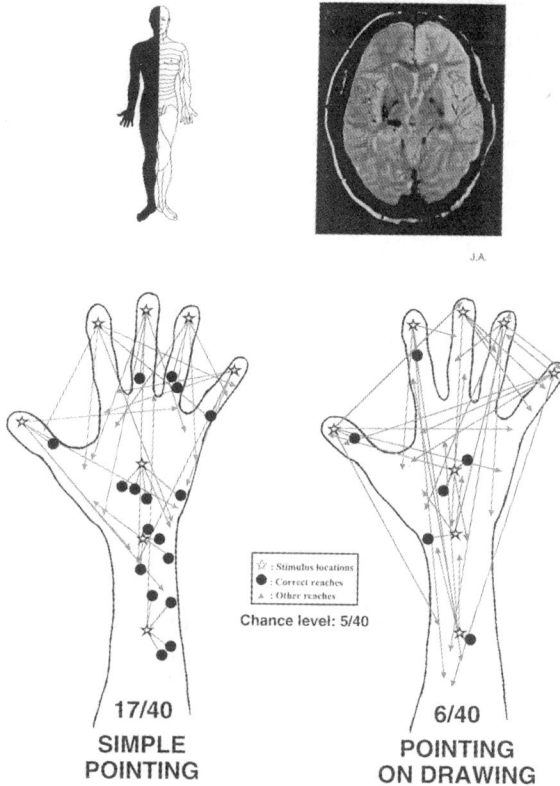

Figure 6.4. Numb-sense: a tactile equivalent to blindsight.
Patient J.A. was fully deprived of right-side tactile and proprioceptive sense after a left thalamus VL-VPL lesion (higher panel). Clinically, no tactile stimulus could be detected or located on his right body. He was then blindfolded and instructed to perform a pointing movement with the left normal hand toward the locus stimulated (stars) on the right hand. To assess his performance, we assigned the value one or zero to each trial, respectively for correct and incorrect responses i.e. inside *vs.* outside the stimulated territory (e.g. the whole finger in the case of fingertip stimulation). Although his errors (arrows) were much greater than in normals, he could strikingly perform well above chance level (p < 0.001) (lower panel). By contrast, when he had to make a similar pointing movement toward a picture of his right arm (the right arm being hidden from his view), he performed at chance level. In addition to the dichotomous correct-incorrect evaluation of the performance, the distance between each stimulus and the corresponding responses was measured, and significantly increased in the pointing-on-drawing condition. *(from Rossetti, Rode, Boisson, 1995)*

could not be simply explained by a conscious-unconscious dissociation, since verbal guesses were made randomly.

Second, we compared pointing responses made in different conditions, which involved distinct levels of representation. In order to test whether the somatic sensation was processed only for motor interaction with the stimulus or if it had also a value for proper location perception, we used another pointing response which was not directed to the stimulus. In this experiment, a drawing of an arm (scale 1) was placed on the table 20 cm left of his hidden, stimulated right arm. J.A. was then asked to point on the drawing to the point matching the location of the stimulus applied to the arm. In this condition, he had to know 'Where' to point instead of simply knowing 'How' to point, as was the case in the pointing-on-arm condition. Comparison of the two conditions clearly showed that the patient could not perform accurately in the pointing-on-drawing condition (see Figure 6.4).

In the same way that a neuroanatomical basis has been proposed to account for blindsight (cf. Bullier et al., 1994; Girard, 1995; Milner and Goodale, 1995), it is interesting to consider the possible pathways that are compatible with numb-sense (see Figure 6.5). Besides the main pathway from the ventroposterolateral nucleus of the thalamus to the primary somatosensory cortex, another pathway links the posterior nucleus of the thalamus to the posterior parietal cortex of the monkey (see Jones, 1985; Martin, 1985). The parietal opercular region (second somatosensory area) would be responsible for "object touch", whereas the posterior parietal areas would mediate "spatial touch", and can be considered as analogs of the visual inferotemporal and posterior parietal areas respectively (Mishkin, 1979; Ettlinger, 1990). Interestingly, these two regions are also highly interconnected (Pandya and Seltzer, 1982; Neal et al., 1987). When the main pathway is lesioned like in patient J.A., the other pathway may still provide information to the areas processing spatial information and projecting to the premotor cortex.

2.3 Numb-sense: The sense of proprioception

Patient J.A. was also tested for his ability to process proprioceptive information (Rossetti, Rode, Boisson, 1995). As for touch, J.A. was blindfolded. A tablet was used above which the patient's right fingertip was positioned. Figure 6.6 shows the results obtained in the pointing condition (The pointing + verbal condition depicted in Figure 6.6 will be described in the following).

Blind-sight   Numb-sense

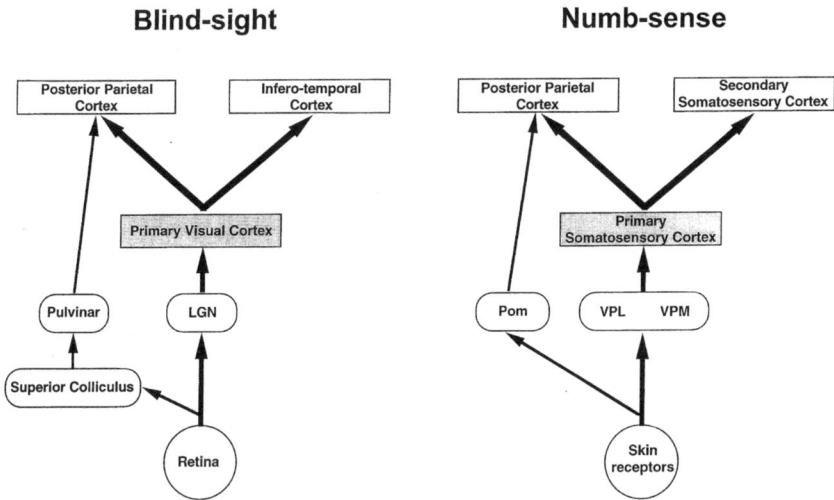

Figure 6.5. Anatomical pathways by-passing primary sensory areas for vision and touch. Blind-sight and numb-sense observations raise the interest for finding sensory pathways by-passing the primary areas. Such pathways can be isolated within the central sensory networks described for both visual and somatic systems. The lateral geniculate nucleus (LGN), the medial region of the posterior nucleus (Pom), the ventrolateral nucleus (VL), and the ventroposterior lateral nucleus (VPL), and the ventroposterior medial nucleus (VPM) are thalamic nuclei. Square boxes indicate cortical areas. Thickness of the arrows reflects the probable importance of projection for driving neurons in the target area in the absence of lesion.

The main thalamic projection to the primary sensory areas can be by-passed by a subcortical-cortical projection to the posterior parietal cortex for vision, and a projection from Pom to posterior parietal cortex for touch. Although the temporal cortex may play a role in object touch (Keating and Horey, 1971), there are similarities between the properties of inferotemporal cortex in vision and SII cortex in touch (Mishkin 1979, Ettlinger 1990). Patients P.J.G. and N.S. described in this chapter had a lesion of the primary visual cortex, whereas patient J.A. had a VPL lesion. The ability of these patients to perform an action toward an undetected stimulus may be sustained by these secondary pathways. *(Drawn from data found in Jones (1985: chap. 11), Martin (1985), Garraghty et al. (1991), Bullier et al. (1994)).*

The right, numb arm was manipulated by the investigator in such a way so as to place the right index fingertip on one of two locations. In order to avoid interference between tactile and proprioceptive information processing, attention was paid to provide the arm with as little tactile stimulation as possible. Since no conscious processing of arm proprioception was available to J.A., guessing responses were evaluated. In one session, he was asked to point underneath the tablet to the point corresponding to his right target-fingertip location. In another session, J.A. was asked to guess verbally whether his

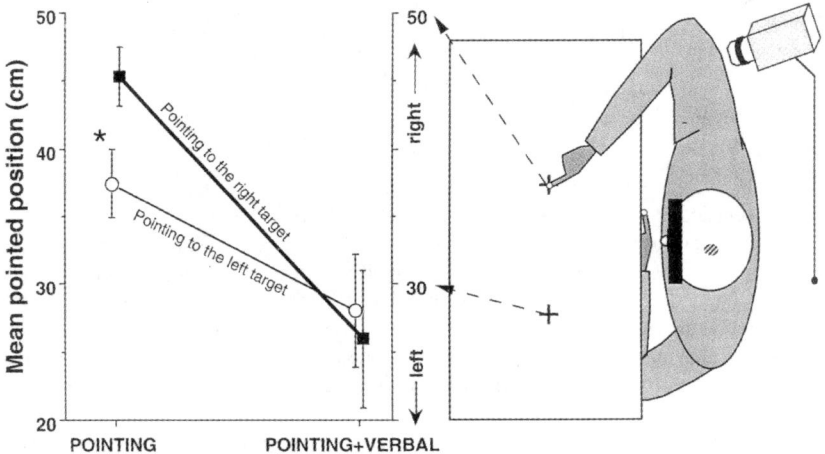

Figure 6.6. Interference between pragmatic and semantic representations in numb-sense.
Patient J.A. was tested for his ability to process proprioceptive information. He was asked to locate proprioceptively defined target location encoded with the left arm. Two locations were tested. When asked to produce a forced-choice verbal response (left vs. right) about the target, he performed at chance level.

When J.A. was asked to perform a pointing movement with the right hand underneath the table, he performed above chance, and could discriminate between the two positions (POINTING condition). Despite of the high variability found in the pointing responses, the difference between the pointings made toward the left target and toward the right target reached statistical significance. When he was required simultaneously to point to the target and to provide a verbal response, his performance was dropped to chance level, and he could no more discriminate between the two loci (POINTING + VERBAL condition). Simultaneous activation of the pragmatic and the semantic representations thus produced a deleterious effect on the motor performance. *(from Rossetti, Rode, Boisson, 1995).*

target-fingertip was on the right or on the left location. J.A. was significantly influenced by the target-finger locus in the pointing task only (Figure 6.6: pointing condition). In contrast, the distribution of verbal forced-choice responses was not significantly different from a random distribution.

2.4 Conclusion

The absence of significant performance in the verbal forced choice condition shows that sensory information (visual, tactile, proprioceptive) may be processed not only implicitly but specifically for motor purposes. Therefore the present result can be interpreted as a dissociation between a motor system responsible for the stimulus driven pointing and a semantic system responsible for the verbal depiction of the same stimulus location (see Perrig and Hofer, 1989). A similar kind of dissociation has been previously proposed between a 'What' system, responsible for a semantic processing, and a 'How' system, responsible for a pragmatic processing (Goodale and Milner, 1992; Jeannerod and Rossetti, 1993; Milner and Goodale, 1993; Jeannerod, 1994a). It is now accepted that the posterior parietal cortex is primarily involved in visual processing for action purposes (see also Jeannerod, 1994b; Rizzolatti et al., 1994; Sakata and Taira, 1994). By contrast, the dissociation observed in the present case holds for stimulus location (i.e. Where vs. How) instead of stimulus intrinsic qualities (i.e. What vs. How), and can be described for both tactile and proprioceptive stimuli. This view is strengthened by the results obtained when J.A. pointed on the arm drawing. In this case, he had to produce the same pointing movement, but with a more elaborated representation of where the stimulus was applied, and consequently his performance was reduced to chance level. The dissociation observed here would therefore not result from the difference in the response provided (pointing vs. verbal), but from the difference between the representations underlying the responses (How vs. Where). This dissociation fits with the more general description of sensorimotor and representational modes of spatial information processing that would respectively use a body-centered and an external frame of reference (see Paillard, 1991; Desmurget et al., 1998). Following Paillard's hypothesis, J.A. appears unable to process the tactile information at levels higher than a direct sensorimotor system, i.e. at more symbolic levels, as was the case in the matching task used with Goodale et al.'s patient D.F. and Perenin and Rossetti's Blindsight patients. The following section will further demonstrate that attempts to make use of a more elaborated representation of the stimulus will disrupt these patients implicit sensorimotor ability.

3. Interaction between conscious perception and action in brain-damaged patients

All the neuropsychological data provided in the previous section support the hypothesis of a dissociation between two streams of sensory processing, respectively devoted to perception and action. However, the degree of anatomical segregation between the dorsal and the ventral pathways remains questionable. Indeed, neuroanatomical connections have been described between the occipitoparietal and the occipitotemporal pathways (e.g. Morel and Bullier, 1990 (see Figure 6.1: lower panel), Knierim and Van Essen, 1992). In addition to these direct or indirect anatomical links, the two visual streams converge onto the superior temporal sulcus (Watson et al., 1994) and onto several cortical areas of the premotor cortex, which results in two "interconnected networks" (Ungerleider, 1995; Bullier et al., 1996; Boussaoud et al., 1996). The above hypothesis of two dissociated representations of space may "imply that the cortical mechanisms for object recognition and for object oriented action are selectively activated by the task in which the subject is involved" (Jeannerod and Rossetti, 1993: 445). Indeed, attention has been continuously focused on dissociation rather than interaction between the two modes of visual processing (e.g. Goodale and Milner, 1992). If this hypothesis holds true, then only one of the two types of representation may be activated at a single time. Alternatively, one may attempt to explore the possible functional interaction between pragmatic and the semantic representations, that could be allowed by the anatomical cross-connections between the dorsal and the ventral streams (Rossetti et al., 1994b). In this context, it becomes interesting to underline the specific conditions in which each of the two types of processing is involved in order to better understand to what extent they are dissociated (see Rossetti and Revonsuo, 2000). For 20 years, research on blindsight has been continuously exploring the extent to which stimuli can be perceived implicitly (position, movement, color, etc.). Because the discovery of new residual abilities in these patients was stressed over this period, the limiting factors of the implicit processing received less attention. Only a few attempts have been made to explore the limitation of the motor processing performed by patients deprived from the ability of identifying objects. One main limiting factor seems to be the time constraints attached to the pragmatic representation. Another crucial factor may be found in experiments trying to simultaneously activate both types of representations (motor and conscious).

3.1 *Time constraints*

An interesting observation was made on a blindsight patient (N.S.) performing task at different paces (Rossetti et al., 1995). It was observed that N.S. performance was first at chance level as she was reacting slowly to the stimuli. A significant performance appeared when movement latency decreased from about 500 ms (0 trial "in the slot" + 1 correctly oriented out of 40 trials) to about 300 ms (7 trials "in the slot" + 2 correctly oriented). This result suggest that sensory information responsible for blindsight in action is available only during a short period following stimulus presentation. An inverse relationship between the latency of response and performance was also reported in another blindsight patient (G.Y.) in an experiment comparing several types of response with several delays (Marcel, 1993). It was found that several detection reports made to identical trials could be dissociated. An eye blink response (latency about 290 ms) provided more accurate detection guesses than a button press (latency about 365 ms). In addition, a speeded condition produced better performance for both types of motor responses. A similar observation was also made on the agnosic patient D.F. presented above. Although she was able to preshape her hand in-flight, her grip size was no longer related to object size when a delay between object viewing and movement initiation was imposed (Goodale et al., 1994).

The same effect of time was also observed for touch and proprioception. Patient J.A. demonstrated that latencies up to 1 s for tactile stimuli and up to about 4 s for proprioceptive stimuli were compatible with above-chance performance in motor tasks, but longer delays completely disrupted his performance (see Figure 6.7). There is thus converging evidence arising from three sensory modalities that the pragmatic representation can only be expressed within a short delay following stimulus presentation (see Pisella and Rossetti, 2000).

These results may lead to a reinterpretation of the data obtained in the verbal and the matching task. In spite of encouragement to perform faster, it took more time for subjects to respond in the verbal task or in the matching

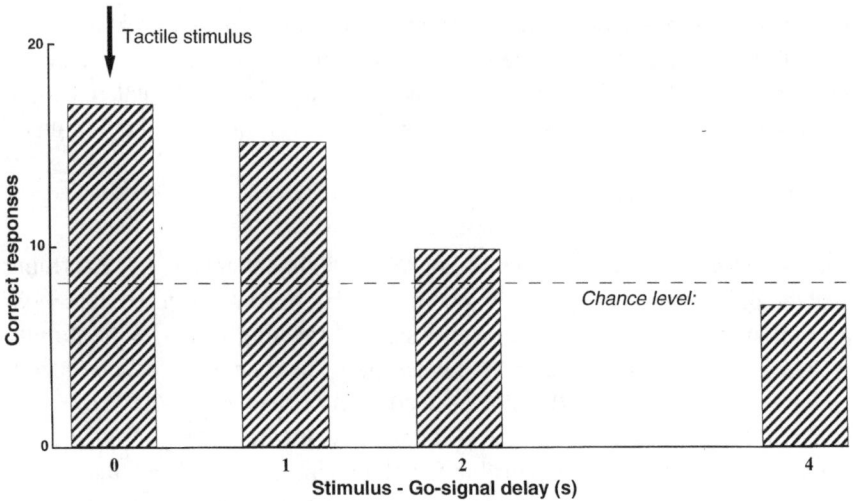

Figure 6.7. Time and implicit perception: short-lived representations in numb-sense.
This experiment was performed with the same methodology as in Figure 4. Six locations of stimulus were used on the patient's left hand. Each stimulus was presented 8 times in a random order. Since J.A. was never aware of the stimulus, a delay could be added between stimulus application and the go signal provided to the patient. No information was provided to the patient about his performance during the experiment. The effect of four delays were investigated in separate sessions. The patient was able to perform above chance level for delays up to 1 s, but his pointing was not influenced by stimulus location for longer delays.

task (in some trials between 1 and 2 s) than it took for the reaching and grasping tasks (see Perenin and Rossetti, 1996). It could therefore be argued that time is the decisive variable for explaining the difference between the fast reaching and the matching task. However, results obtained with numb-sense are not compatible with this interpretation. Indeed, J.A. performed at chance level when asked to point on the arm drawing, although the latencies observed in these cases were shorter than 2 s. This result suggests that the effect of verbalization cannot only be due to a problem with time. Experiments presented in section 5 will show that similar effects of verbalization and time can also be observed in healthy subjects.

3.2 Semantic-pragmatic interference

The deleterious effect of time on motor representations suggests that delayed actions may be based on more cognitive representations. Experiments were specifically designed to search for such a possibility. The logic of this paradigm, in contrast with the previous experiments, consisted of coactivating the two types of representation and then looking for effects on the patient's performance. In this experimental condition, patients were instructed to simultaneously produce a movement toward the stimulus and a verbal forced-choice response about the same stimulus. This task could be performed easily after little training, and the verbal response generally occurred during the second half of the arm movement. Three predictions could be put forward: 1) if there is a complete independence of the pragmatic and the semantic systems, then the verbal and motor responses will be performed without any modification of their respective performances. 2) if a transfer of information is possible from the pragmatic to the semantic stream, then the verbal response will gain accuracy. 3) if a transfer of information is possible from the semantic to the pragmatic stream, then the previous performance observed for the arm movement will disappear. Surprisingly, responses provided by the patients showed that the simultaneous task totally suppressed the ability to process implicitly the sensory information.

Patient N.S. could introduce the postcard in the slot although she could not perceive its orientation. When required to perform the same task, but to guess aloud the orientation of the slot during the movement, her motor performance was considerably deteriorated (1 vs. 9 trials in the slot) (Rossetti et al., 1995). In addition, no facilitation of the verbal response by the simultaneous pointing movement was observed. An observation possibly related to the competition between the two representations was also reported by Weiskrantz about blindsight (1989: 379): "...it was actually better to use less salient stimuli to improve performance by switching the subject into an 'implicit' rather than an 'explicit' mode, in which [the patient] depended upon his real but non-veridical experiences".

The effect of semantic-pragmatic coactivation was also investigated for touch and proprioception. In the tactile modality, there was a congruency between the pointing and the verbal responses when J.A. was required to produce the two responses at the same time. However, J.A.'s responses were not significantly influenced by stimulus location (correct trials: 6/40 for 6 possible locations). In addition, the distance between the stimulus and the response increased up to similar values as when pointing on the arm drawing (Rossetti, Rode, Boisson, 1995).

When the verbal and the pointing responses were produced simultaneously in the proprioceptive task (see Figure 6.6), they always were congruent, but the pointing performance was reduced to random (17/40, p > 0.10). This was confirmed by the mean locations reached for the two locations explored: the mean pointing toward the left location was located right of the mean pointing to the right location. As for touch, activation of a semantic representation of where the target finger was (required for the verbal response) disrupted J.A.'s ability to point to this finger. Section five (5.2.) will provide evidence that the interference effect does not appear when the verbal response is not specific to the representation of the goal.

3.3 Conclusion

When the pointing was delayed or associated with verbal responses about the stimulus for either visual, tactile or proprioceptive targets, the pointing performance was reduced to random. These findings first suggest that the implicit processing observed in these patients is specifically observed during aiming movements rapidly and directly oriented toward the stimulus. They also confirm that attempts to elaborate a semantic representation of the stimulus location can have detrimental effects on the relatively intact sensorimotor processing. Whether this may be due to interconnections or convergence between the dorsal and the ventral stream will be discussed further.

4.   Relations between conscious perception and action in healthy subjects

Results obtained with visual illusions have raised the problem of the reference frame used to perform an action. It was suggested that the egocentric reference frame could only be used during a restricted delay following stimulus presentation. Observations made on patients also revealed that the representation of space at work shortly after target presentation was likely to have a rather limited life-span. Action requires an encoding of metric properties of objects. In particular, object location must be encoded relative to the body. In the external frame of reference used by the perceptual system however the same point has to be encoded, but as a part of the visual context. It is this visual context that can induce illusory perception (see also Bridgeman, 1999). Consequently, we may apply the experimental paradigms used by the neuropsychological approach to healthy subjects to seek dissociation and interaction of the two frames of reference that can be used in action.

Relevant experiments performed by Graves and colleagues should be mentioned here (Meeres and Graves, 1990; Graves and Jones, 1992). Interestingly, these experiments were aimed at describing a possible analogue of blindsight in normal subjects. Short-duration masked patterns were presented tachistoscopically to subjects who were asked to produce three verbal responses indicating detection, identification and localization of the stimulus. Their results strikingly showed that undetected stimuli could be localized by the verbal guesses. Although it is not directly related to action, this observation shows that attempts can be made to seek "neuropsychological phenomena" in normals, and that threshold for unconscious localization may be lower than for unconscious detection (cf. Price, chapter 2 of this book). It may be hypothesized that the threshold for locating a target for an action purpose could even be lower.

4.1 *Time constraints*

Effects of the delay between stimulus viewing and movement onset have been repeatedly reported in the present chapter. In particular, the work performed on visual illusions demonstrated that motor behavior can be affected by perceptual illusions when the response is delayed by a few seconds (Bridgeman, 1991; Gentilucci et al., 1995b). Explanation of why a slight, but significant effect of the visual illusion could be observed in the Aglioti et al. (1995) and the Gentilucci et al. (1995b) experiments may be found in the retinal component of the illusion (see Gentilucci et al., 1995b). But it may also be found in the several seconds delay used between stimulus appearance and movement onset.

The results obtained with visual illusions suggest that healthy subjects may exhibit an effect of time on their natural aiming movements, as patients did. Several experiments were performed to seek such an effect.

Goodale et al. (1994) applied the delay paradigm used with their patient D.F. to a group of healthy subjects. They reported that many parameters of the grasping movement were affected by a 2 s delay introduced between stimulus viewing and movement onset. In particular, the opening and closure of the finger grip was altered and maximal grip size was reduced as compared to normal movements. Strikingly, movements delayed by 30 s and pantomimed movements performed beside the object were similar to those observed after 2 s. Allowing a good comparison with experiments performed on patients, this study further supported the view that brain mechanisms underlying perceptual representations are quite independent of those activated during action, and stressed the necessity for motor representations to have an on-line access to target parameters.

Grasping movements are by nature dependent on the metric properties of the object. Because they impose less constraint on the final posture, pointing movement may provide a better tool to address the problem of frame of reference. Following this idea, we studied pointing movements to memorized targets. The first series of experiment was carried out with visual targets briefly presented on a computer screen (Rossetti et al., 1994c). Subjects were required to point accurately to the target location when a go signal was provided, i.e. between 0 and 8 s following target presentation (see Figure 6.8). The results clearly showed that both constant and variable error parameters were strongly affected when the delay reached about 1 s, but then followed a plateau. Analysis of the pointing distributions observed with two experimental setups suggested that endpoints obtained at the shortest delay were coded using a reference frame centered on the starting position. In contrast, pointing distributions obtained for longer delays suggested that endpoints were encoded in an external frame of reference, that was mainly based on the target array used in the current experimental session (see Figure 6.8). It resulted that movements aimed to the same physical target could be affected by different biases according to the delay and the experimental set-up. As in Bridgeman's experiment (1991), it seems that the target was encoded as part of the visual context in the delay condition.

Experiments in monkey have shown that saccades to a memorized target were much less accurate when the delay became longer than 400 ms (White et al., 1994). The time course of saccadic errors measured with several delays is comparable to that obtained in our experiment. This may further support the idea that two distinct systems can be activated during eye movements as well as arm movements. Alternatively, this alteration of eye movements might also explain the results obtained for arm movements. A control experiment was thus realized on a subject performing an eye fixation during the memory

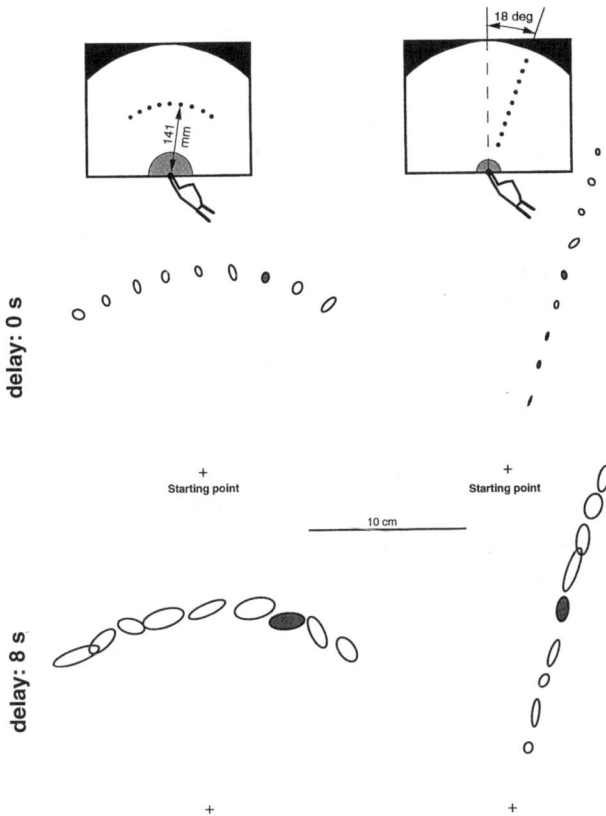

Figure 6.8. Time and the representations in action in healthy subjects.
Subjects had to point to visual stimuli briefly presented on a computer screen (300 ms). The delay between stimulus offset and the go signal provided to the subject (change in color) was randomly varied from 0 to 8 s within each session (0, 0.5, 1, 2, 4, or 8 s). Two different arrays of 9 targets (arc and line) were used in different sessions. Ellipses presented on the figure are confidence ellipse (60%) of the pointing scatter obtained for each target for a representative subject. The shaded ellipses correspond to scatters obtained for the same physical point in space. It can be observed that ellipse size is increased by the delay. The most interesting observation can be made about the ellipse orientation. With the 0 s delay, ellipses tend to be aligned with movement direction. With the 8 s delay (and other delays longer than about 1 s), comparison of the results obtained for the two target arrays shows that ellipse orientation tended to be aligned with the target array, and thus became dependent on the visual context provided by the experimental design. These results suggest that different frames of reference were used according to the delay. Movements are likely to be encoded in an egocentric reference frame for immediate responses, whereas they may be encoded in the extrapersonal space after the delay. (From Rossetti et al., 1994c; 2000).

delay. Pointing scatters obtained in this condition were larger, but they were elongated as in the previous experiment.

Our findings may result from the strong capacity of vision to process information about several spatial targets in parallel. Although only one of the targets forming the visual context appeared in each trial, they could each be located relative to each other using the additional reference provided by the screen border. A similar experiment was designed to determine whether the above results were dependent on the sensory modality. In order to provide the subject with a minimal side-information during each target location encoding, the proprioceptive modality was chosen. For a wider generalization of the results obtained with vision, targets were proprioceptively defined in the subjects sagittal plane, by a passively-guided movement of the left arm, the index finger being briefly kept on the target. Subjects were then required to point with their right index to the memorized target location. Again, two different arrays of 6 targets were tested in separate sessions, with two delays (0 and 8 s). As for vision, the effect of the context was strongly apparent only after the 8 s delay, so that pointing scatters became elongated in the same direction as the target array (Rossetti and Régnier, 1995; Rossetti et al., 1995; Rossett and Procyk, 1997). In a similar link between memory and extraction of auditory subjective patterns, MEG experiments have shown that subjective auditory features, such as pitch, are likely to be extracted from objective tone parameters before storing acoustic information in memory (Winkler et al., 1995). Also, recent studies have found that the visual processing of a dorsal attribute is performed faster than that of a ventral attribute (Pisella et al., 1998; Rossetti et al., 1999)

These experiments strongly supported the existence of two distinct ways of encoding spatial information for action. As it has been suggested earlier by Bridgeman (1991), an immediate sensorimotor system would depend on an egocentric frame of reference, whereas a second, slower system would represent the target within an external context. As a function of the delay between target encoding and the motor response, the result of the action would exclusively reflect one type of organization (Pisella and Rossetti, 2000). Alternatively, and according to Gentilucci et al. (1995b), the effect of time shown in Figure 6.2 suggests that the two systems can gradually interact. The experiment reported in the next section attempted to coactivate the two types of representations in healthy subjects.

4.2 *Semantic-pragmatic interference*

The effect of simultaneous activation of semantic and pragmatic representations was tested in healthy subjects pointing to memorized proprioceptive targets (Rossetti et al., 1995). Prior to the session, subjects were instructed to associate a number with each of the six target positions from the arc array. As in patients, they were then required to speak aloud the number corresponding to the pointed target during each movement (target number condition). In a control condition, subjects were required to count backward aloud (from 6 to 1 and so forth), so that an utterance (without spatial content) would be provided during each movement (number condition). The orientation of the pointing scatters was analyzed as in the experiments reported in the previous section. Distributions of these orientations are shown in Figure 6.9. It can be seen that the "target number condition" affected only the distribution obtained for the 0 s delay, so that no difference was observed between the two delays. Indeed, results of the "number condition" were comparable to those previously obtained without verbal response (see 5.1.) and with a less specific verbal response (Rossetti and Régnier, 1995). This finding suggests that activation of a semantic representation of target position had the same effect as the memory delay. Indeed, similar distributions were observed after a delay but without verbalization and without delay but with verbalization of target position. In other words, it is likely that the specific verbalization forced the motor system to immediately use the same frame of reference as was normally used after the delay, namely an external frame dependent on the target array.

4.3 *Conclusion*

These experiments performed on healthy subjects were aimed at replicating the situations found to activate (sequentially or simultaneously) a motor and a more cognitive representation of the goal. A great convergence is found when results of the present section are compared to those presented in section four. Patients could perform accurately only in the "natural" condition, in which they neither delayed the response, nor attempted to represent the stimulus at a higher cognitive level (matching, pantomiming, verbalizing). The performance observed in healthy subjects in the same "natural" condition may also suggest that a motor representation was used, whereas the representation used after the delay or the verbalization became contingent upon the external

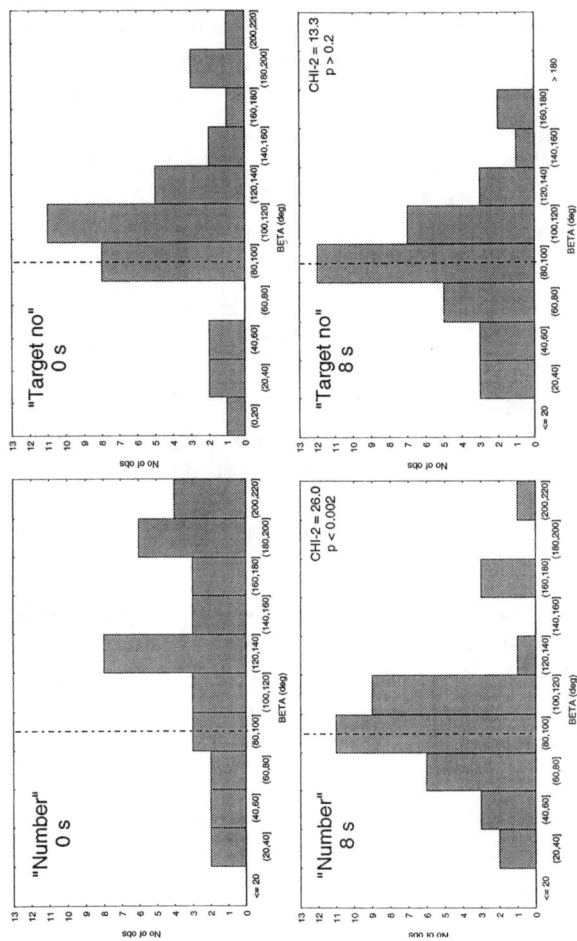

Figure 6.9. Simultaneous activation of pragmatic and semantic representations in healthy subjects.
This experiment involved a similar methodology as is described in Figure 8, unless the target location was defined in the sagittal plane through a movement of the left arm guided by the investigator and left in place for about 300 ms. Histograms show the distribution of ellipse orientations relative to movement direction obtained in 6 subjects for 6 target locations located on an arc array. In addition to the experiment presented on Figure 8, the subject was required to perform a verbal task during each pointing movement. When the verbal response consisted in counting backward from 6 to 1 repeatedly, providing a single number during each movement, confidence ellipses orientation varied with the delay: as for visual targets, they tended to be aligned with the target arc (i.e. angle = 90°) only for the longer delay. By contrast, when the number provided by the subject expressed target location (1 for the nearest to 6 to the farthest), ellipse orientation were clustered around 90° for both delays. This result strongly suggests that building a semantic representation about target location is responsible for a change in the frame of reference used by the motor system, which becomes dependent upon the target array. *(drawn from Rossetti and Régnier, 1995).*

context. The effect of the simultaneous verbalization was crucial to demonstrate that segregation of the two representations is based not only on the response delay.

5. General conclusions

Considered together, empirical data reviewed here show that the implicit processing of sensory information for action is not an anecdotal curiosity, nor restricted to some very limited control process operating at a low level of motor organization. They rather provide evidence for two distinct representations of space and suggest that these sensory representations can interfere before motor output, which allows us to better understand spatial functions involved in action. Automatic processing has long been thought to result from a bottom-up processing and unrelated to current intention (see Imanaka and Abernethy, 1992: 690). Examples provided here demonstrate that this is not always the case. None of the patients or healthy subjects would respond to the stimulus unless they have been trained to do so or are instructed to perform a guessing action. This major top-down component suggests that the function studied here can be affiliated to the representations own by the 'cognitive unconscious' (Kihlstrom, 1987, 1993).

A considerable amount of evidence for short-lived motor representations can be found in the various experimental fields reviewed in this chapter. These examples of dissociation between behavior and awareness share the feature that a stimulus can affect an action in a way that is not congruent with the manner in which it is consciously represented. In this respect, all these examples depart from the naive conception of a linear process leading from sensors to muscles, such as depicted by Descartes (cf. Figure 6.1). Other common features can be listed below: (1) Dissociations reviewed here support a segregation of sensory processing between two streams respectively devoted to identification and action. (2) Sensory information can be processed at an implicit level to perform an object oriented action. This implicit pragmatic processing seems to be faster than the explicit one. (3) Although the two types of space representation can be dissociated under particular circumstances, other conditions provide evidence for a possible interaction between the two systems (see Rossetti and Revonsuo, 2000). (4) Pragmatic representations are at work in conditions where the subject performs an immediate goal directed movement directly aimed to a stimulus (see Pisella and Rossetti, 2000). (5) Any attempt to elaborate a higher level representation of the spatial goal to be achieved apparently disrupts the pragmatic processing of the goal. Demonstration of similar dissociation between the two systems in normal subjects as well as in brain damaged patients makes these distinctions all the more pronounced.

These common features allow us to outline a specific representation of space, whose limitations are coherent with its implication in action. First, and as emphasized by Milner and Goodale (1993), performing an action requires an on-line treatment of pertinent goal characteristics. Thus there may be no need for long-lived representations in such a process. Second, action is primarily concerned with virtually all metric qualities of objects in space that are useful to guide an action, and does not require a binding of all object properties into a conscious unified percept. Thus only a partial specification of the goal is requested in pragmatic representations (cf. Jeannerod and Rossetti, 1993).

The rather restricted function of pragmatic representations makes it a system highly specific to a simple action performed toward a physical goal. This partial representation is built up faster than identification, but it allows less general and less flexible types of response (cf. Lahav, 1993, 1996). In contrast to visual awareness (Milner, 1995), the sensory capacity of the motor system may be restricted to a single or few features of the goal at a single time. For example, when the location of J.A.'s hand was unexpectedly modified between two trials of a tactile numb-sense experiment, he consistently pointed to the previous location of his hand, even though he could discriminate between two hand positions during proprioceptive numb-sense sessions. In the same way, when Goodale et al.'s agnosic patient was required to perform a task in which she had to insert a T-shaped object into an appropriate slot, she failed even though she was able to perform the task with a simpler object characterized by a single orientation (Goodale et al., 1994). It may well be that such a complex task cannot be performed in a purely egocentric frame of reference (Milner, 1995). The interference observed between pragmatic and semantic representations also relates to different frames of reference. The verbal requirement may accelerate the elaboration of the context-dependent representation built in an external frame of reference, or increase its influence on brain structures responsible for movement control.

Discussion of the interference in terms of coactivation or competition should include anatomical data. Recent neuroanatomical data obtained in monkey provide a support for an extension of the segregation between the two visual systems in their frontal connections. A vision-for-action pathway would project from the dorsal visual system directly to the dorsal and ventral premotor cortex (Watson et al., 1994; Ungerleider, 1995; Jeannerod et al., 1995; Tanné et al., 1996). By contrast, projections from the ventral visual system to the premotor cortex are less direct and involve prefrontal areas. In the same way, projections from the two visual systems to frontal areas involved in eye movements can be distinguished (Bullier et al., 1996). Convergence of inputs in the premotor cortex may be more relevant than the interconnections found between the two streams. The anatomical organization of the projections to the premotor cortex may provide a basis for understanding the timing difference observed between motor reaction to a stimulus and identification of the same stimulus. Indeed, visual activation of neurons in area MT (or V5) (a major input to the dorsal stream) appears earlier than in V4 (a major input to the ventral stream) (see Bullier and Nowak, 1995; Nowak and Bullier, 1997). There seems to be no evidence in the literature, however, that the density of premotor projections from the ventral stream is higher than that from the dorsal stream. As far as eye movements are concerned, projections from the dorsal stream onto FEF (frontal eye field) are even more numerous than projections from the ventral stream (Schall et al., 1995). Two alternative hypotheses may then be put forward to explain why pragmatic representations are supplanted by semantic ones as soon as the latter have been elaborated. First, an ephemeral activation of the dorsal stream could account for short lived-motor representations. Second, a hierarchy of the frontal inputs may then be evoked. Indeed afferents from the ventral stream may project on more anterior frontal regions (prefrontal) than does the direct dorsal afferents. If one assumes that the more anterior areas of the frontal cortex are involved in higher level of action control (e.g. Norman and Shallice, 1986), ventral projections may well dominate dorsal projections after a short delay. It is also worth noticing that prefrontal cortex is a key structure for short-term (working) memory (Ungerleider, 1995: 774). Then it may be hypothesized that prefrontal cortex would be involved in forming representations in an external frame of reference, such as can be observed in the many delayed tasks described in this chapter.

6. Speculations

Lesions restricted to primary sensory areas seem to be compatible with the sparing of many activities performed in the whole sensorimotor system (involving sensory and motor areas from the parietal to the frontal cortex) (see also Stoerig and Cowey, 1993). It is remarkable that the most patent deficit observed after such lesion is the loss of conscious perception reported by these patients. In line with the current thinking about connections within the visual system, blindsight raises the question of the possible implication of V1 connections in perceptual awareness (see Milner, 1995). Synchronization has been considered as essential for conscious awareness (see Picton and Stuss, 1994). For example, synchronization of neuronal activity can be observed between V1 and V2, and V2 and MT, and may involve other areas (see Bullier and Nowak, 1995). Such a mechanism may theoretically explain why many functions of vision are partially spared after a V1 lesion (because of subcortical projections) while conscious awareness is lost. Although the current experimental data do not allow to speculate further, it should be mentioned that similar thinking has been evoked to explain conscious touch: primary component of S1 response recorded from human somatosensory evoked potentials (Libet et al. 1979) and backward projections from S2 to S1 reported in monkeys (Cauller and Kulics, 1991) could provide such physiological correlates of consciousness. As for vision, implication of several sensory areas including the primary area could explain the dissociation between the loss of consciousness and the sparing of other functions.

   If it is true that understanding the cognitive unconscious may provide a basis for exploring conscious events (Rossetti, 1992), then time factors should be worth considering in future investigations. However, one should be cautious not to amalgamate the simple processes reviewed in this chapter with the complexity of consciousness. Several discrepancies may be emphasized to illustrate possible misunderstanding. First, most of the phenomena reported here are subjected to statistical analyses and are endowed with a probabilistic nature. They can be observed and described only as such 'probable tendencies', that obviously contrast with the usual winner-takes-all decisions made about environmental stimuli in conscious perception, which provides a basis for global and integrated responses (see also Perrig and Hofer, 1989; Norman and Shallice, 1986; Merickle, 1992; Lahav, 1993, 1997; Cabanac, 1996). In addition, it should be again noted that consciousness may refer to different concepts. Let us consider for example Block's recent distinction between access consciousness and phenomenal consciousness (Block, 1995). Access-consciousness may be considered as a continuous stream of serial, unlinked sensory experience of the surrounding. Such phenomenon may share mechanisms with the short-lived representations described here and involved in specific responses. It is indeed questionable whether it actually belongs to consciousness at all (Revonsuo, 1995). However, phenomenal consciousness may well not be linked to a specific time nor restricted in duration (cf. Dennett and Kinsbourne, 1992). With this respect, it might relate more to the binding processes involved in other non-motor types of representations (see Rossetti and Revonsuo, 2000). If one considers that consciousness may have evolved from sensation (Cabanac, 1996), the distinction defended in this chapter between two types of perception may become crucial for understanding conscious states.

Notes

I am most indebted to Marc Jeannerod and Claude Prablanc for having inspired and encouraged this work. I wish to thank Jean Bullier, Yann Coello, Michel Desmurget, Isabelle Faillenot, Laurent Goffart, Peter Grossenbacher, Patrick Mazoyer, Denis Pélisson, Céline Régnier, Gilles Rode, Peter Weiss for their valuable collaboration, and thoughtful comments on the ideas presented in this chapter, Isabelle Faillenot, Jean Bullier and Maurizio Gentilucci for their authorization to make use of their figures or data. This work has been supported by grants from Région Rhône-Alpes.

References

Adams, A.J., Bodis-Wollner, I., Enoch, J.M. Jeannerod, M., Mitchell, D.E. 1990. Normal and abnormal mechanisms of vision. In: L. Spillmann and J.S. Werner (eds). *Visual perception: The neurophysiological foundations*. San Diego, Academic Press: 381-416.
Aglioti, S., DeSouza, J.F.X., Goodale, M.A. 1995. Size-contrast illusions deceive the eye but not the hand. *Current Biology* 5 (6): 679-685.
Becker, W., Fuchs, A.F. 1969. Further properties of the human saccadic system: eye movements and correction saccades with and without visual fixation points. *Vision Research* 9: 1247-1258.
Block, N., 1995. On a confusion about a function of consciousness. *Behavioral and Brain Sciences* 18 (2): 227- 287.
Bock, G.R., Marsh, J. (eds). 1993. *Experimental and theoretical studies of consciousness* (Ciba Foundation Symposium 174). Chichester: Wiley

Boussaoud, D., di Pellegrino, G., Wise, S.P. 1996. Frontal lobe mechanisms subserving vision-for-action versus vision-for-perception. *Behavioural Brain Research* 72, 1-15.

Bridgeman, B. 1991. Complementary cognitive and motor image processing. In: Obrecht, G. and Stark, L.W. (eds), *Presbyopia Research: From molecular biology to visual adaptation.* New York: Plenum Press, 189-198.

Bridgeman, B. 1992. Conscious vs unconscious processes: The case of vision. *Theory & Psychology* 2 (1) 73-88.

Bridgeman, B. 1999. Interactions between vision for perception and vision for behavior. In Y. Rossetti & A. Revonsuo (Eds.), *Beyond dissociations: Interactions between dissociated implicit and explicit processing.* In press.

Bridgeman, B., Hendry, D., Stark, L. 1975. Failure to detect displacement of the visual world during saccadic eye movements. *Vision Research* 15: 719-722.

Bridgeman, B., Kirch, M., Sperling, A. 1981. Segregation of cognitive and motor aspects of visual function using induced motion. *Perception & Psychophysics* 29: 336-342

Bridgeman, B., Lewis, S., Heit, F., Nagle, M. 1979. Relation between cognitive and motor-oriented systems of visual perception. *Journal of Experimental Psychology: Human Attention and Performance* 5: 692-700.

Bridgeman, B., Peery, S., & Anand, S. 1997. Interaction of cognitive and sensorimotor maps of visual space. *Percept Psychophys,* 59(3), 456-469.

Brochier, T., Habib, M., Brouchon, M. 1994. Covert processing of information in hemianesthesia: A case report. Cortex 30: 135-144.

Bullier, J., Girard, P., Salin, P.-A. 1994. The role of area 17 in the transfer of information to extrastriate visual cortex. In: A. Peters and K.S. Rockland S. (eds). *Cerebral Cortex, Volume 10.* New york, Plenum Press, 301-330.

Bullier, J., Nowak, L. 1995. Parallel versus serial processing: New vistas on the distributed organization of the visual system. *Current Opinion in Neurobiology* 5: 497-503.

Bullier, J., Schall, J.D., Morel, A. 1996. Functional streams in occipito-frontal connections in the monkey. *Behavioural Brain Research* 76, 89-97.

Cabanac, M. 1996. On the origin of consciousness, a postulate and its corollary. *Neuroscience and Biobehavioral Reviews* 20 (1): 33-40.

Campbell, F.G., Wurtz, R.H. 1978. Saccadic ommission: Why we do not see a grey-out during a saccadic eye-movement. *Vision Research* 18: 1297-1303.

Castiello, U. Paulignan, Y., Jeannerod, M. 1991. Temporal dissociation of motor responses and subjective awareness. A study in normal subjects. *Brain* 114: 2639-2655.

Castiello, U., Jeannerod, M. 1991 Measuring time to awareness. *Neuroreport* 2: 797-800.

Cauller, L.G., Kulics, A.T. 1991. The neural basis of the behaviorally relevant N1 component of the somatosensoty-evoked potential in SI cortex of awake monkeys: Evidence that backward cortical projections signal conscious touch sensation. *Experimental Brain Research* 84: 607-619.

Cohen, N.J., Squire, L.R. 1980. Preserved learning and retention of pattern-analysing skill in amnesia: Dissociation of knowing how and knowing that. *Science* 210: 207-210.

Cowey, A., Stoerig, P. 1991. Reflections on blindsight. In: A.D. Miler and M.D. Rugg (eds). *The neuropsychology of consciousness.* London, Academic Press. 11-37.

Dennett, D., Kinsbourne, M. 1992. Time and the observer: the where and the when of consciousness in the brain. *Behavioral and Brain Sciences* 15: 183-220.

Desmurget, M., Pélisson, D., Rossett, Y., & Prablanc, C. 1998. From eye to hand: Planning goal directed movements. *Neuroscience and Biobehavioral, 22*(6), 761-788.

Desmurget, M., Prablanc, C., Arzi, M., Rossetti, Y., Paulignan, Y., Urquizar, C. 1996. Integrated control of hand transport and orientation during prehension movements. *Experimental Brain Research* (in press).

Desmurget, M., Rossetti, Y., Prablanc, C. Stelmach, G., Jeannerod, M. 1995a. Integration of sensory cues about hand position used for a goal directed movement. *Canadian Journal of Physiology and Pharmacology* 73 (2): 262-272.

Dodge, R. 1900. Visual perception during eye movement. *Psychological Review* 7: 454-465.

Elliott, D., Carson, R.G., Goodman, D., Chua, R. 1991. Discrete vs. continuous control of manual aiming. *Human Movement Science*, 10, 4, 393-418.

Ettlinger, G. 1990. "Object vision" and "spatial vision": The neuropsychological evidence for the distinction. *Cortex*, 26: 319-341.

Farah, M.J. 1994. Visual perception and visual awareness after brain damage: A tutorial review. In: C. Umilta, M. Moscovitch (eds) *Attention and Performance XV Conscious and nonconscious information processing*. Cambridge: MIT Press, 37-76.

Fehrer, E., Biederman, I. 1962. A comparison of reaction time and verbal report in the detection of masked stimuli. *Journal of Experimental Psychology* 64: 126-130.

Garraghty P.E., Florence, S.L., Tenhula, W.N., Kaas, J.H. 1991. Parallel thalamic activation of the first and second somatosensory areas in prosimian primates and tree shrews. *Journal for Comparative Neurology* 311: 289-299.

Gentilucci, M., Chieffi, S., Daprati, E. 1995b. Visual illusion and action. *Neuropsychologia.*

Gentilucci, M., Daprati, E., Toni, I., Chieffi, S., Saetti, M.C. 1996. Unconscious updating of grasp motor program. *Experimental Brain Research*, 105, 291-303.

Gentilucci, M., & Negrotti, A. 1996. Mechanisms for distance reproduction in perceptual and motor tasks. *Experimental Brain Research, 108*, 140-146.

Girard, P. 1995. Bases anatomo-physiologiques de la vision résiduelle après lésion de l'aire visuelle primaire. *Revue Neurologique (Paris)* 151 (8-9)): 457-465.

Goodale M.A. 1983. Neural mechanisms of visual orientation in rodents: Target vs places. in Hein, A. & Jeannerod, M. (eds) Spatially oriented behavior, pp 35-62. Springer Verlag, New York.

Goodale, M.A., Jacobson, L.S., Milner, A.D., Perret, D.I., Benson, P.J., Hietanen, J.K. 1994. The nature and limits of orientation and pattern processing supporting visuomotor control in a visual form agnosic. *Journal of Cognitive Neuroscience* 6 (1): 46-56.

Goodale, M.A., Milner, A.D. 1992. Separate visual pathways for perception and action. *Trends in Neurosciences* 15 (1): 20-25.

Goodale, M.A., Milner, A.D., Jacobson, L.S., Carey, D.P. 1991. A neurological dissociation between perceiving objects and grasping them. *Nature*, 349, 154-156.

Goodale, M.A., Pélisson, D., Prablanc, C. 1986. Large ajustments in visually guided reaching do not depend on vision of the hand or perception of target displacement. *Nature* 320, 6064, 748-750.

Graves, R.E., Jones, B.S. 1992. Conscious visual perceptual awareness vs. non-conscious visual spatial localisation examined with normal subjects using possible analogues of blindsight and neglect. *Cognitive Neuropsychology* 9 (6): 487-508.

Haffenden, A., & Goodale, M.A. 1998. The effect of picturial illusion on prehension and perception. *Journal of Cognitive Neuroscience, 10*, 122-136.

Hallett, P.E., Lightstone, A.D. 1976. Saccadic eye movements towards stimuli triggered by prior saccades. *Vision Research* 16: 99-106.

Hommel, B. 1999. The prepared reflex: Automaticity and control in stimulus-response translation. In S. Monsell & J. Driver (Eds.), *Attention and Performance*. Cambridge MA: MIT Press. In press.

Honda, H. 1990. Eye movements to a visual stimulus flashed before, during, or after a saccade. In M. Jeannerod (ed), *Attention and Performance XIII: Motor representation and control*. Hillsdale NJ, Lawrence Erlbaum Ass Publ, 567-582.

Humphreys, N.K., Weiskrantz, L. 1967. Vision in monkey after removal of striate cortex. *Nature* 215: 595-597.

Imanaka, K., Abernethy, B. 1992. Cognitive strategies and short-term memory for movement distance and location. *The Quarterly Journal of Experimental Psychology* 45A (4): 669-700.

Jakobson, L.S., Archibald, Y.M., Carey, D.P., Goodale, M.A. 1991 A kinematic analysis of reaching and grasping movements in a patient recovering from optic ataxia. *Neuropsychologia* 29, 803-809.

Jakobson, L.S., Goodale, M.A. 1989 Trajectories of reaches to prismatically-displaced targets: evidence for "automatic" visuomotor recalibration. *Experimental Brain Research* 78: 575-587.

Jeannerod M. 1981. Specialized channels for cognitive responses. Cognition 10, 135-137.

Jeannerod, M. 1986. The formation of finger grip during prehension. A cortically mediated visuomotor pattern. *Behavioural Brain Research* 19: 99-116.

Jeannerod, M. 1988. *The neural and behavioural organization of goal-directed movements*. Oxford: Oxford University Press.

Jeannerod, M. 1994a. The representing brain: Neural correlates of motor intention and imagery. *Behavioral and Brain Sciences* 17 (2): 187-245.

Jeannerod, M. 1994b. The hand and the object: The role of posterior parietal cortex in formaing motor representations. *Canadian Journal of Physiology and Pharmacology* 72: 535-541.

Jeannerod, M., Arbib, M.A., Rizzolatti, G., Sakata, H. 1995. Grasping objects: the cortical mechanisms of visuomotor transformation. *Trends in Neurosciences* 18 (7): 314-320.

Jeannerod, M., Decety, J., Michel, F. 1994 Impairment of grasping movements following a bilateral posterior parietal lesion. *Neuropsychologia* 32 (4): 369-380.

Jeannerod, M., Rossetti, Y. 1993. Visuomotor coordination as a dissociable function: Experimental and clinical evidence. In: C. Kennard (ed.) *Visual Perceptual Defects*, Baillière's Clinical Neurology, International Practise and Research, Baillière Tindall Ltd/Saunders, 439-460.

Jones, E.G. 1985. *The Thalamus*. New York: Plenum Press.

Keating E.G., Horel, J.A. 1971. Somatosensory deficits produced by parietal-temporal cortical disconnection in the monkey. *Experimental Neurology* 33: 547-565.

Kihlstrom, J.F. 1987. The cognitive unconscious. *Science* 237: 1445-1452.

Kihlstrom, J.F. 1993. The psychological unconscious and the self. In: *Experimental and theoretical studies of consciousness* (Ciba Foundation Symposium 174). Chichester:

Wiley, 147-167.

Knierim, J.J., Van Essen, D.C. 1992.Visual cortex: Cartography, connectivity, and concurrent processing. *Current Opinion in Neurobiology* 2: 150-155.

Komilis, E., Pelisson, D., Prablanc, D. 1994. Error processing in pointing at randomly feedback-induced double-step stimuli. *Journal of Motor Behavior* 25 (4): 299-308.

Lahav, R. 1993. What neuropsychology tells us about consciousness. *Philosophy of Science* 60: 67-85.

Lahav, R. 1997. The conscious and the unconscious. In M. Carrier & P. Machamer (Eds.), *Mindscapes, Philosophy, Science and the Mind*. Pittsburgh: Pittsburgh University Press.

Libet, B., Wright, E.W., Feinstein, B., Pearl, D.K. 1979. Subjective referral of the timing for a conscious sensory experience: a functional role for the somatosensory specific projection system in man. *Brain* 102: 191-222.

Marcel, A. 1993. Slippage in the unity of consciousness. In: *Experimental and theoretical studies of consciousness* (Ciba Foundation Symposium 174). Chichester: Wiley, 168-186.

Martin, J.H. 1985. Anatomical substrates for somatic sensation. In: Kandel E.R. & Schwartz J.H. (eds), *Principles of neural science, 2nd edition*. New York: Elsevier, 301-315.

Matin, E., Clymer, A.B., Matin, L. 1972. Metacontrast and saccadic suppression. *Science* 178: 179-182.

Matthews, T.C., & Kennard, C. 1993. Residual vision following geniculostriate lesions. In C. Kennard (Ed.), *Visual perceptual defects*. (pp. 227-241). London: Ballière Tindall.

Meeres, S.L., Graves, R.E. 1990. Localization of unseen visual stimuli by humans with normal vision. *Neuropsychologia* 28 (12): 1231-1237.

Merickle, P.M. 1992. Perception without awareness. Critical issues. *American Psychologist* 47 (6), 792-795.

Milner, A.D. 1995. Cerebral correlates of visual awareness. *Neuropsychologia* 33 (9): 1117-1130.

Milner, D.A., Goodale, M.A. 1993. Visual pathways to perception and action. In: T.P. Hicks, S. Molotchnikoff and Y. Ono (eds) *Progress in Brain Research vol. 95*. Elsevier Science Publishers. 317-337.

Milner, Goodale, M.A. 1995. *The visual brain in action* (Oxford Psychology Series 27). Oxford: Oxford University Press. 248p.

Mishkin, M. 1979. Analogous neural models for tactual and visual learning. *Neuropsychologia* 17: 139-151.

Mishkin, M., Ungerleider, L.G., Macko, K.A. 1983 Object vision and spatial vision: Two cortical pathways. *Trends in Neuroscience* 6: 414-417.

Morel, A., Bullier, J. 1990. Anatomical segregation of two cortical visual pathways in the macaque monkey. *Visual Neuroscience* 4: 555-578.

Nandrino, J.L., El Massioui, F. (1995) Etude au moyen des potentiels évoqués tardifs de l'organisation des étapes sensorimotrices chez l'homme. *Neurophysiologie Clinique* 25: 96-108.

Neal, J.W., Pearson, R.C.A., Powell, T.P.S. 1987. The cortico-cortical connections of area 7b, PF, in the parietal lobe of the monkey. *Brain Research* 419: 341-346.

Neely, J.H. 1991. Semantic priming effects in visual word recognition: a selective review of current findings and theories. In: D. Besner, G.W. Humphreys (Eds), *Basic processes in reading. Visual word recognition*. Hillsdale N.J.: Lawrence Erlbaum Ass., 264-336.

Neumann, O., Klotz, W. 1994. Motor responses to nonreportable, masked stimuli : Where is the limit of direct parameter specification?. In: C. Umilta, M. Moscovitch (eds) *Attention and Performance XV Conscious and nonconscious information processing*. Cambridge: MIT Press, 123-150.

Norman, D.A., Shallice, T. 1986. Attention to action. Willed and automatic control of behavior. In: R.J. Davidson, G.E. Schwartz and D. Shapiro (Eds), *Consciousness and self regulation, vol. 4*, New York: Plenum, 1-18.

Nowak, L., & Bullier, J. 1997. The timing of information transfer in the visual system. In J. Kaas, K. Rochland, & A. Peters (Eds.), *Extrastriate cortex in primates*. pp. 205-241.

Paillard, J. 1987. Cognitive versus sensorimotor encoding of spatial information. In: Ellen, P. and Thinus-Blanc (Eds) *Cognitive processes and spatial orientation in animal and man*. Dordrecht: Nijhoff, 43-77.

Paillard, J. 1991 Motor and representational framing of space. In: J. Paillard (ed) *Brain and space*. Oxford: Oxford University Press, 163-181.

Paillard, J., Michel, F. Stelmach, G. 1983. Localization without content. A tactile analogue of 'Blindsight'. *Archives of Neurology* 40: 548-551.

Pandya, D.N., Seltzer, B. 1982 Intrinsic connections and architectonics of posterior parietal cortex in the rhesus monkey. *Journal of Comparative Neurology*, 204: 196-210.

Paulignan, Y., Jeannerod, M. 1991. Selective perturbation of visual input during prehension movements. 2. The effect of changing object size. *Experimental Brain Research* 87: 407-420.

Paulignan, Y., MacKenzie, C.L., Marteniuk, R.G., Jeannerod, M. 1991. Selective perturbation of visual input during prehension movements. 1. The effect of changing object position. *Experimental Brain Research* 83: 502-512.

Pélisson, D., Prablanc, C., Goodale, M.A., Jeannerod, M. 1986. Visual control of reaching movements without vision of the limb. II. evidence of fast unconscious processes correcting the trajectory of the hand to the final position of a double-step stimulus. *Experimental Brain Research* 62: 303-311.

Perenin, M-T, Jeannerod, M. 1975. Residual vision in cortically blind hemifields. *Neuropsychologia* 13: 1-7.

Perenin, M-T, Jeannerod, M. 1978. Visual function within the hemianopic field following early cerebral hemidecortication in man. I. Spatial localisation. *Neuropsychologia* 16: 1-13.

Perenin, M.-T. 1991. Discrimination of motion direction in perimetrically blind fields. *Neuroreport* 2: 397-400.

Perenin, M.-T., Rossetti, Y. 1993. Residual grasping in a hemianopic field. *European Brain and Behaviour Society* (EBBS), september 1993.

Perenin, M.-T., Rossetti, Y. 1995. Saisir sans voir dans un champs hémianopsique: Un autre exemple de dissociation entre perception et action. Annual meeting of the French Neuroscience Society, Lyon May 1995.

Perenin, M.-T., Rossetti, Y. 1996. Grasping in an hemianopic field. Another instance of dissociation between perception and action. *Neuroreport* 7(3), 793-797.

Perenin, M.T., Vighetto, A. 1988 Optic ataxia: a specific disruption in visuomotor mechanisms. I. Different aspects of the deficit in reaching for objects. *Brain* 111: 643-674.

Perrig, W.J., Hofer, D. 1989. Sensory and conceptual representations in memory: motor images that cannot be imaged. *Psychological Research* 51: 201-207.

Picton, T.W., Stuss, D.T. 1994. Neurobiology of conscious experience. *Current Opinion in Neurobiology* 4: 256-265.

Pisella, L., Arzi, M., & Rossetti, Y. 1998. The timing of color and location processing in the motor context. *Experimental Brain Research, 121*(3), 270-276.

Pisella, L., Rossetti, Y. 2000. Interaction between conscious identification and non-conscious sensori-motor processing: Temporal constraints. In: Rossetti & Revonsuo (Eds) *Beyond dissociation: Interaction between dissociated implicit and explicit processing.* Amsterdam: Benjamins, pp. 129-151.

Pöppel, E., Held, R., Frost, D. 1973. Residual visual function after brain wounds involving the central visual pathways in man. *Nature* 243: 295-296.

Prablanc, C., Echallier, J.F., Jeannerod, M., Komilis, E. 1979. Optimal response of eye and hand motor systems in pointing at a visual target. II Static and dynamic visual cues in the control of hand movement. *Biological Cybernetics* 35: 183-187.

Prablanc, C., Jeannerod, M. 1975. Corrective saccades: dependence on retinal reafferent signals. *Vision Research* 15: 465-469.

Prablanc, C., Martin, O. 1992. Automatic control during hand reaching at undetected two-dimensional target displacements. *Journal of Neurophysiology* 67 (2): 455-469.

Price, M. 2000. Now you see it, now you don't: Preventing consciousness with visual masking. In: P. Grossenbacher (Ed) *Finding Consciousness in the Brain: A Neurocognitive Approach.* Amsterdam: John Benjamins. 25-57.

Revonsuo, A. 1995. Conscious and nonconscious control of action. *Behavioral and Brain Sciences* 18 (2): 265-266.

Rizzolatti, G., Riggio, L., Sheliga, B.M. 1994. Space and selective attention. In: Umilta, C. & Moscovitch, M. (Eds) *Attention and Performance XV.* Cambridge: MIT Press: 231-265.

Rossetti, Y. 1992. Multidisciplinary approach to consciousness: from mind-brain problem to conscious-unconscious processing. *Trends In NeuroSciences* 12: 467-468.

Rossetti, Y. 1998. Implicit short-lived motor representations of space in brain damaged and healthy subjects. *Consciousness and Cognition, 7,* 520-558.

Rossetti, Y. 1999. In search of immaculate perception: evidence from motor perception of space. In: Stuart Hameroff, Al Kaszniak and David Chalmers (Eds) *Towards a science of consciousness.* MIT Press, pp. 141-148.

Rossetti, Y., Desmurget, M., Prablanc, C. 1995. Vectorial coding of movement: vision, proprioception or both? *Journal of Neurophysiology* 74 (1): 457-463.

Rossetti, Y., Gaunet, F., & Thinus-Blanc, C. 1996. Early visual experience affects memorization and spatial representation of proprioceptive targets. *Neuroreport, 7*(6), 1219-1223.

Rossetti, Y., Koga K. 1996. Visual-proprioceptive discrepancy and motor control: modification of fast-pointing trajectories during prismatic displacement of vision. Unpublished manuscript.

Rossetti, Y., Koga, K., Mano, T. 1993. Transient changes in eye-hand coordination during early exposure to wedge-prism. *Perception & Psychophysics* 54 (3): 355-364.

Rossetti, Y., Lacquaniti, F., Carrozzo, M., Borghese, A. 1994. Errors of pointing toward memorized visual targets indicate a change in reference frame with memory delay. unpublished.

Rossetti, Y., Procyk, E. 1997. What memory is for action: The gap between percepts and concepts. *Behavioral and Brain Sciences, 20*(1), 34-36.

Rossetti, Y., Régnier, C. 1995. Representations in action: pointing to a target with various representations. In: B.G. Bardy, R.J. Bootsma, Y. Guiard (Eds) *Studies in perception and action III*, Lawrence Erlbaum Ass., Mahwah N.J., 233-236.

Rossetti, Y., Régnier, C., Perenin, M.-T., Rode, G., Lacquaniti, F., Boisson, D. 1995. Actions et représentations: influence de la mémorisation et de la verbalisation du but sur les mouvements de patients et de sujets sains. Annual meeting of the French Neuroscience Association (Société des Neurosciences), Lyon May 1995.

Rossetti, Y., Pisella, L., Pélisson, D. 2000. Eye blindness and hand sight: Temporal aspects of visuo-motor processing. *Visual Cognition, 7,* in press.

Rossetti, Y., Revonsuo, A. (Eds). 2000. *Beyond dissociation: Interaction between dissociated implicit and explicit processing.* Amsterdam: Benjamins.

Rossetti, Y., Rode, G., Boisson, D. 1995. Implicit processing of somesthetic information: a dissociation between Where and How? *Neuroreport* 6: 3, 506-510.

Rossetti, Y., Rode, G., Boisson, D. 2000. Numbsense: A case study and implications. In: De Gelder, B., De Haan, E., Heywood, C. (Eds) *Varieties of unconscious processing.* Oxford University Press, in press.

Rossetti, Y., Rode, G., Perenin, M.-T. 1994. Dissociations and interactions between implicit and explicit sensory processing. Annual meeting of the *European Society for Philosophy and Psychology*, Paris, September 1994.

Rossetti, Y., Stelmach, G., Desmurget, M., Prablanc, C., Jeannerod, M. 1994. The effect of viewing the static hand prior to movement onset on pointing kinematics and variability. *Experimental Brain Research* 101: 323-330.

Sakata, H., Taira, M. 1994. Parietal control of hand action. *Current Opinion in Neurobiology* 4: 847-856.

Schall, J.D., Morel, A., King, D.J., Bullier, J. 1995. Topography of visual cortex connections with frontal eye field in macaque: convergence and segregation of processing streams. *Journal of Neuroscience* 15 (6): 4464-4487.

Shacter, D.L. 1987. Implicit memory: History and current status. *Journal of Experimental Psychology: Learning, Memory and Cognition* 13: 501-518.

Soechting, J.F., Lacquaniti, F. 1983. Modification of trajectory of a pointing movement in response to a change in target location. *Journal of Neurophysiology* 49 (2): 548-564.

Stoerig, P., Cowey, A. 1992. Wavelength discrimination in blindsight. *Brain* 115: 425-444.

Stoerig, P., Cowey, A. 1993. Blindsight and perceptual consciousness: Neuropsychological aspects of striate cortical function. In: B. Gulhas, D. Ottoson, P. Roland (Eds), *Functional organization of the human visual cortex.* Oxford: Pergamon Press, 181-193.

Stoerig, P., & Cowey, A. 1997. Blindsight in man and monkey. *Brain, 120,* 535-559.

Stoerig, P., Hübner, M., Pöppel, E. 1985. Signal detection analysis of residual vision in a field defect due to a post-geniculate lesion. *Neuropsychogia* 23 (5): 589-599.

Tanné, J., Boussaoud, D, Boyer-Zeller, N., Rouiller, E. 1995. Direct visual pathways for

reaching movements in the macaque monkey. *Neuroreport*: 7(1), 267-272.
Taylor, T.L., McCloskey, D. 1990. Triggering of preprogrammed movements as reactions to masked stimuli. *Journal of Neurophysiology* 63: 439-446.0
Trevarthen, C.B. 1968. Two mechanisms of vision in primates. *Psychologishe Forschung* 31: 299-227.
Ungerleider L. 1995. Functional brain imaging studies of cortical mechanisms for memory. *Science* 270: 769-775.
Ungerleider, L., Mishkin, M. 1982. Two cortical visual systems. In Ingle DJ, Goodale MA & Mansfield RJW (Eds) *Analysis of motor behavior*, Cambridge MIT Press: 549-586.
Velay, J.-L., Roll, R., Rossetti, Y. 1995. Proprioceptive encoding of initial hand position demonstrated by tendon vibration. Unpublished manuscript.
Volpe, B.T., Ledoux, J.E., Gazzaniga, M.S. 1979. Spatially oriented movements in the absence of proprioception. *Neurology* 29 (9): 1309-1313.
Watson, R.T., Valenstein, E., Day, A., Heilman, K.M. 1994. Posterior neocortical systems subserving awareness and neglect. *Archives of Neurology* 51: 1014-1021.
Weiskrantz, L. 1989 Blindsight. In: F. Boller, J. Grafman (Eds), *Handbook of Neuropsychology Volume 2*, Amsterdam: Elsevier, 375-385.
Weiskrantz, L. 1991. Some contributions of neuropsychology of vision and memory to the problem of consciousness. In: A.J. Marcel and E. Bisiach (Eds) *Consciousness in contemporary science*. Oxford: Oxford Science Publications. 184-199.
Weiskrantz, L., Warrington, E.K., Sanders, M.D., Marshall J. 1974. Visual capacity in hemianopic field following a restricted occipital ablation. *Brain* 97: 709-728.
Weiskrantz, L., Zhang, D. 1987. Residual tactile sensitivity with self-directed stimulation in hemianaesthesia. Journal of Neurology, Neurosurgery, and Psychiatry 50: 632-634.
White, J.M., Sparks, D.L., Stanford, T.R. 1994. Saccades to remembered target locations: An analysis of systematic and variable errors. *Vision Research* 34 (1): 79-92.
Winkler, I., Tervaniemi, M., Huotilainen, M., Ilmoniemi, R., Ahonen, A., Salonen, O., Standertskjöld-Nordenstam, C.-G., Näätänen, R. 1995. From objective to subjective: Pitch representation in the human auditory cortex. *Neuroreport* 6, 2317-2320.
Wong, E., Mack, A. 1981 Saccadic programming and perceived location. *Acta Psychologica* 48: 123-131.

Section III

Frame of Mind

Introduction

This section approaches the *subjectivity* inherent in conscious experience by examining the subjective regard which frames conscious content. The ever-renewing sequence of unfolding phenomenal experience which is subjectively known to a person often has an obvious emotional cast to it. One feels imbued with more or less energy, and an experience can either be enjoyed or suffered through. This section explains some of the most powerful influences on frame of mind which are known to affect the experience of conscious content.

Hypothetically, you could compare for yourself how you feel after a full night's sleep versus at the end of a long day. And without resorting to hypotheticals, you may compare your *recollection* of such occasions with how you feel right now.

||••||

Does *your* frame of mind change over time? Objectively, most of us conclude that somebody else's subjective state does change over time. But how can we really tell what is going on inside somebody else's nervous system? Consider the person who says, "I feel down, I am too tired to think this hard." Would you believe this statement if it were made while the speaker was smiling and walking at a brisk pace? Language is only one channel of interpersonal communication. The living brain passes through frames of mind which affect both conscious experience and bodily activity. Body posture is an objective source of clues as to how a person might be feeling subjectively (Grossenbacher, Potts, Liotti & McQuaid, 1994). As evident in subsequent chapters, neuroscience offers amazingly useful techniques for directly observing changes in brain activity which underlie the changes in subjective experience and overt behavior.

People often use the term "consciousness" to distinguish being awake from being asleep, but this rather stark dichotomy fails to capture the full range of neurophysiological states of consciousness. Medical doctors have taken a step in right direction by categorizing "levels of consciousness" ranging from alert wakefulness through lethargy, obtundation, and stupor to coma (Dorland, 1988). This clinical operationalization of neurophysiological state of consciousness is convenient because it is easily discerned through simple interpersonal interaction (talking with, looking at, and touching the patient) without requiring elaborate methods or sophisticated technology. Though appropriate for many medical scenarios, this kind of subjective, interpersonal appraisal oversimplifies the complex neurophysiological processes which enable and constrain the patient's conscious experience. Roger Whitehead and Scott Schliebner, in Chapter 7, explain the neuromodulatory systems which control the various kinds of wakefulness we experience at different times. Their discussion of arousal provides an explanatory bridge between conscious experience and the brain systems which distribute neurotransmitters known to affect degree of wakefulness.

Along with variations of wakefulness, the emotional tone which colors experience also depends on neurochemistry. In Chapter 8, Douglas Derryberry reports on the complex neurophysiology which underlies emotionally charged experience. His discussion of the brain systems involved in motivation shed light on the nuanced control of attention exerted by emotional relevance.

At first blush it seems simple to distinguish between the current conscious *content* (for example, the percept of the page you now hold in your hand) and the current *frame of mind*, that ensemble of motivations, attitudes and feelings which now shapes your regard for this page. But it may not always be so easy to neatly separate *what* we are aware of and *the way* we are aware of it. When you become aware of your emotional feeling, for example, current conscious content may include internal percepts of abdominal tension or negative affect, as well as conscious recognition of these feelings as "belonging to you," that is, as reflecting your current frame of mind. Awareness of emotional feelings *as such* exemplifies one way in which the content and frame are intimately bound together.

Even without that sort of sensitive meta-awareness, the current frame of mind (the way content is regarded) can influence upcoming content. To the extent that ongoing thought and behavior are shaped by one's current frame of mind (the way content is regarded) can influence upcoming content. To the extent that ongoing thought and behavior are shaped by one's current frame of mind, the representations available for consciousness are fundamentally constrained by frame of mind. Consider the case of someone who is experiencing paranoia. This person is likely to be perceiving threat (the content of consciousness) and be operating in a cautious, jittery mode (the frame of mind). In this example, the *content* bears a kinship with the *frame* within which the content is held.

The interdependence between content and frame works both ways: the emotional and energetic framing of current contents can also depend on the current and previous content of consciousness. This happens whenever a mood gets triggered by percepts or thoughts. In short, frame of mind and phenomenological content mutually influence each other. But this mutual influence is a tricky business — despite ones own intentions, moods often persist with a momentum not easily swayed by attempts to quickly think of something else!

References

Dorland, W.A.N. 1980. *Dorland's Illustrated Medical Dictionary*. (27th ed.). Philadelphia, PA: W.B. Saunders.

Grossenbacher, P.G., Potts, G.F., Liotti, M., & McQuaid, J.R. 1994. Identification of emotion through postural cues. In N.H. Frijda (Ed.), *Proceedings of the VIIIth Conference of the International Society for Research on Emotions*, (pp. 260-263). Storrs, CT: ISRE Publications.

CHAPTER 7

Arousal

Conscious experience and brain mechanisms

Roger Whitehead
University of Colorado at Denver

Scott Schliebner
University of Utah

Bridge jumping is a pastime quite similar to bungee jumping, except that the rope doesn't stretch as much, and there is a horizontal, as well as a vertical component to the jump. The jump described below was made in Annecy in the South of France, where two parallel bridges span a very deep gorge. One end of a rope was tied to the center of one bridge, and the other end was tied to the jumper who stood in the middle of the other bridge. A large loop of rope separating the bridges descended into the gorge. Following is the jumper's account:

> I knew I couldn't stand on the edge of the bridge. Straight over and off — that's what I told myself. All my senses seemed so finely tuned. I could feel my heart racing. Details on small objects were as clear as could be. Concentrate now; just hold the rope in front of you, keep it from tangling on your legs. Go off straight, don't twist or you'll risk falling out of your harness. Don't wait on the other side of the railings. Over and off, over and off.
> My heart upped the tempo another notch as I climbed over the railings. Go Go Go!! Do it now! The pressure in my chest was overwhelming. Jump! One last big breath. The pressure increased again. It was released with a scream. I grabbed the rope in front of me, held it away from my legs, and let go. Dropping, dropping, still accelerating.... I could see the ropes snaking

down in front of me. Still building speed, the trees were closing in, coming up so fast. No way I should be falling this far. The ropes must have come loose. Oh God, this is it... At last I felt the forward tug and the swing began. Sweet relief. Still accelerating, but not straight down.

The transition to horizontal motion was sheer bliss. The wind was deafening through the bottom of the swing, skimming the trees at nearly terminal velocity, but the real danger was over. Coming up the other side, the wind eased, quieter and still quieter. Zero gravity at the end of the swing, perfectly quiet. From an impossible position, way out to the side of the bridge, I could see my friends on the bridge in front of me, not above. Just a touch of slack in the ropes and then back down again — stretch out straight, loving the speed now and pushing for more. This was more like it!

Mark followed my technique of not hanging around on the edge. Over a couple of glasses of glorious cool beer, we relived every wonderful moment. Grinning like maniacs and feeling like gods. No matter how many would come after us we were it! We did a quick encore by tightrope-walking the cables of the suspension bridge — it was easy. We'd already learnt to fly.

Reading Hesse on the drive back to Paris, Mark shoved a passage in front of me: "And already I was falling, I plunged, I leaped, I flew; wrapped in a cold vortex, I shot, blissful and palpitating with ecstatic pain, down through infinity....." He leaned back in his seat with a smug grin. "Hesse must have done it too!"

This passage illustrates some important characteristics of 'arousal'. First, arousal manifests itself in different forms. It can, for example, be experienced negatively, as a component of anxiety or terror, or positively, as thrill or elation. Also, we can become aware of our arousal as a result of changes in either our emotional or physiological state. Finally, arousal impacts upon conscious awareness, for example, by making one's senses seem "finely tuned". Each of these themes will be developed more fully in the following section of this chapter, in which we will consider the nature of arousal, especially as it influences conscious experience. We will then turn to a discussion of the neurophysiological mechanisms that are responsible for arousal, and which mediate its influence.

1.   Arousal and conscious experience

1.1 *One "arousal" or many?*

Our first task is to decide what we mean by 'arousal'. Perhaps the easiest way to think of arousal is as neurological and physiological excitation. When we are anxious, as opposed to relaxed, we are more excited. When we are enthused, as opposed to bored, we are more excited. Emotions such as apprehension, fear, panic, terror, euphoria, bliss, and ecstasy all contain, to varying degrees, a component of excitation or arousal. What appears to distinguish these emotions from one another is the cognition that accompanies the excitement. It may be possible to claim that terror and ecstasy involve an equivalent amount of arousal, but that they differ on a dimension of pleasure, or on our appraisal of whether we can cope with an existing situation. This idea promotes the argument that arousal is non-specific. It is simply the amount of energy (or potential energy) in the system at any point in time. For many years, this notion of non-specific or unidimensional arousal was held to be true by psychologists and physiologists alike. In 1949, Moruzzi and Magoun discovered an area of the brainstem, the reticular formation, which seemed to be wholly responsible for arousal. Lesions of the reticular formation resulted in somnolence, or even coma, in cats, while electrical stimulation often produced hyperactivity. It was assumed on the basis of such evidence, that arousal constitutes the intensity dimension of behavior, and that it is manifest on a unidimensional continuum ranging from deep sleep to extreme excitement.

Since the early 1960s, however, evidence has accumulated to support the contrary view; that arousal is most profitably thought of as being multidimensional. Consider for a moment the following three scenarios:

> You have become lost in a foreign city. In your attempts to find your way back to your hotel, you have strayed into a "bad part of town", and you are now aware that you are being followed by a gang. You imagine that you will be mugged.

> You have just met a member of the opposite sex, whom you find extremely attractive. This person appears to enjoy your company, compliments your physical attractiveness, and after a few minutes of conversation, suggests that you should meet again, so that you can "spend more time together". It becomes obvious that you have both been flirting with each other, and that an "intimate acquaintance" is a very real possibility.

> You are working on a difficult crossword. The final clue is not leading you to the answer. You rack your brains, and then realize that the clue can be interpreted differently. Suddenly there is no doubt that you have the answer. You fill it in, and sit back in your chair feeling quite pleased with yourself.

In each of these scenarios, arousal is present, but in different forms. Suppose that in each case, you had the cognitive resources (and the inclination) to monitor your emotional and physiological state as the scenario progressed. You would probably detect some fairly specific changes. In the first scenario, your cognitive appraisal of the situation might lead you from mild frustration or worry to acute anxiety or even panic. Accompanying would be physiological changes — an increase in the secretion of adrenaline, faster heart rate, deeper breathing, and perhaps a dry mouth. This pattern of activity, known as the "fight or flight" response, is almost wholly mediated by the sympathetic branch of the autonomic nervous system (ANS). In the second scenario, detached interest might be replaced by a feeling of intrigue or high anticipation. Your physiological state might be characterized by increased alertness and sexual arousal — pupil dilation, blood pressure increases, and possibly genital excitation. This pattern of physiological activity would probably involve a moderate level of activity in the sympathetic branch of the ANS, but may be accompanied (in the case of a sexual response) by increased parasympathetic ANS activity. The third scenario might take you from an emotional state of frustration to one of self-satisfaction. Simultaneously, you would probably exhibit high neurological excitement, and increased blood flow in the word association areas of the brain, which would decrease once the final answer had been obtained. Clearly, in each case, there are different levels of activity in a variety of systems. Thus, it seems that it is overly simplistic to consider arousal to be unidimensional, rather, we should be able to account for the fact that we become aroused in different ways.

This intuitive idea has received a great deal of empirical investigation. For example, Lacey (1967) argued that if we consider arousal to be unidimensional, then correlations between indices of autonomic nervous activity should be high (for example, between heart rate, skin conductance and pupil dilation). In practice however, conditions assumed to increase autonomic activity generally yield low to modest correlations between such indices. Lacey demonstrated that not only are the correlations merely modest, but also that different patterns of autonomic activity are found in different situations. He termed this effect "situational stereotypy" and argued for the dissociation between activation systems, claiming the existence of at least three: behavioral, cortical and autonomic.

Many other authors have also provided evidence for the multidimensionality of arousal. Later in this chapter, we will devote considerable attention to brain mechanisms of arousal, and in particular toward the multi-component arousal models of Gray (1975), and of Pribram and McGuinness (1975).

Suffice it to say for the present, that these authors have provided compelling evidence for the multidimensionality of arousal, implicating in particular the reticular and limbic systems, the basal ganglia, and in the case of Robbins (1986), for differentiation among ascending neurotransmitter systems arising in the reticular formation.

Our position regarding the definition of arousal is, that it is generally appropriate to equate arousal to excitation; but, that in order to consider the complexities of human behavior, it is necessary to consider which (of several) systems are becoming excited.

1.2 Arousal, attention and the contents of consciousness

In the passage introducing this chapter, the bridge-jumper claims that his "senses seemed so finely tuned", and that "details on small objects were as clear as could be". These statements suggest that arousal influences the mechanisms by which we obtain information for conscious processing. Empirical research (conducted predominantly by investigators of human performance) strongly suggests that this is the case. However, the consensus in the literature is that arousal does not affect the contents of consciousness directly, but does so through modulating the effects of attention.

Interest in the relationship between arousal, attention and performance was greatly promoted by Easterbrook's (1959) claim that in conditions of increased emotional arousal, "the use of peripheral (occasionally or partially relevant) cues" is reduced while "the use of central and immediately relevant cues" is maintained (p. 183). Human performance theorists argued that suboptimal performance in conditions of underarousal (e.g., as a result of fatigue) was due to the processing of too many task-irrelevant cues. With increased arousal, more of these would be excluded to the point at which only task-relevant cues would be processed, that is, at the point of optimum performance. Further increases in arousal (e.g., those elicited by extreme competitive anxiety) would then serve to exclude task-relevant cues leading to performance impairment.

Although Easterbrook emphasized reduced range of cue utilization in terms of task relevancy, several authors explored the possibility that the phenomenon would extend to reduced awareness of events occurring in the visual periphery. Cornsweet (1969) found that under stressful conditions (fear of electric shock) subjects actually showed increased use of cues in the visual periphery when these were task relevant. Similarly, Hockey (1970a, 1970b) found that in a dual task situation, loud noise increased the detection of targets in central but not peripheral locations, but only as a function of their higher subjective probability of occurrence. When the majority of targets were presented in the periphery, attention to the periphery was increased by loud noise. His conclusion was that the increased selectivity due to loud noise was more a function of the allocation of resources according to task priorities than to reduced sensitivity in the visual periphery. Bacon (1974) concurred with Hockey in claiming that increased arousal, due to threat of electric shock, serves to narrow the range of cue utilization by reducing responsiveness to those aspects of the situation initially attracting a lesser degree of attentional focus. Moreover, he also claimed that this effect is due to arousal effects upon capacity limitations and attentional processes rather than upon the initial sensory impressions of stimuli.

Wachtel (1967) described three dimensions of perceptual narrowing, and argued that arousal and anxiety differentially affect these dimensions. Wachtel adopted the searchlight analogy of attention in which focus is represented by the width of the beam, and scanning by the amount of movement of the beam around the perceptual field. Selectivity, on the other hand, describes the phenomenon of subjects selectively attending to certain aspects of the environmental display while ignoring others. Wachtel proposed that perceptual focus narrows under conditions of increased arousal but, in trait anxious subjects, both focus and scanning are affected; focus being narrowed while scanning is simultaneously increased. Wachtel also argued that stress increases selectivity, but emphasized that "the range of attentional deployment may be somewhat under the control of the individual" (p. 422).

Let us summarize what this means in terms of conscious awareness. First, as arousal increases, we do not necessarily loose awareness of visually peripheral events, or become 'tunnel-visioned'. Second, our attentional mechanisms are affected by arousal, but our sensory apparatus is not, and as arousal increases, we tend to sacrifice attention to less important aspects of a situation in order to concentrate more fully on those we consider to be most relevant. Finally, the perceptual narrowing described above might be due to effects upon attentional focus, scanning or selectivity.

When processing external sources of information, we appear to direct our attention increasingly toward task-relevant, or dominant aspects of a situation as arousal increases. This increased selectivity appears to also hold true for internally generated sources of information. Eysenck (1975, 1976) extended the earlier work of Bacon (1974) to describe similar arousal effects on memory as well as perception. Where Bacon suggested that arousal reduces responsiveness still further to cues initially attracting a lesser degree of attentional focus, Eysenck showed that noise at output facilitates the recall of high association items, but impairs that for low association items. Hamilton, Hockey and Quinn (1972) also found facilitated processing of task relevant input (as shown by improved recall) concurrent with reduced awareness of task irrelevant cues. Hence, the effects of arousal (mediated by attention) upon the contents of consciousness seem to generalize across inputs from both perception and memory.

Returning for a moment to the experiences of the bridge jumper, we are able to reassess his experiences in the light of empirical study. Although he claimed that his "senses seemed so finely tuned", it is likely that his senses were not in the least affected by his being highly aroused. However, it may well have *seemed* to him that they were, because his attention was very likely to have been affected. In all probability, peripheral information that would normally have become a part of his conscious experience, would have been selectively filtered out. Instead, his conscious processing would probably have been of task-relevant information. Again, it may well have been that he found "details on small objects as clear as could be" because he had an extremely narrow attentional focus. Unfortunately, we do not know which small objects he was attending to, but, on the basis of the studies we reviewed above, we would suppose that they were again task relevant, a piece of his equipment perhaps.

Our consideration of the effects of arousal upon consciousness thus far, presupposes that information is being actively processed. It is also possible to investigate the effects of arousal upon the subsequent, as well as the present contents of consciousness. Posner (1978) argued for the treatment of one aspect of arousal, alertness, as a subcomponent of attention itself. He viewed alertness in the restricted sense of being a pre-stimulus input state responsible for maintaining preparedness to process incoming signals. That is, that alertness is an attentional subsystem which serves to support subsequent processing. In Posner's paradigm, subjects are told that they must identify a target as quickly as possible after its onset. Shortly prior to the presentation of the target, a warning signal is presented which puts the subjects in a state of high phasic alertness. The physiological concomitants of this state have been well documented, and consist of alpha desynchronization, a slow negative drift in

EEG, heart rate deceleration, pupil dilation and a general state of sympathetic ANS activity (Kahneman, 1973). Experientially, a state of high preparatory alertness is viewed as being an "empty-headed" state in which ongoing conscious processing is suspended in anticipation of the imperative target. Posner (1978) found that a state of high alertness does not influence the rate of accumulation of information in sensory systems but that it does affect the rate at which attention can respond to the stimulus. High phasic alertness was associated with more rapid, but less accurate responding. In other words, in a highly alert state, one is able to more rapidly detect, and respond to the presence of information in consciousness, but if processing of that information is required prior to the response, that processing is more likely to be incomplete, and errors may result.

1.3 Self-awareness of arousal

How are we able to monitor our arousal? There would be appear to be three types of indicator. First, we can monitor our behavior. For example, if you have ever attempted to drive a car while feeling drowsy, you may recognize these attempts to combat sleepiness — turning up the stereo, singing out loud, and rolling down the window. We are also able to monitor our emotional state. As we saw in the bridge jumping passage, the writer was able, presumably accurately, to report rapid emotional transitions through states of anxiety, terror, relief, exhilaration, and euphoria. Third, we can derive information about our arousal levels from physiological information. Again for example, in the bridge jumping passage, the writer is aware of the pounding of his heart and an overwhelming pressure in his chest.

Despite the apparent self-evidence of the preceding examples, relatively little direct empirical study has been conducted on our ability to consciously self-monitor arousal.

How is it that we are able to know that we are too tired to take on a difficult task, or that we are too excited to perform at our best? A relatively simple, though not particularly enlightening explanation, would posit that we make our best cognitive appraisal of the energetical demands of the task, and assess our physiological, emotional and behavioral states in regard to the perceived demands. Such an analysis might lead to the following conclusions: "I'm too tired to work in the yard right now", "I'm too wound up to think straight", or, "OK, take a deep breath, relax, and don't loose your temper".

The question still remains regarding the mechanism(s) by which we are able to assess our current state. Although we are currently unable to specify exactly how this may take place, several lines of inquiry have relevance for further study.

Psychophysiologists have long been interested in the ability to detect and control changes in autonomic functions. Largely because of technical ease, most of the work in this area has been conducted with heart rate. For example, Brener and Jones (1974) found that subjects could learn to discriminate between vibratory stimuli that were either contingent or non-contingent upon heart beats. Ashton, White and Hodgson (1979) found that subjects were able to decide in which of two adjacent periods of time their heart rate was the highest. Interestingly, in each of these studies, discrimination in the absence of feedback was greatly enhanced by prior training with biofeedback . Other researchers have obtained similar results for skin conductance (Stern, 1972), blood pressure (Greenstadt, Shapiro & Whitehead, 1986), and pulse transit time (Martin, Epstein & Cinciripini, 1980). An unfortunate impediment to the interpretation of these discrimination studies however, is the lack of certainty that the subjects were only discriminating, as opposed to controlling their autonomic functions.

In the realm of cognitive psychology, researchers have long recognized the utility of incorporating "arousal monitors" in their models. Broadbent (1971) proposed a model incorporating two independent arousal mechanisms: a lower mechanism, posited to be involved in the execution of well-established decision processes, and an upper mechanism which monitors and adjusts the lower mechanism in an attempt to maintain optimum performance.

Sanders (1983) proposed that three mechanisms of energetical supply support human performance; arousal, activation and effort. These were hypothesized to supply the linear stages (described below) of feature extraction, motor adjustment and response choice, respectively. The energetical constructs were borrowed from the work of Pribram and McGuinness (1975), in which arousal is defined as a phasic response to input, and activation as a tonic readiness to respond. The effort mechanism is regarded as a higher level executive, responsible for the coordination of arousal and activation.

Sanders viewed the linear stages of the model as being closely associated with components of attention previously identified by Posner and Boies (1971). Motor adjustment is associated with Posner's alertness in the sense that efficiency in motor adjustment is maximized by preparatory processes through which motor adjustment is preset as close as possible to the "motor action limit" (Näätänen & Merisalo, 1977). In the case where feature extraction is an active rather than an automatic process, for example in the case of degraded stimuli, or where a stimulus requires recognition in a complex display, Sanders views these active processes as corresponding to the "selective attention" of Posner and Boies (1971). The response choice stage, in which perception and action are linked, and in which decision rules and reasoning are involved, Sanders relates to Posner's (1978) "conscious processing". Sanders described five patterns of stress, each of which arises as a result of the effort mechanism failing to accomplish some energetical adjustment. Namely, when it fails to correct for under- or over-arousal, for under- or over-activation, or when effort fails to supply sufficient energetical resources to the response choice stage. Let us consider each of these failures from an experiential perspective. To be under-aroused would mean that attention is not being sufficiently selective, in essence, that we are processing too much irrelevant information. To be over-aroused, in Sanders's terminology, would mean that attention is being excessively selective, that is, that we are so restrictive in our analysis of the environment that we fail to process task-relevant cues. Under-activation would mean that we are not sufficiently alert, and we would be slow to respond to an imperative event. Over-activation, on the other hand would mean that we are setting ourselves too close to the motor action limit, as for example when a sprinter makes a false start. Finally, if sufficient resources are not supplied to conscious processing, we are unable to make an appropriate response selection from an array of alternatives. Clearly, conscious processing is more demanding for complex, rather than simple tasks; hence, the effort mechanism is more likely to fail under such circumstances.

It can be seen from the preceding review that although cognitive models are able to consider various subtleties of the arousal-performance relationship, they do not directly explain how one's state of arousal is assessed. A few additional points are worthy of brief mention before leaving the issue of self-assessment of arousal — these center upon the study of anxiety, rather than arousal per se. Anxiety has received a good deal of investigation from several different psychological perspectives. For example, clinical psychologists have researched the topic in order to better assist those suffering from anxiety disorders, or phobias for example. Derryberry (this volume) addresses this issue in his section on "Orienting to interoceptive information". Sport psychologists have also devoted their attention toward anxiety. To a certain degree, sport and clinical psychologists share a common characteristic of being more concerned with the control of anxiety, especially when it is maladaptive, than with a detailed analysis of the mechanisms of self-assessment. Nevertheless, in both disciplines, considerable efforts are being devoted to the development of sophisticated self-assessment tools.

1.4 *Summary*

Our discussion to this point has highlighted various aspects of arousal as it relates to conscious experience. Although arousal can be thought of as being roughly equivalent to excitation, it is more appropriate to consider excitation in specific systems when analyzing the effects of arousal on conscious experience. Arousal can contribute to both positive and negative emotions, and we can detect changes in our state of arousal on the basis of both emotional and physiological evidence. Arousal exerts an influence upon the contents of consciousness through its modulation of attention. The most common finding in this regard is that increased arousal serves to bias attentional selectivity toward dominant aspects of memory, environmental displays, and cognitive tasks at the expense of task-irrelevant or peripheral information.

In regard to self-awareness of arousal, psychophysiological research indicates that we are able to detect changes in autonomic functions, especially when trained (through biofeedback) to do so, and cognitive psychologists agree that an "arousal monitor" is an essential component for a model of the arousal-performance relationship. Despite this, and the fact that we are clearly able to detect at least gross changes in our own arousal levels, the mechanism by which we do so is not clearly understood.

2. Neurophysiology of arousal

In this section, we will first discuss two multiple-component models of arousal, each of which has received a good deal of empirical support. Second, we will discuss the action of three 'arousing' neurotransmitter systems, each arising in sub-cortical centers and potentially capable of modulating cortical activity. Our discussion will then focus upon one additional neurotransmitter system, the noradrenergic system, and we will review the evidence that suggests that this system in particular exerts an influence on attention. The section will conclude with a discussion of cortical influence on subcortical structures.

2.1 *Multiple-component models of arousal*

Several authors have followed Lacey's (1967) lead in proposing multiple component arousal systems in which two components are proposed to be mutually antagonistic. Two models in particular, those of Gray (1975) and Pribram and McGuinness (1975), have received a great deal of scrutiny and empirical support.

Gray's model can be thought of as having three major components:

1.   A physiological arousal system (NAS) centered on the reticular formation which is responsible for the intensity dimension of behavior.

2.   A behavioral activating system (BAS) centered on the medial forebrain bundle, lateral hypothalamus, rostral septal area and amygdala, responsible for the selection and organization of all active behavioral responses, and which responds to reward- conditioned stimuli.

3.   A behavioral inhibition system (BIS) which is centered on the hippocampus and medial septal area, and which attempts to suppress all ongoing operant behavior whenever the organism is faced by punishment-conditioned or novel stimuli.

The BAS and BIS are mutually antagonistic, but both have direct positive input to the physiological arousal system. Imagine yourself walking down the sidewalk, and finding some cash. In Gray's terminology, cash is a reward-conditioned stimulus, and would serve to energize the BAS. You would be likely to exhibit approach behavior, that is, pick up the cash. As a result of the BAS activity, you would also receive a positive input to the NAS, that is, there would be some physiological excitement (no doubt dependent upon the amount of cash). Now, imagine yourself walking along, turning a corner, and finding a snarling dog in front of you. In this case, you would be faced with a punishment-conditioned stimulus, and your behavior (walking) would be likely to be inhibited. You would again receive positive input to the NAS, resulting in physiological excitement (in this case probably dependent upon the size of the dog).

Pribram and McGuinness (1975) have also proposed a model having three energetic components, these being:

1.   An arousal system located in the reticular formation and anterior hypothalamus, which is controlled by the amygdala. This system controls all phasic

physiological responses, and in particular is responsible for the orienting reflex.

2. An activation system which controls the organism's tonic readiness to respond, and which is located in the medial forebrain bundle, lateral hypothalamus and basal ganglia.

3. A coordinating system which demands effort on the part of the organism, and which is centered on the hippocampal and septal areas. This system is responsible for the coordination of arousal and activation in establishing the more difficult relationships between perception and action.

Hence, according to Pribram and McGuinness (1975), the effort mechanism both monitors and controls performance. It "asks questions" such as "Am I too excited or too fatigued to perform this task?", "Am I being distracted by irrelevant information?", "Am I locking on to one solution, and failing to consider all possible alternatives?", and "Am I trying as hard as I can?". Depending upon the answers to those questions, resources are reallocated accordingly.

From the preceding descriptions of these two models, we can see that the chief difference between them lies in the roles assigned to the amygdala and hippocampal circuit. For Pribram and McGuinness, the amygdala is primarily responsible for the controlling of all phasic responses to input, while Gray contends that the amygdala is involved in the selection and organization of behavioral responses, that is, that it is more closely tied to activation rather than arousal in Pribram and McGuinness's terminology. On the role of the hippocampus, Gray proposes it to have an inhibitory responsibility. Gray's argument is based upon a demonstrated association between inhibitory behavior and hippocampal theta (Gray, 1977), and the finding of response specificity in hippocampal theta (Gray, 1978). Pribram and McGuinness agree with Gray in ascribing an inhibitory role to the hippocampus, in the sense that it performs a coordinating role, but claim that it does so by suppressing inappropriate stimulus-response relationships.

Despite the idiosyncrasies of the models described above, a common characteristic is that both associate the reticular system with influence on physiological arousal responses. Since the identification of the reticular formation however, evidence has accumulated to specify a degree of neurochemical differentiation among ascending neurotransmitter systems. This has led to the notion that it is now perhaps more profitable to focus attention upon the location and type of influence exerted by these pathways, rather than to consider the potential influence of the undifferentiated reticular structure (Robbins, 1986).

2.2 Arousing neurotransmitter systems

Robbins has suggested that investigation of the multiplicity in ascending neurotransmitter pathways may illuminate the multidimensional nature of arousal. Our discussion here will focus upon the serotonergic (5-HT), cholinergic (ACh), and dopaminergic (DA) systems. The noradrenergic (NA) system will be discussed in greater detail separately.

2.2.1 Serotonin

The central serotonin (also called 5-hydroxytryptamine or 5-HT) projections arise in the dorsal and medial raphe nuclei (Azmitia, 1978). Terminal innervation, which is inhibitory, targets frontal cortex and neocortex, septal and hippocampal regions and the basal ganglia. With the exception of the latter, this topography is strikingly similar to that of the noradrenergic system. Indeed, it has been suggested (Zhang et al., 1995) that there is reciprocity in 5-HT and NA pathways in the support of alertness and responsivity to perceptual input. Similarly, electrophysiological activity in 5-HT neurons in the mesencephalon is directly related to behavioral arousal, a finding which is also common to NA (Robbins, 1986). Furthermore, as is the case with NA, the 5-HT system is responsive during stress (Feldman, Conforti & Weidenfeld, 1995), and has been implicated in behavioral inhibition resulting from anxiety-producing situations (Iversen, 1983). Interestingly, Gray (1982) emphasizes a role for NA in behavioral inhibition. There is one further commonality between the systems — in respect to the control of sleep. First, it has been demonstrated that lesions of the raphe (the central source of 5-HT) produce insomnia. Using cats, Jouvet and Renault (1966) destroyed 80 to 90 percent of the raphe, and observed complete insomnia for 3 to 4 days. Minimal restoration of slow wave sleep, but not REM sleep subsequently occurred. Jouvet (1968) further demonstrated that raphe lesions deplete cortical 5-HT, and that the amount of 5-HT present in the brain was positively correlated with the amount of time the animals spent sleeping. In addition, it has been found that administration of PCPA, which acts to limit 5-HT synthesis, also suppresses sleep (Mouret, Bobillier & Jouvet, 1968). NA mechanisms have also been implicated in the control of sleep, in particular, in regulating the cyclicity of REM sleep (Hobson, McCarley & Wyzinski, 1975) and in the control of waking mechanisms (Jouvet, 1972).

2.2.2 Acetylcholine

There are three primary sources of central ACh. The neocortex receives an extrinsic projection from the basal forebrain, while a projection to the hippocampus arises in the medial septum (Mesulam, Mufson, Levey & Wainer, 1983). In addition, intrinsic cortical innervation arises from cell bodies in cortical layers. Mesulam and Mufson (1984) have also described the afferents to the basal forebrain. In addition to subcortical inputs, they also found reciprocal pathways from the entorhinal cortex, the medial temporal cortex and the orbitofrontal cortex. Importantly, Robbins (1986) argues that these regions may be able to alter the activity of the ACh they receive and ACh innervation of the entire cortex.

ACh has been demonstrated to exert both excitatory and inhibitory effects upon the evoked potentials of many cortical regions, and appears to produce a relatively long-lasting effect on target cells, wherein the response of the target cell to its other inputs is exaggerated (Stone, 1972). ACh may also be involved in regulating discrimination performance in humans and other mammals (Bartus, 1980). ACh receptors are classified as being either nicotinic or muscarinic, and hence, it is of further interest to note that nicotine has been shown to improve attention (Wesnes & Warburton, 1983), while scopolamine (a muscarinic antagonist) has been shown to detrimentally affect vigilance, attention and learning (Kopelman, 1985). Robbins (1986) argues that in combination, the findings reported above may be interpreted as suggesting a role for ACh in maintaining cortical arousal at an optimal level for cue discrimination. Robbins however also notes the work of Stanes, Brown and Singer (1976), which found extreme dose (anticholinesterase) specificity in the effects of ACh, suggesting that performance improvements occur within a narrow range of ACh activity.

In addition to being widespread in the cortex, ACh is also prevalent in the amygdala, where it appears to contribute to aggression. Hernandez-Peon, O'Flaherty and Mazzuchelli-O'Flaherty (1967) elicited both rage and flight by applying ACh to the amygdala. The affective reaction depended upon the precise locus of ACh application. On a final note, recent reports (LaBerge, 1995; Steriade, McCormick & Sejnowski, 1993) have implicated ACh pathways in the governance of slow-wave sleep, and possibly also in the consolidation of information acquired during the waking state.

2.2.3 Dopamine

Considerable overlap can be seen in the projection areas of NA, 5-HT and ACh. While dopaminergic projections bear some similarity to those of ACh, they are grossly dissimilar to those of NA and 5-HT. The primary central dopamine projection arises in the ventral tegmental region (Robbins & Everitt, 1982) and specifically targets the frontal cortex (the only cortical region to receive a substantial ascending DA input) and those elements of the limbic system having connections to the basal ganglia. A functional similarity to the ACh system may exist in that cortical feedback loops in the DA system could regulate mesencephalic DA activity (Nauta & Domesick, 1984).

There is some controversy in regard to the role of DA as a mediator of cortical arousal. Whereas Jacobs (1984) found that DA activity in the substantia nigra does not covary with changes in the waking state, and DA does not appear to play a major role in mediating activity in sensory systems (Ljungberg & Ungerstedt, 1976), other researchers (e.g., Trampus, Ferri, Adami & Ongini, 1993) have argued that DA is involved in the control of tonic cortical arousal, and as a mediator of stress (Imperato, Puglisi-Allegra, Casolini & Angelucci, 1991).

No such controversy exists in regard to the importance of the DA system as a regulator of motor activation. It has long been known that the motor symptoms of Parkinson's disease result from the degeneration of dopaminergic neurons in a pathway from the substantia nigra to the caudate nucleus. Lesions of this pathway in rats have resulted in a response and postural bias and paw preference ipsilateral to the side of the lesion (Carli, Evenden & Robbins, 1985; Evenden & Robbins, 1985). Since the right hemisphere of the brain controls the left side of the body, and vice versa, the interpretation of this finding is that the ipsilateral preference results from a contralateral deficit.

2.3 The Noradrenergic System

We have chosen to consider the noradrenergic system in far greater detail than the 5-HT, ACh and DA systems. The reason for this being that a considerable amount of diverse evidence has converged on the idea of the NA system having specific effects upon arousal and attention.

Within the central nervous system, NA cells are distributed in the medulla of the hindbrain, and more densely in the locus ceruleus of the pons. Two primary ascending pathways originate in the locus ceruleus, the dorsal pathway innervating the neocortex, hippocampus, thalamus, cerebellum, certain portions of the hypothalamus and limbic system, and the ventral pathway which projects mainly to the limbic system and hypothalamus. Inputs to the locus ceruleus originate primarily in the visceral centers of the medulla, the reticular formation and the limbic system, including the septum. Locus ceruleus cells appear to respond to polymodal stimuli as a function of their intensive rather than spatial or temporal properties (Watabe, Nakai & Kasamatsu, 1982). For example, they respond to novel light flashes rather than to lines at particular orientations (as is the case with simple and complex cortical visual cells).

The wide-ranging influence of the dorsal pathway provides NA with diffusion sufficient to exert a modulatory effect on many diverse regions simultaneously. This is a topographic organization that would be expected from a system with some general function such as arousal. However, there is more precise evidence that associates NA with such a function.

The influence of NA in its terminal regions is to inhibit spontaneous activity in those areas, leading to an increase in the signal to noise ratio for inputs to a target cell, thereby resulting in a greater evoked potential to a sensory event (Segal, 1985). However, Woodward, Moises, Waterhouse, Hoffer and Freedman (1979) have provided strong evidence to suggest that the modulatory function of NA is to increase the effects of other inputs such that the current activity of a cell, whether facilitatory or inhibitory, is further accentuated. This view was originally proposed by Kety (1970) and has since been substantiated in a number of studies (e.g., Foote, Aston-Jones & Bloom, 1980; Kasamatsu & Heggelund, 1982).

In tests of NA influence on the performance of certain tasks, the most frequent conclusion has been an association of NA with attentional and arousing functions. Carli, Robbins, Evenden and Everitt (1983) found that an 84% reduction in cortical NA (due to 6-OHDA lesioning of the dorsal pathway) had no effect on rats' performance in spatial and brightness discrimination tasks when stimuli were presented slowly and regularly, but that performance was impaired when stimuli were presented at faster more unpredictable rates in the presence of white noise. The authors interpreted their results as implicating the ceruleo-cortical NA system in the preservation of discriminative accuracy under conditions of elevated arousal.

Feldman, Conforti and Weidenfeld (1995) have documented mechanisms by which NA may mediate the stress response. Robbins and Everitt (1982) have also reviewed evidence indicating that NA turnover increases following an organism's exposure to a variety of stressors. However, they emphasized that NA depletion occurs in rats faced with inescapable electric shock, that is in conditions of learned helplessness, but no such depletion is found when the shock is escapable (Anisman, Kokkindis & Sklar, 1981). Hence, the activity of the central NA system does not simply seem to be driven by environmental input, but is affected by efforts to cope on the part of the organism.

Aston-Jones (1985) has argued that the locus ceruleus system is responsible for the control of vigilance. This argument was based on a study of electrophysiological activity of NA neurons in the locus ceruleus of unanesthetized rats. In addition, McCormick (1989) has reviewed evidence demonstrating that NA is at least partially responsible for the disruption of rhythmic oscillations in thalamocortical activity (which are associated with drowsiness) and for their replacement with a state of excitability that is consistent with cognition. In the following two sections we will consider studies of hemispheric specialization that lend further support to the argument that attention may be influenced by noradrenaline.

2.3.1 Hemispheric specialization of attention
In experimental tasks employing warning signals, subjects display a number of characteristic physiological responses to the warning signal. Among these are heart rate deceleration, a galvanic skin response, and a slow negative EEG shift, termed contingent negative variation (CNV). It has been shown that right hemisphere (RH) lesions may result in the disruption of these responses. For example, Yokoyama, Jennings, Ackles, Hood and Boller (1987) have shown that heart rate deceleration in RH patients is less pronounced than in left hemisphere (LH) patients and controls. Similarly, Heilman, Watson and Valenstein (1985) showed that RH lesions in both humans and monkeys disrupt normal galvanic skin responses.

Other physiological data suggest a right hemisphere specialization for the sustaining of attention. Deutsch, Papanicolaou, Bourbon and Eisenberg (1987) reported greater blood flow to the right rather than to the left frontal regions during conditions of sustained attention, as did Cohen et al. (1987). A similar finding was reported by Pardo, Fox and Raichle (1990) who also found increased blood flow to the right but not left superior parietal cortex in both visual and somatosensory vigilance conditions. Hemispheric asymmetry in alpha desynchronization has also been observed. Heilman and Van Den Abell (1980) found right parietal alpha desynchronization following lateralized warning signals presented to either hemisphere, but left parietal desynchronization only after a lateralized warning signal had been presented to that hemisphere.

Behavioral data also suggest the right hemisphere's importance in conditions of sustained attention. Wilkins, Shallice and McCarthy (1987) demonstrated that right, but not left frontal patients were impaired in their ability to voluntarily sustain attention in a monotonous signal detection task. A similar result was obtained by Coslett, Bowers and Heilman (1987). Also, in studies of split brain patients, Dimond (1979) found that tactual, visual and auditory vigilance performance was substantially better when stimuli were presented to the right rather than to the left hemisphere.

Behavioral data from normals have also revealed hemispheric asymmetries. In a signal detection task, Dimond and Beaumont (1973) found that when targets were presented to the left hemisphere, vigilance deteriorated over the course of an eighty minute experimental session. Right hemisphere performance, although at a lower level (fewer signals detected) showed no such decrement. It should be noted however, that during the course of a session, targets were presented only to one hemisphere or the other. The results were interpreted as suggesting that the left hemisphere is the more sensitive "watchkeeper" but that its relatively high level of performance is susceptible to exhaustibility. The right hemisphere however, is more able to sustain vigilance, albeit at a lower level.

Whitehead (1991a) provided further behavioral evidence for the notion of right hemisphere superiority in maintaining the alert state. In a series of studies, normal subjects received a warning signal followed by a stimulus onset asynchrony of between 3 and 30 seconds. When stimuli were presented following delays of 12 seconds or more, subjects responded significantly faster to stimuli presented to their left rather than to their right visual field. During further investigation (Whitehead, 1991b), it was found that an external alerting stimulus (presented at the same time as the visual target) eliminated this effect, suggesting a common pathway between voluntary sustained attention and the more automatic alerting effects of external stimuli. It was also found that the act of sustaining alertness interacted with the covert orienting of attention. Specifically, when alertness was sustained for a long, as opposed to a short period, attentional engagement in the left visual field was increased. This result was interpreted as indicating some commonality between the structures affected by the act of sustaining alertness, and those responsible for the covert orienting of attention. Direct physiological evidence regarding this issue will be presented later.

Heilman and Van Den Abell (1979) have shown that warning stimuli projected to the right hemisphere reduce reaction times of the right hand to a greater extent than left hemisphere warning stimuli reduce left hand reaction times, and more importantly, to a greater extent than left hemisphere warning stimuli reduce right hand reaction times. Their interpretation was that although each hemisphere can mediate its own activation (Heilman & Valenstein, 1979), the right hemisphere is better able to activate the left hemisphere than the reverse. Subsequent to this study however, Heilman argued that the findings suggest that attention is entirely mediated by the right hemisphere. In support of his argument, he cited evidence indicating that neglect of the contralateral visual hemifield is more prevalent following right than left parietal lesions (Heilman & Van Den Abell, 1980). Weintraub and Mesulam (1987) adopted a similar position in claiming that right hemisphere lesions are more likely to lead to both contralateral and ipsilateral neglect, whereas left hemisphere lesions are more often associated with only contralateral neglect. Also, contralateral neglect is often less severe when the left rather than the right hemisphere is lesioned.

Although Heilman's data permit either an activational or attentional explanation, the majority of the extant data, including Heilman's, indicates right hemisphere dominance only when the task requires the sustaining of attention for long periods, either in situations where continuous processing is involved, or where subjects must remain alert in anticipation of a signal. When highly phasic attentional effects are studied, as for example in Derryberry's (1989) study, left hemisphere superiority may be evident. Also, even when vigilance is required, as was the case with Dimond and Beaumont's (1973) study, the left hemisphere displayed superior performance (albeit with impairment over time) in conditions where only one hemisphere received stimulation, that is, when there was little likelihood of the right hemisphere activating the left. One further piece of evidence is particularly damaging for Heilman's idea of attention itself being a right hemisphere function. Luck, Hillyard, Mangun and Gazzaniga (1989) have demonstrated quite convincingly that in split brain patients, visual search of each hemifield is performed independently by the contralateral hemisphere.

In summary, it would appear most plausible that demonstrations of hemispheric asymmetries of attention are not due to the fact that the visuo-spatial attention system resides solely or even primarily within the right hemisphere. Rather, some more general influence serves to sustain the activity of right hemisphere attentional mechanisms in conditions requiring continuous activity or vigilance, and that this influence is either less pronounced or is more phasic in its effect upon the left hemisphere.

2.3.2 Hemispheric specialization in noradrenergic pathways

In addition to findings of hemispheric specialization in the control of attention, several lines of research have indicated hemispheric specialization in neurotransmitter systems. For example, Tucker and Williamson (1984) have proposed that lateralization in neurotransmitter pathways may be responsible for the production of subjectively meaningful affective states. Specifically, left lateralized dopaminergic innervation is associated with fluctuations in anxiety, and right lateralized NA pathways are responsible for variations in mood level.

The demonstration of a right hemisphere affinity for noradrenaline is of particular interest in view of the association between NA and arousal and attention, and demonstrations of right hemisphere specialization for sustained attention. Several lines of evidence converge on the supposition of right lateralization in NA pathways.

The dorsal NA pathway described earlier, in its innervation of the cortex is known to enter the cortex at the frontal pole, to ascend to the superficial layers, and to pass through a horizontal layer parallel to the outer layer of the cortex as it traverses toward posterior regions (Emson & Lindvall, 1979). Consistent with the notion of right hemisphere bias in this pathway, it has been shown that lesions of the right but not left hemisphere lead to depletions of NA bilaterally (Robinson, 1979). Furthermore, this effect is more pronounced when lesions are close to the frontal pole (Robinson, 1985). A behavioral consequence of these lesions is the production of spontaneous hyperactivity in rats. Oke, Lewis and Adams (1980) have shown that higher NA levels are found in the right rather than in the left thalamus of rats, and in humans, post mortem studies have shown a similar lateralization (Oke, Keller, Mefford & Adams, 1978).

Further indirect evidence contributes to the belief of right lateralized NA. It is thought that the effects of electro-convulsive therapy (ECT) on mood are mediated through the action of NA. Hence, the finding that right hemisphere activation is particularly facilitated by ECT (Kronfol, Hamsher, Digre & Waziri, 1978) is of relevance. Similarly, tricyclic antidepressant medication is NA mediated; following administration of this medication to children, it was found that improvements resulted in right but not left hemisphere cognitive performance (Brumback, Staton & Wilson 1980).

Since NA innervation of the cortex and thalamus is provided by the dorsal NA pathway, and since both display right lateralized NA concentrations, a parsimonious explanation of this distribution would be to suggest right lateralization in this pathway. Within the cortex, evidence suggests that NA innervation is provided by a branch of the dorsal pathway entering at the frontal pole, and which arborizes as it moves toward posterior regions.

If it is indeed the case that findings of right lateralization of attention and arousal are related, and that both are mediated by right lateralization of NA pathways, then it would be reasonable to expect an effect of NA manipulation on attention shifting. In a study of this question, Clark, Geffen and Geffen (1989) investigated the effects of an NA blocker, clonidine, on responses to visual targets which had been preceded by a directional cue (valid, neutral or invalid). They found that clonidine increased reaction times in general, but reduced the cost of invalid cueing — in essence, it made the subjects more distractible. This result does indeed suggest that NA exerts a specific action on the visuo-spatial attention system. In this study laterality effects were not investigated, however, using the same paradigm, Posner, Inhoff, Friedrich and Cohen (1987) did study hemispheric specialization in this system. They found that if a warning signal is omitted before a target, then right parietal patients are greatly slowed in their ability to respond to targets, whereas left parietal patients are not. This finding again supports the notion that the right hemisphere contains the mechanism responsible for the sustaining of alertness. When a patient who has damage to that mechanism must sustain alertness without an external aid (the warning signal), then he or she is poor at detecting the target. In contrast, left parietal patients suffer no comparable loss.

If one is to suppose that the shifting of attention is influenced by NA, then one would expect evidence of NA innervation of the structures involved. This has been shown to be the case. As mentioned earlier, Oke et al. (1978) have shown dense NA innervation of the thalamus. Of greater interest however, are the findings of Morrison and Foote (1986), wherein each of the structures of the visuo-spatial attention system (the posterior parietal lobe, superior colliculus and lateral pulvinar of the thalamus) was shown to be densely innervated by NA, while much weaker NA innervation was found in the geniculo-striate and ventral pattern recognition pathways.

Finally, if NA is hypothesized to be responsible for the maintenance of the alert state, one would expect to find an inter-relationship between the physiological indices of that state and NA manipulation. Tackett, Webb and Privitera (1981) have indeed shown that following the release of NA, heart rate deceleration occurs, and Walker and Sandman (1979) have shown that evoked potentials are higher in the right than in the left cortical region during periods of heart rate deceleration.

2.4 Interactions between cortical and subcortical systems

Our discussion thus far has reviewed models of arousal which incorporate multiple component systems, and it has emphasized the diversity in potential neuromodulation among different ascending neurotransmitter systems. While these systems have been categorized as ascending, it is not the case that their influence on cortical structures and functions is simply unidirectional. Rather, there are many lines of research indicating reciprocity between cortical and subcortical systems in determining the ultimate influence exerted upon cognitive processing. Our review of this research will be less than comprehensive; instead, we will attempt to provide examples of the more concrete findings in this area. For a more complete treatment of this issue, the interested reader is referred to two excellent reviews by Tucker and his colleagues (Tucker & Williamson, 1984; Tucker & Derryberry, 1990).

The idea of executive cortical control over sub-cortical arousal mechanisms is not a new one, dating back at least to the work of Pribram and McGuinness (1975) who argued for "the involvement of the amygdala and related frontal cortical structures in the attentional control of the core brain arousal systems" (p. 119). Pribram and McGuinness based their argument upon extensive lesion evidence (e.g., Bagshaw & Benzies, 1968; Bagshaw, Kimble & Pribram, 1965) indicating that a frontolimbic circuit is involved in the control of the orienting response. Moreover, their evidence indicated some differentiation between a facilitatory system involving the dorsolateral frontal cortex, and an inhibitory system related to the orbitofrontal cortex.

Evidence also indicates a high degree of response specificity in amygdala neurons, in particular in their selection for reward- and punishment- conditioned stimuli (Ono, Tamura, Nishijo, Nakamura & Tabuchi, 1989) and in attributing motivational significance to incoming signals (Sarter & Markowitsh, 1985). The amygdala has also been ascribed a role in the mediation of anxiety, especially as a contributor to the fight or flight response (Sarter & Markowitsch, 1985). In addition, Applegate, Kapp, Underwood & McNall (1983) found that electrical stimulation of the amygdala results in feelings of anxiety in humans, and Davis, Hitchcock & Rosen (1987) eliminated conditioned fear responses through amygdaloid lesions.

The amygdala and orbitofrontal cortex share several characteristics in regard to responsivity. In addition to being implicated by Pribram and McGuinness in a system controlling the orienting response, cells in the orbito-frontal cortex, like those in the amygdala have been shown to respond to cues for reward and punishment (Thorpe, Rolls & Madison, 1983), to influence autonomic control (Mesulam & Mufson, 1982) and to be excessively active in patients with anxiety disorders (Baxter, Phelps, Mazziotta & Guze, 1987). Several authors (e.g., Nauta, 1971; Tucker & Derryberry, 1990) have argued that the orbital cortex has an executive role in anticipating the expected significance of events and in developing intentionality. Since the orbital cortex is reciprocally connected with the limbic structures, and also projects efferents to the amygdala via temporal regions, its potential sphere of influ-ence is very extensive. For example, from the amygdala alone, influence may be exerted upon forebrain and brainstem circuits, temporal and occipital cortices, and (via the basal nucleus of Mynert) upon all cortical acetylcholine projections.

A second paralimbic circuit consists of a complex interaction between the posterior cingulate, entorhinal and parahippocampal cortices, the anterior thalamus, septum and hippocampus. As discussed earlier, Gray (1982) has implicated several of the components of this circuit in facilitating passive avoidance and behavioral inhibition, and also in modulating cortical and autonomic arousal. Gray views the Behavioral Inhibition System as a monitor-ing mechanism, one which responds to conditioned stimuli signifying forth-coming punishment or non-reward. In the event that such signals are detected, ongoing behavior is inhibited, cortical and autonomic arousal are adjusted and attentional resources are directed toward the avoidance of punishment. Hence, this system can also be characterized as exhibiting reciprocity between corti-cal and subcortical activation. In order for the BIS to perform its function, some evaluation of what constitutes a cue for punishment must have preceded, and in response to the evaluation, the BIS is apt to influence the adjustment of both the intensity and direction of behavior. In terms of brain activity, the progression may involve evaluation at the cortical level, BIS activity at the level of the limbic forebrain, arousal through the action of the reticular formation (with subsequent cortical repercussions), and a new direction of behavior, mediated again by cortical structures.

The control of motor actions constitutes another example of behavior being regulated by an interaction of cortical and subcortical systems. The association between dopaminergic pathways involving the basal ganglia and motor control is well established. There is massive interconnectivity between the frontal cortex and the basal ganglia, an area well established as contributing to the initiation of movements (Kornhuber, 1974). These connections take the form of multiple parallel loops (Groenewegen, 1988) in which, most importantly, activity may be modulated at both the cortical and subcortical level by ascending dopaminergic projections from the ventral tegmental region (Taber, Das & Fibiger, 1995). Since dopamine has also been associated with higher order motivational processes (Bunney & Aghajanian, 1977), and response readiness (Tucker & Williamson, 1984), we are again presented with a scenario in which cortical arousal, ostensibly arising in subcortical systems, is mediated not simply in a bottom-up fashion, but as the result of activity in cortico-subcortical feedback loops.

The control of attention is another area in which cortical and subcortical mechanisms interact. Mangun et al. (1994) have produced a compelling argument in favor of the idea that the right hemisphere's specialization for the control of attention is mediated in part by subcortical pathways. These researchers obtained findings suggesting an executive role for the right hemisphere in the control of attention, and importantly, did so in split-brain patients. Since the cerebral hemispheres of these patients are separated at the cortical level, one must suppose that the right hemisphere's influence over the left hemisphere is mediated, at least in part, by subcortical pathways. Furthermore, the cortical mechanisms of attention appear to involve both frontal and posterior mechanisms, with the former being characterized as having a regulatory control over the latter. For example, in a study of scalp potentials, Deeke, Kornhuber, Lang and Schreiber (1985) required subjects to make responses to visual targets. They documented a temporal progression of activity wherein subjects exhibited first frontal negativity, followed by occipital negativity and the resolution of the frontal activity, and finally, once a response was made, resolution of the occipital negativity. The implication here being that the frontal areas possess executive control, which in this task, was passed to the occipital cortex. In terms of personal experience, this would equate to the subjects deliberately shutting down on-going cognitive processing, since it might distract them from detection of the target, and assuming an 'empty-headed' state, in which attentional resources are directed only toward target detection. Once the target has been detected, conscious processing resumes, and activity in target detection mechanisms subsides. Similarly, Posner, Petersen, Fox and Raichle (1988) have found that in anticipation of a target, blood flow to the anterior cingulate gyrus is depleted, but once the target is presented and requires further processing, blood flow to the anterior cingulate is elevated above baseline. In addition, studies of event-related potentials (ERPs) in frontal patients further the supposition that an attentional mechanism resides within the frontal cortex. Frontally lesioned patients do not show the normal enhancement of temporal ERPs to attended auditory stimuli, but do exhibit abnormally large ERPs to unattended stimuli (Knight, Hillyard, Woods & Neville, 1981). The implication here is that the unattended stimuli are not being adequately screened out.

Finally, there is evidence suggesting that the frontal cortex is able to selectively inhibit posterior perceptual mechanisms. Ascending pathways from the thalamus to the posterior cortex are affected by activity in the reticular nucleus of the thalamus. The reticular nucleus has been argued to serve a central gating function (Skinner & Yingling, 1977; Yingling & Skinner, 1977). More importantly, the frontal cortex exerts an excitatory control over the reticular nucleus. Hence, frontal activation of the reticular nucleus encourages its gating function, resulting in inhibitory effects on posterior cortex. LaBerge (1995) has extended the Yingling and Skinner model to suggest that thalamo-cortical circuits operate as an attentional enhancement mechanism, wherein information (internally or externally generated) passing through the thalamus is enhanced for further higher-order processing. LaBerge elaborates on the earlier work of Yingling and Skinner by ascribing specific roles to each of the thalamic nuclei, and by detailing the anatomical connections between these nuclei and other brain areas.

2.5 *Summary*

We have made several references to the experiences of the bridge jumper in order to introduce various characteristics of arousal. By way of providing a final example of our understanding of the brain mechanisms of arousal, let us now speculate upon the bridge jumper's experiences from a neurophysiological perspective

As he stood on the bridge contemplating what he was about to do, he perceived cues for both punishment (threat of injury or death) and reward (anticipation of thrill). According to Gray's theory, these cues encouraged both behavioral inhibition and activation, hence energizing the hippocampus and medial septum (inhibition), and the medial forebrain, lateral hypothalamus, and amygdala (activation). As these two systems fought for control over the direction of behavior (to jump or not) each innervated the non-specific arousal system in the reticular formation. As a result of the amygdaloid, hypothalamic and reticular activity, the sympathetic branch of the ANS achieved dominance over the parasympathetic branch. Hence, the jumper exhibited a fight or flight response — heart rate, blood pressure and respiration increased, digestion was suspended. Also as a function of reticular activity, the locus ceruleus increased noradrenergic supply to other areas of the brain; in particular to attentional mechanisms. The jumper's mode of processing information became highly selective. As a result of thalamic activity, task-relevant information was engaged and enhanced. In the example we employed, the jumper did actually jump — therefore activation overcame inhibition. Once the jumper realized that he was out of danger, his anxiety diminished, but his still high state of autonomic arousal found a new outlet, in thrill or euphoria.

In conclusion, early views of arousal as unidimensional have been superseded by multidimensional conceptualizations. Multi-component structural models of arousal have become necessary in accounting for findings of situational specificity in human and animal behavior. Earlier views of the effect of arousal on information processing emphasized a one-way causal mechanism, wherein arousal was thought to impact upon higher-order processing but not the reverse. More recent thinking has emphasized reciprocity between higher- and lower-order processing, a view strengthened by anatomical findings of interactivity between cortical and subcortical structures. Neurophysiological studies of neurotransmitter systems, the NA system in particular, have further aided our understanding of energetical effects on behavior, and in particular, have helped us to accommodate arousal within a rapidly developing field — the neuroscientific study of human behavior.

Notes

Correspondence regarding this chapter should be addressed to: Roger Whitehead, Department of Psychology, University of Colorado at Denver, Campus Box 173, P.O. Box 173364, Denver, CO 80217-3364. Electronic mail may be sent to RWHITEHEAD@CASTLE.CUDENVER.EDU via Internet.
    We wish to thank Jeanne Jackson and Don Tucker for their suggestions regarding the manuscript. We also acknowledge the earlier work of Trevor Robbins, Don Tucker, Peter Williamson, and Doug Derryberry. Their earlier reviews of parts of the literature that we have summarized greatly assisted us in the writing of this chapter.

References

Anisman, H., Kokkindis, L. and Sklar, L. S. 1981. "Contribution of neurochemical change to stress induced behavioral deficits". In S. J. Cooper (ed.), *Theory in psychopharmacology*. London: Academic Press, 65-102

Applegate, C.D., Kapp, B. S., Underwood, M.D. and McNall, C. L. 1983. "Autonomic and somatomotor effects of amygdala central nucleus stimulation in awake rabbits". *Physiology and Behavior* 31: 353-360.

Ashton, R., White, K. D., and Hodgson, G. 1979. "Sensitivity to heart rate: A psychophysiological study". *Psychophysiology* 16: 463-466

Aston-Jones, G. 1985. "Behavioral functions of locus coeruleus derived from cellular attributes". *Physiological Psychology* 13: 118-126.

Azmitia, E. C. 1978. "The serotonin-producing neurons of the midbrain median and dorsal raphe nuclei". In L. L. Iversen, S. D. Iversen and S. H. Snyder (eds), *Handbook of psychopharmacology: Vol. 9.* New York: Plenum Press.

Bacon, S.J. 1974. "Arousal and the range of cue utilization". *Journal of Experimental Psychology* 102: 81-87.

Bagshaw, M.H. and Benzies, S. (1968). "Multiple measure of the orienting reaction and their dissociation after amygdalectomy in monkeys". *Experimental Neurology* 20: 175-187.

Bagshaw, M.H., Kimble, D.P. and Pribram, K.H. 1965. "The GSR of monkeys during orienting and habituation after ablation of the amygdala, hippocampus and inferotemporal cortex". *Neuropsychologia* 3: 111-119.

Bartus, R.F.T. 1980. "Cholinergic effects on cognition and memory in animals". In L.W. Poon (ed), *Aging in the 1980s: Psychological issues.* Washington DC: American Psychological Association.

Baxter, L.R., Phelps, M. E., Mazziotta, J.C. and Guze, B.H. 1987. "Local cerebral glucose metabolic rates in obsessive-compulsive disorder: A comparison with rates in unipolar depression and in normal controls". *Archives of General Psychiatry* 44: 211-218.

Brener, J. and Jones, J.M. 1974. "Interoceptive discrimination in intact humans: Detection of cardiac activity". *Physiology and Behavior* 13: 763-767.

Broadbent, D.E. 1971. *Decision and Stress*. London: Academic Press.

Brumback, R.A., Staton, R.D. and Wilson, H. 1980. "Neuropsychological study of children during and after remission of endogenous depressive episodes". *Perceptual and Motor Skills* 50: 1163-1167.

Bunney, B.S. and Aghajanian, G.K. 1977. "Electrophysical studies of dopamine-innervated cells in the frontal cortex". *Advances in Biochemical Psychopharmacology* 16: 65-70.

Carli, M., Evenden, J. and Robbins, T.W. 1985. "Depletion of unilateral striatal dopamine impairs initiation of contralateral actions and not sensory attention". *Nature* 313: 679-682.

Carli, M., Robbins, T.W., Evenden, J. and Everitt, B.J. 1983. "Effects of lesions to ascending noradrenergic neurons on performance of a 5-choice serial reaction task in rats: Implications for theories of dorsal noradrenergic function based on selective attention and arousal". *Behavioural Brain Research* 9: 361-380.

Clark, M., Geffen, G.M. and Geffen, L.B. 1989. "Catecholamines and the covert orienting of attention". *Neuropsychologia* 27: 131-139.

Cornsweet, D.M. 1969. "Use of cues in the visual periphery under conditions of arousal". *Journal of Experimental Psychology* 80: 14-18.

Coslett, H.B., Bowers, D. and Heilman, K.M. 1987. "Reduction in cerebral activation after right hemisphere stroke". *Neurology* 37: 957-962.

Davis, M., Hitchcock, J.M. and Rosen, J.B. 1987. "Anxiety and the amygdala: Pharmacological and anatomical analysis of the fear-potentiated startle paradigm". In G. Bower (ed), *The psychology of learning and motivation*. New York: Academic Press, 263-305.

Deeke, L., Kornhuber, H.H., Lang, W. and Schreiber, H. 1985. "Timing function of the frontal cortex in sequential motor and learning tasks". *Human Neurobiology* 4: 143-154.

Derryberry, D. 1989. "Hemispheric consequences of success-related emotional states: Roles of arousal and attention". *Brain and Cognition,* 11, 258-274.

Deutsch, G., Papanicolaou, A.C., Bourbon, T. and Eisenberg, H.M. 1988. "Cerebral blood flow evidence of right cerebral activation in attention demanding tasks". *International Journal of Neuroscience* 36: 23-28.

Dimond, S.J. 1979. "Performance by split-brain humans on lateralized vigilance tasks". *Cortex* 15: 43-50.

Dimond, S.J. and Beaumont, J.G. 1973. "Differences in the vigilance performance of the right and left hemispheres". *Cortex* 9: 259-265.

Easterbrook, J.A. 1959. "The effect of emotion on cue utilization and the organization of behavior". *Psychological Review* 66: 183-201.

Emson, P.C. and Lindvall, O. 1979. "Distribution of putative neurotransmitters in the cortex". *Neuroscience* 4: 1-30.

Evenden, J. and Robbins, T.W. 1985. "Effects of unilateral 6-hydroxydopamine lesions of the caudate-putamen on skilled forepaw use in the rat". *Behavioural Brain Research* 14: 61-68.

Eysenck, M.W. 1975. "Effects of noise, activation level and response dominance in retrieval from semantic memory". *Journal of Experimental Psychology: Human Learning and Memory* 104: 143-148.

Eysenck, M.W. 1976. "Arousal, memory and learning". *Psychological Bulletin* 83: 389-404.

Feldman, S., Conforti, N. and Weidenfeld, J. 1995. "Limbic pathways and hypothalamic neurotransmitters mediating adrenocortical responses to novel stimuli". *Neuroscience and Beahvioral Reviews* 19: 235-240.

Foote, S.L., Aston-Jones, G. and Bloom, F.E. 1980. "Impulse activity of locus coeruleus neurons in awake rats and squirrel monkeys is a function of sensory stimulation and arousal". *Proceedings of the National Academy of Sciences* 77: 3033-3037.

Gray, J.A. 1975. *Elements of a two-process theory of learning.* London: Academic Press.

Gray, J.A. 1977. "Drug effects on fear and frustration: Possible limbic site of action of minor tranquilizers". In L.L. Iverson, S D. Iverson and S.H. Snyder (eds), *Handbook of psychopharmacology: Vol. 8.* New York: Plenum Press, 433-529.

Gray, J.A. 1978. "The neuropsychology of anxiety". *British Journal of Psychology* 69: 417-434.

Gray, J.A. 1982. *The neuropsychology of anxiety.* New York: Oxford University Press.

Greenstadt, L., Shapiro, D. and Whitehead, R. 1986. "Blood pressure discrimination". *Psychophysiology* 23: 500-509.

Groenewegen, H.J. 1988. "Organization of the afferent connections of the mediodorsal thalamic nucleus in the rat, related to the mediodorsal prefrontal topography". *Neuroscience* 24: 379-431.

Hamilton, P., Hockey, G.R.J. and Quinn, J.G. 1972. "Information selection, arousal and memory". *British Journal of Psychology* 63: 181-189.

Heilman, K.M. and Valenstein, E. 1979. "Mechanisms underlying spatial neglect". *Annals of Neurology* 5: 166-170.

Heilman, K.M. and Van Den Abell, T. 1979. "Right hemispheric dominance for mediating cerebral activation". *Neuropsychologia* 17: 315-321.

Heilman, K.M. and Van Den Abell, T. 1980. "Right hemisphere dominance for attention: The mechanism underlying hemispheric asymmetries of inattention". *Neurology* 30: 327-330.

Heilman, K.M., Watson, R.T. and Valenstein, E. 1985. "Neglect and related disorders". In K.M. Heilman and E.Valenstein (eds), *Clinical neuropsychology.* New York: Oxford, 243-293.

Hernandez-Peon, R., O'Flaherty, J.J. and Mazzuchelli-O'Flaherty, A.L. 1967. "Sleep and other behavioral effects induced by acetylcholine stimulation of basal temporal cortex and striate structures". *Brain Research* 4: 243-267.

Hobson, J.A., McCarley, R.W. and Wyzinski, P.W. 1975. "Sleep cycle oscillation: Reciprocal discharge by two brainstem neuronal groups". *Science* 189: 55-58.

Hockey, G.R.J. 1970a. "Signal probability and spatial location as possible bases for increased selectivity in noise". *Quarterly Journal of Experimental Psychology* 22: 37-42.

Hockey, G.R.J. 1970b. "Effect of loud noise on attentional selectivity". *Quarterly Journal of Experimental Psychology* 22: 28-36.

Imperato, A., Puglisi-Allegra, S., Casolini, P and Angelucci, L. 1991. "Changes in brain dopamine and acetylcholine release following stress are independent of the pituitary-adrenocortical axis". *Brain Research* 538: 111-117.

Iversen, S.D. 1983. "Where in the brain do benzodiazepines act"? In M. Trimble (ed), *Benzodiazepines divided*. Chichester: Wiley.

Jacobs, B.L. 1984. "Single unit activity of brain monoaminergic neurons in freely moving animals: A brief review". In R. Bandler (ed*)*, *Modulation of sensorimotor activity during alterations in behavioral states*. New York: Liss, 601-646.

Jouvet, M. 1968. "Insomnia and decrease of cerebral 5-hydroxy-tryptamine after destruction of the raphe system in the cat". *Advances in Pharmacology* 6: 265-279.

Jouvet, M. 1972. "The role of monamines and acetylcholine containing neurons in the regulation of the sleep-waking cycle". *Ergebnisse der Phyiologie* 64: 166-307.

Jouvet, M. and Renault, J. 1966. "Insomnie persistante aprÈs lÈsions des noyaux du raphÈ chez le chat". *Comptes Rendus Dendus de la SociÈtÈ de Biologie et de ses Filiales* 160: 1461-1465.

Kahneman, D. 1973. *Attention and effort*. Englewood Cliffs NJ: Prentice-Hall.

Kasamatsu, T. and Heggelund, P. 1982. "Single cell responses in cat visual cortex to visual stimulation during iontophoresis of noradrenaline". *Experimental Brain Research* 45: 317-327.

Kety, S.S. 1970. "The biogenic amines in the central nervous system: Their possible roles in arousal, emotion and learning". In F.O. Schmitt (ed*)*, *The neurosciences second study program*. New York: Rockefeller University Press, 324-335.

Knight, R.T., Hillyard, S.A., Woods, D.L. and Neville, H.J. 1981. "The effects of frontal cortex lesions on event-related potentials during auditory selective attention". *Electro-encephalography and Clinical Neurophysiology* 52: 571-582.

Kopelman, M.D. 1985. "Multiple memory deficits in Alzheimer-type dementia; implications for pharmacotherapy". *Psychological Medicine* 15: 527-541.

Kornhuber, H.H. 1974. "Cerebral cortex, cerebellum, and basal ganglia: An introduction to their motor functions". In F.O. Schmitt and F.G. Worden (eds), *The neurosciences: Third study program*. Cambridge, MIT Press.Kronfol, Z., Hamsher, K., Digre, K. and Waziri, R. 1978. "Depression and hemisphere functions: Changes associated with unilateral ECT". *British Journal of Psychiatry* 132: 560-567.

LaBerge, D. 1995. *Attentional processing*. Harvard University Press: Cambridge, MA.

Lacey, J.I. 1967. "Somatic response patterning and stress: Some revisions of activation theory". In M.H. Appley and R. Trumbull (eds), *Psychological stress: Issues in research*. New York: Appleton-Century-Crofts, 14-44.

Ljungberg, T. and Ungerstedt, U. 1976. "Sensory inattention produced by 6-hydroxy-dopamine-induced degeneration of ascending dopamine neurons in the brain". *Experimental Neurology* 53: 585-600.

Luck, S.J., Hillyard, S.A., Mangun, G. R. and Gazzaniga, M.S. 1989. "Independent hemispheric attentional systems mediate visual search in split-brain patients". *Nature* 342: 543-545.

Mangun, G.R., Luck, S.J., Plager, R., Loftus, W., Hillyard, S.A., Handy, T., Clark, V.P. and Gazzaniga, M.S. 1994. "Monitoring the visual world: Hemispheric asymmetries and subcortical processes in attention". *Journal of Cognitive Neuroscience* 6: 267-275.

Martin, J.E., Epstein, L.H. and Cinciripini, P.M. 1980. "Effects of feedback and stimulus control on pulse transit time discrimination". *Psychophysiology* 17:431-436.

McCormick, D.A. 1989. "Cholinergic and noradrenergic modulation of thalamocortical processing". *Trends in Neurosciences* 12: 215-221.

McGuinness, D. and Pribram, K. 1980. "The neuropsychology of attention: Emotional and motivational controls". In M.C. Wittrock (Ed.), *The brain and psychology*. New York: Academic Press, 27-70.

Mesulam, M.M. and Mufson, E.J. 1984. "Neural inputs into the nucleus basalis of the substantia inominata (Ch4) in the rhesus monkey". *Brain* 107: 253-274.

Mesulam, M.M., Mufson, E.J., Levey, A.,I., and Wainer, B.H. 1983. "Cholinergic innervation of cortex by the basal forebrain: cytochemistry and cortical connections of the septal area, diagonal band nuclei, nucleus basalis (substantia inominata) and hypothalamus in the rhesus monkey". *Journal of Comparative Neurology* 214: 170-197.

Morrison, J.H. and Foote, S.L. 1986. "Noradrenergic and serotonergic innervation of cortical, thalamic and tectal visual structures in old and new world monkeys". *Journal of Comparative Neurology* 243: 117-128.

Moruzzi, G. and Magoun, H.W. 1949. "Brain stem reticular formation and activation in the EEG". *Electroencephalography and Clinical Neurophysiology* 1: 455-473.

Näätänen, R. and Merisalo, A. 1977. "Expectancy and preparation in simple reaction time". In S. Dornic(ed), *Attention and performance VI*. Hillsdale: Erlbaum, 115-138.

Nauta, W.J.H. 1971. "The problem of the frontal lobe: A reinterpretation". *Journal of Psychiatric Research* 8: 167-187.

Nauta, W.J.H. and Domesick, V.B. 1984. "Afferent and efferent relationships of the basal ganglia". In D. Evered and M. O'Conner (eds), *Functions of the basal ganglia*. London: Pitman, (pp. 167-187).

Oke, A., Keller, R., Mefford, I. and Adams, R. 1978. "Lateralization of norepinepherine in human thalamus". *Science* 200: 1411-1413.

Oke, A., Lewis, R. and Adams, R. 1980. "Hemispheric asymmetry of norepinepherine distribution in rat thalamus". *Brain Research* 188: 269-272.

Ono, T., Tamura, R., Nishijo, H., Nakamura, K. and Tabuchi, E. 1989. "Contribution of amygdalar and lateral hypothalamic neurons to visual information processing of food and nonfood in monkey". *Physiology and Behavior* 45: 411-421.

Pardo, J.V., Fox, P.T. and Raichle, M.E. 1990. "A heteromodal neural network activated in humans during sustained attention to sensory stimuli". *Unpublished manuscript*, Washington University, St. Louis, Missouri.

Posner, M.I. 1978. *Chronometric explorations of mind*. Hillsdale: Erlbaum.

Posner, M.I. and Boies, S.J. 1971. "Components of attention". *Psychological Review* 78: 391-408.

Posner, M.I., Inhoff, A., Friedrich, F.J. and Cohen, A. 1987. "Isolating attentional mechanisms: A cognitive anatomical analysis". *Psychobiology* 15: 107-121.

Posner, M.I., Petersen, S.E., Fox, P.T. and Raichle, M.E. 1988. "Localization of cognitive operations in the human brain". *Science* 240: 1627-1631.

Pribram, K.H. and McGuinness, D. 1975. "Arousal, activation and effort in the control of attention". *Psychological Review* 82: 116-149.

Pribram, K.H., Reitz, S., McNeil, M. and Spevack, A.A. 1974. "The effect of amygdalectomy on orienting and classical conditioning". In *Mechanisms of formation and inhibition of conditional reflex*. Moscow: Publications Office, Academy of Sciences of USSR.

Robbins, T.W. 1986. "Psychopharmacological and neurobiological aspects of the energetics of information processing". In G.R.J. Hockey, A. Gaillard and M.G. Coles (eds), *Energetics and human information processing*. Amsterdam: Martinus Nijhoff, (pp. 71-90).

Robbins, T.W. and Everitt, B.J. 1982. "Functional studies of the central catecholamines". *International Review of Neurobiology* 23: 303-365.

Robinson, R.G. 1985. "Lateralized behavioral and neurochemical consequences of unilateral brain injury in rats". In S. G. Glick (ed), *Cerebral lateralization in nonhuman species*. Orlando: Academic Press, 135-156.

Robinson, R.G. 1979. "Differential behavioral and biochemical effects of right and left hemispheric cerebral infarction in the rat". *Science* 205: 707-710.

Sanders, A.F. 1983. "Toward a model of stress and human performance". *Acta Psychologia* 53: 61-97.

Sarter, M. and Markowitsch, H.J. 1985. "Involvement of the amygdala in learning and memory: A critical review, with emphasis on anatomical relations". *Behavioral Neuroscience* 99: 342-380.

Segal, M. 1985. "Mechanisms of action of noradrenaline in the brain". *Physiological Psychology* 13: 172-178.

Skinner, J.E. and Yingling, C.D. 1977. "Central gating mechanisms that regulate event-related potentials and behavior. A neural model for attention". In J. E. Desmedt (ed), *Attention, voluntary contraction and event-related cerebral potentials*. Basel: Karger.

Stanes, M.D., Brown, C.P. and Singer, G. 1976. "Effect of physostigmine on Y-maze retention in the rat". *Psychopharmacologia* 46: 269-276.

Steriade, M., McCormick, D.A. and Sejnowski, T.J. 1993. "Thalamocortical oscillations in the sleeping and aroused brain". *Science* 262: 679-685.

Stern, R.M. 1972. "Detection of one's own spontaneous GSRs". *Psychonomic Science* 29: 354-356.

Stone, T.W. 1972. "Cholinergic mechanisms in the rat somatosensory cortex". *Journal of Physiology* 225: 485-499.

Taber, M.T., Das, S. and Fibiger, H.C. 1995. "Cortical regulation of subcortical dopamine release: Mediation via the ventral tegmental area". *Journal of Neurochemistry* 65, 1407-1410.

Tackett, R.L., Webb, J.G. and Privitera, P.J. 1981. "Cerebroventricular propanolol elevates cerebrospinal fluid norepinepherine and lowers blood pressure". *Science* 213: 911-913.

Thorpe, SJ., Rolls, E.T. and Madison, S. 1983. "The orbitofrontal cortex: Neuronal activity in the behaving monkey". *Experimental Brain Research* 49: 93-115.

Trampus, M., Ferri, N., Adami, M. and Ongini, E. 1993. "The dopamine D_1 receptor agonists A68930 and SKF 38393 induce arousal and suppress REM sleep in the rat". *European Journal of Pharmacology* 235: 83-87.

Tucker, D.M. and Derryberry, D. 1992. "Motivated attention: Anxiety and the frontal executive functions". *Neuropsychiatry, Neuropsychology and Behavioral Neurology* 5: 233-252.

220   ROGER WHITEHEAD AND SCOTT SCHLIEBNER

Tucker, D.M. and Williamson, P.A. 1984. "Asymmetric neural control systems in human self-regulation". *Psychological Review* 91: 185-215.
Wachtel, P.L. 1967. "Conceptions of broad and narrow attention". *Psychological Bulletin* 68: 417-429.
Walker, B.B. and Sandman, C.A. 1979. "Human visual evoked responses are related to heart rate". *Journal of Comparative and Physiological Psychology* 93: 717-729.
Watabe, K., Nakai, K. and Kasamatsu, T. 1982. "Visual afferents to norepinepherine-containing neurons in cat locus coeruleus". *Experimental Brain Research* 48: 66-80.
Weintraub, S. and Mesulam, M.M. 1987. "Right cerebral dominance in spatial attention". *Archives of Neurology* 44: 621-625.
Wesnes, K. and Warburton, D.M. 1983. "Stress and drugs". In G. R. J. Hockey (ed), *Stress and fatigue in human performance*. Chichester: Wiley.
Whitehead, R. 1991a. "Hemispheric specialization in sustained visuo-spatial attention". *Unpublished doctoral dissertation*, University of Oregon, Eugene.
Whitehead, R. 1991b. "Right hemisphere processing superiority during sustained visual attention". *Journal of Cognitive Neuroscience* 3: 329-334.
Wilkins, A.J., Shallice, T. and McCarthy, R. 1987. "Frontal lesions and sustained attention". *Neuropsychologia* 25: 359-366.
Woodward, D.J., Moises, H.C., Waterhouse, B.D. and Hoffer, B.J. 1979. "Modulatory actions of norepinepherine in the central nervous system". *Federal Proceedings* 38: 2109-2116.
Yingling, C.D. and Skinner, J.E. 1977. "Gating of thalamic input to cerebral cortex by nucleus reticularis thalami". In J.E. Desmedt (ed), *Attention, voluntary contraction and event-related cerebral potentials*. Basel: Karger.
Yokoyama, K., Jennings, R., Ackles, P., Hood, P. and Boller, F. 1987. "Lack of heart rate changes during an attention demanding task after right hemisphere lesions". *Neurology* 37: 624-630.
Zhang, Z., Yang, S., Chen, J., Xie, Y., Qiao, J. and Dafny, N. 1995. "Interaction of serotonin and norepinepherine in spinal antinociception". *Brain Research Bulletin* 38: 167-171.

CHAPTER 8

Emotion and Conscious Experience
Perceptual and attentional influences of anxiety

Douglas Derryberry
Oregon State University

In recent years, approaches within the cognitive and neurosciences have come a long way toward understanding the perceptual and cognitive processes contributing to conscious awareness. However, similar progress has not been made concerning the contributions of emotional and motivational processes. This is of course not surprising, because emotion is among the most elusive and problematic aspects of consciousness. Nevertheless, emotion is in many ways fundamental, providing the basis for the varied states of consciousness that we experience.

The present paper explores emotional contributions to consciousness by adopting a motivational approach. It is assumed that the brain contains a set of parallel 'motivational systems' that have been shaped through evolutionary history to regulate information processing in light of the adaptive needs of the individual. Within this set, the emphasis will be on the limbic circuitry related to defensive motivation and emotions of fear and anxiety. It will be proposed that the defensive system contributes to consciousness in two general ways. First, the limbic circuits exert descending control over the body's response systems, thereby producing the interoceptive feelings that accompany the anxious state. Second, the limbic circuits exert ascending control over perceptual and conceptual processing within the cortex. This modulation of cortical processing serves to control attention, and thus the selection of information for entry into consciousness. The paper begins with a physiological overview of the relevant neural circuits and connections that contribute to these two functions. It concludes with a review of recent psychological research concerning attentional processes during anxious states.

1. Neural systems and anxiety

The neural circuitry related to fear and anxiety consists of a complex system distributed across the orbital and medial regions of the frontal cortex, limbic structures such as the amygdala, hippocampus, bed nucleus of the stria terminalis, and hypothalamus, and brainstem regions such as the periaqueductal gray and ventrolateral medulla (Bandler & Shipley, 1994; Heimer, de Olmos, Alheid & Zaborszky, 1991; Morgan & LeDoux, 1995). Within this distributed system, however, it appears that the amygdala, particularly the central nucleus and its connections with the basolateral and basomedial nuclei, plays a central role (Davis, 1992; LeDoux, 1995). The central amygdaloid nucleus has been found to respond to fear-related stimuli, such as painful stimuli (Bernard, Huang & Besson, 1992) and conditioned auditory tones that predict shock (LeDoux, 1995). In addition, lesion studies have demonstrated that damage to the central nucleus impairs the behavioral components of fear, such as heart rate conditioning and potentiation of the startle reflex (Davis, Hitchcock & Rosen, 1987). These lesion studies are complemented by demonstrations that electrical stimulation of the central nucleus produces behavioral (freezing) and autonomic (increased respiratory frequency) components of fear (Applegate, Kapp, Underwood & McNall, 1983). Thus, evidence from recording, lesion, and stimulation converges on the central nucleus as central to the brain's defensive circuitry.

1.1 Afferent inputs of the amygdala

Anatomical evidence is also supportive of an amygdaloid involvement in fear. In particular, the central amygdala receives afferent inputs from many parts of the brain, which support the many types of stimuli that can promote a fear reaction. The simplest inputs involve direct projections of partially processed sensory information from brainstem regions. For example, interoceptive pain signals are received from the brainstem parabrachial nucleus (Bernard, Alden

& Besson, 1993). Conditioned auditory signals are received from the thalamus, and engage the amygdala with short latencies under 50 ms (LeDoux, 1987, 1995). These direct inputs are consistent with evidence that fear can be elicited by partially processed stimulus features, at times without (or prior to) conscious awareness (Ohman & Soares, 1993). The short latencies are important in that they allow the amygdala to adjust attentional mechanisms (discussed below) at a very early stage of processing, and thereby to regulate subsequent processing within the cortex.

While the amygdala is capable of responding rapidly given partially processed information, it is also responsive to more extensively processed information delivered from cortical association areas. For example, the amygdala receives input from unimodal association areas within the temporal lobe that are dedicated to visual or auditory information, and also from polymodal areas (e.g., perirhinal, orbital frontal, insular, temporal pole) that are responsive to multiple types of information. LeDoux (1995) suggests that whereas the thalamic input may elicit a fear reaction based on simple sensory 'features', the unimodal and polymodal inputs may initiate fear based on the processing of perceptual 'objects' and 'concepts', respectively. Furthermore, the amygdala receives 'contextual' information from the hippocampus, allowing fear to be conditioned to general characteristics of the environment (Phillips & LeDoux, 1995). The hippocampus also appears involved in detecting mismatches between expected and actual events, promoting anxiety and inhibition given novel or unexpected events (Gray, 1982).

Finally, it should be noted that in addition to these informational inputs, the central amygdala is influenced by variety of modulatory substances such as the monoaminergic projections from the brainstem reticular nuclei and glucocorticoid hormones from the adrenal gland. These modulatory inputs appear to regulate the general reactivity of the defensive system in response to featural, object-based, conceptual, and contextual information. For example, the monoamine serotonin may serve to constrain activity within the defensive circuits (Spoont, 1992), whereas glucocorticoids appear to potentiate fear reactivity (Shulkin, McEwen & Gold, 1994).

1.2 *Efferent outputs of the amygdala*

While the amygdala's afferents help to clarify the many causes of anxiety, it is the efferent outputs that are most helpful in viewing the consequences for consciousness. Upon activation, the amygdala circuitry employs widespread efferent projections to promote an adaptive neural state distributed throughout the brain. These efferent effects can be divided into two general kinds. Response effects are promoted via descending connections to peripheral muscular and organ systems, whose activation generates the interoceptive perceptual content (i.e., feelings) accompanying the anxiety. Attentional effects are promoted by ascending connections to the cortex, which modulate processing in terms of the adaptive needs of the state.

1.3 The generation of anxious feelings

The response effects of fear are orchestrated through amygdaloid projections to endocrine, autonomic, and motor circuits. The endocrine influence involves projections to the hypothalamic paraventricular nucleus, which in turn regulates the secretion of ACTH from the anterior pituitary, which in turn regulates the release of cortisol from the adrenal gland (Dunn & Berridge, 1990; Stansbury & Gunnar, 1994). Circulating cortisol has a number of complex energizing effects on the body's organ systems that are adaptive given the behavioral mobilization required during defensive states. Its initial release appears to be associated with feelings of energy and concentration (Stansbury & Gunnar, 1994), while more prolonged secretion may give rise to feelings of distress (Dienstbier, 1989).

The autonomic adjustments are mediated by amygdaloid projections to hypothalamic and brainstem centers that regulate peripheral organs via sympathetic and parasympathetic influences. An initial increase in the relative sympathetic (i.e., energy mobilizing) influence results primarily from a withdrawal of the antagonistic parasympathetic influence (Porges, Doussard-Roosevelt & Maiti, 1994). For example, projections to the dorsal motor nucleus of the vagus underlie fear-related changes in heart rate and blood pressure, while projections to the parabrachial nucleus contribute to respiratory changes (Davis et al., 1987; Loewy & Spyer, 1990). If the emotion is prolonged, direct activation of sympathetic cell groups sustains the mobilized state. Such influences are mediated by amygdaloid and hypothalamic projections to medullary cell groups such as the nucleus paragigantocellularis and the Kolliker-Fuse nucleus (Loewy & Spyer, 1990; Van Bockstaele, Pieribone & Aston-Jones, 1989).

The motor adjustments are mediated through general and specific mechanisms. General effects are implemented by amygdaloid efferents to medullary cell groups that send serotonergic and noradrenergic projections throughout the spinal cord. These projections are thought to act as a gain-setting system, enhancing the overall responsiveness of spinal motor neurons (Holstege, 1991). Additional projections from the central amygdala to brainstem cell groups allow more specific control of facial expression (the trigeminal facial motor nucleus), startle excitability (nucleus reticularis pontis caudalis), freezing behavior (periaqueductal gray), and running behavior (pedunculopontine nucleus) (Davis, 1987; Holstege, 1991). It is also worth noting that the basolateral amygdala projects to multiple motor areas of the forebrain, such as the caudate nucleus and frontal cortex, which may allow the fearful state to contribute to high level response selection and planning (McDonald, 1991).

In general, the behavioral consequences of defensive motivation involve a complex pattern of activation across the body's muscular and organ systems. Besides promoting the individual's physical survival, this body pattern generates interoceptive sensory information that can have a profound impact on the individual's state of consciousness. Sensory feedback from receptors in organs, joints, and muscles is relayed through parallel pathways (the spinothalamic, spinoreticular, spinomesencephalic, and spinosolitary tracts) back to the forebrain (Cervero & Foreman, 1990; Loewy, 1990). Although for many years it was thought that visceral input was processed by primarily non-conscious, subcortical circuits (e.g., the hypothalamus and amygdala), it has now become clear that visceral information, like exteroceptive input, is also relayed through the thalamus to the cortex. In particular, visceral sensory areas have been discovered within the insular and orbital regions of the cortex (Cechetto & Saper, 1987; Cechetto & Saper, 1990).

The interoceptive cortical areas lie adjacent to somatosensory areas and appear to represent distinct visceral modalities (e.g., cardiovascular, gastrointestinal). The visceral fields tend to be located in phylogenetically older, "paralimbic" regions of the cortex, which feature a simpler laminar architecture compared to the neocortical fields devoted to visual and auditory inputs. This is consistent with the vague and often diffuse nature of interoceptive percepts. It has been suggested that processing within the visceral and somatosensory fields supports the various hedonic feelings, energetic feelings, and felt action tendencies accompanying ongoing emotional responses (Derryberry & Tucker, 1991). In addition, these fields and feelings may be engaged in the absence of an ongoing bodily reaction, allowing the individual to feel 'as-if' they were having an emotional state (Damasio, 1994; Nauta, 1971). Thus, current models follow James' (1890) classic argument that emotional feelings depend upon bodily reactions, but also suggest that they may at times be centrally generated.

The anxious feelings themselves can be described at several levels. Some researchers have focused on sensations related to specific organs, including muscular tension, trembling, cardiac palpitations, dizziness, weakness in the limbs, breathing difficulties, a dry mouth, and a sinking feeling in the stomach (Amies, Gelder & Shaw, 1983). Others have focused at more general levels, attempting to describe the anxious state in terms of dimensions related to hedonic and energetic experiences. In hedonic terms, anxiety can be described as consisting of high levels of unpleasant negative affect, at times accompanied by reduced positive affect (Clark, Watson & Mineka, 1994). In energetic terms, anxiety involves a form of "tense arousal" that combines muscular inhibition and high energy, and in some instances may also involve a decrease in the "energetic arousal" that accompanies positive states (Thayer, 1989).

A key characteristic of the interoceptive fields is their extensive interconnectivity with cortical areas processing other types of sensory information (Derryberry & Tucker, 1991). Coordinated activity among these exteroceptive and interoceptive regions may allow objects, concepts, and actions to be associated with various hedonic and energetic feelings. By integrating interoceptive with exteroceptive information, the resulting representations can support crucial functions in evaluating objects and guiding behavior. For example, when faced with two food objects, one might elicit a feeling of enhanced energy while the other calls up a twinge of nausea. Or in the case of fearful behavior, a young child might perceive his or her mother accompanied by a stranger, with these two social objects activating feelings of safety and threat. By providing such rapid affective evaluation, the interoceptive feelings can be crucial in the guidance of behavior; i.e., responses can be oriented toward the more positive object and away from the more negative object. Such guidance functions have been discussed in neuropsychological models of frontal lobe function in terms of "affective reference points" (Nauta, 1971) and "somatic markers" (Damasio, 1994). Similar notions have appeared in the developmental literature, such as the association of actions with feelings of fear and empathy in the guidance of moral behavior (Hoffman, 1988).

It can be seen that the defensive circuitry of the amygdala can make substantial, in some ways fundamental, contributions to consciousness by regulating response processes. Not only do the resulting interoceptive feelings contribute to ongoing awareness, but they can also be stabilized as representations that promote future evaluation and guidance. But to truly understand their influence on consciousness, it is important to consider the individual's attentional processes. Attention will be crucial in determining the relative salience of interoceptive and exteroceptive information, and in the integrative processes linking feelings to the world.

1.4 *Attentional regulation of cortical reactivity*

The attentional effects of the amygdaloid circuitry arise from indirect and direct influences upon the cortex. The indirect effects are mediated by connections from the amygdala to the brainstem's reticular subsystems, which send ascending projections to the thalamus and cortex. Many of these subsystems appear to be accessed by the amygdaloid defensive system. The central nucleus regulates ascending cholinergic projections from the pedunculopontine nucleus to the thalamus (Pascoe & Kapp, 1993) and from the nucleus basalis of meynert to the cortex (Alheid & Heimer, 1988). Also influenced are the ascending monoaminergic systems, including the noradrenergic projections from the locus ceruleus, the serotonergic projections from the rostral raphe nuclei, and the dopaminergic projections from the ventral tegmental area (Davis, 1992; Wallace, Magnuson & Gray, 1992). These subsystems project to multiple regions of the cortex, where they regulate processing through a variety of complex neuromodulatory mechanisms. For example, the noradrenergic and cholinergic projections enhance signal to noise ratios in certain areas, whereas the dopaminergic and serotonergic projections may exert more suppressive effects (Foote, Bloom & Aston-Jones, 1983; Sillito & Murphy, 1987; Waterhouse, Moises & Woodward, 1986). Although the detailed patterning of these subsystems during anxious states is not understood, they can be viewed as functioning to set up a general state of reactivity across multiple fields that is adaptive given the processing requirements of threatening contexts (Robbins & Everitt, 1995).

One facet of this general state can be found in Posner's construct of "alertness" (Posner, 1978; Posner & Raichle, 1994). Thought to be related to the noradrenergic projections, alerting has been investigated through the presentation of a non-informative warning signal prior to a target. The warnings speed reaction times to all subsequent targets, but may also lead to decreased accuracy. Such speed-accuracy tradeoffs suggest that alerting does not selectively influence the buildup of information within cortical pathways, but rather, facilitates the cortical orienting system (i.e., the "posterior attentional system") so that it acts faster to select the accumulating information (Posner & Raichle, 1994). This would be adaptive during anxious states, where rapid responses are often essential to survival, and false alarms may not carry a great cost.

Another aspect of the anxious processing state can be found in Tucker's construct of 'tonic activation' (Tucker & Derryberry, 1992; Tucker & Williamson, 1984). This arousal pattern is thought to be established during anxious states by means of dopaminergic projections from the ventral tegmental region to limbic, striatal, and frontal cortical circuits. While promoting the motor readiness adaptive to flight and fight reactions, tonic activation also increases the constancy or redundancy of information processing. This redundancy bias limits the range of information processed, thereby promoting a "focused" mode of attention that integrates events in the immediate past and impending future. This focused, future-oriented mode of processing is adaptive in threatening situations — it focuses attention on current and future threats, prevents distraction by irrelevant stimuli, and promotes a rapid coupling of responses to perceptual objects.

While the reticular projections contribute to the anxious state by promoting faster and more focused attentional selectivity, direct projections from the amygdala to the cortex may underlie more specific attentional functions. As discussed below, these functions involve the facilitation of specific objects and locations that are crucial within the threatening context. The relevant projections arise primarily from the lateral and basolateral nuclei and reach many areas of the cortex. Among the most important are projections to the paralimbic fields such as the orbital frontal and cingulate regions. The primitive paralimbic regions receive inputs from the limbic system, project back upon the more recently evolved neocortical fields, and serve as central links within cortical attentional networks. In an early model, Mesulam (1981) proposed that the cingulate cortex serves to construct a map of "motivational space", which influences attention by modulating sensory and response maps in parietal and frontal regions, respectively. More recently, Posner has suggested that the anterior cingulate cortex constitutes a pivotal component within an executive attentional system. This "anterior attentional system" is thought to serve integrative functions in controlling orienting, working memory, and effortful behavior (Posner & Petersen, 1990; Posner & Raichle, 1994). During anxious states, amygdaloid input to these cingulate regions may serve an important role in guiding attention in relation to threat-related stimuli (Derryberry & Tucker, 1991).

In addition to its paralimbic connections, the amygdala projects directly to multiple neocortical areas. The basolateral nucleus projects to polymodal and unimodal visual areas of the temporal lobe, and even to the earliest sensory areas of the occipital lobe (Iwai & Yukie, 1987). These projections place the amygdala in position to regulate the entire processing stream for visual object information. In addition, the basolateral amygdala projects to the primary visceral sensory area within the insular cortex (Krushel & Van Der Kooy, 1988), allowing it to influence the early processing of interoceptive information. These direct connections from the amygdala to cortical sensory areas are consistent with an early and specific attentional influence, but their relative contributions to processes such as attention, storage, and retrieval remain to be investigated.

In general, however, it appears that the amygdaloid defensive system has rather remarkable access, directly and indirectly, to the networks of the cortex. This is perhaps not surprising given the adaptive value of attention. By promoting a general state involving alertness and focused processing, and by more selectively facilitating relevant sources of danger and safety, the system greatly enhances the individual's capacity to cope with threat. Not only does the motivational system adjust the state of the body, but it also tunes the individual's state of consciousness. Unfortunately, current physiological perspectives can only take us so far in understanding these effects upon consciousness. In the last section of the paper, we turn to psychological studies to examine more specific characteristics of attentional processes during anxiety.

2.   Psychological studies of anxiety and attention

Most psychological studies of anxiety and attention have employed between-subjects designs comparing chronically-anxious and non-anxious individuals. The majority of these studies have focused on individuals suffering from a clinically-diagnosed anxiety disorder, although others have sampled anxiety across a more normal range by comparing individuals who are above and below the median in terms of the personality dimension of 'trait anxiety'. These are reasonable research strategies in that both clinical and trait measures of anxiety are assumed to reflect chronic activation in the brain's defensive circuitry. It should be kept in mind, however, that many of these studies do not directly manipulate the individuals level of state anxiety, and when they do, the manipulations tend to promote relatively mild states of anxiety.

2.1 *Attentional narrowing*

Many of the earliest studies provided evidence that anxiety produces a focused state of attention characterized by impaired processing of peripheral or secondary sources of information (Easterbrook, 1959). For example, when engaged in a central perceptual task, subjects rendered anxious by exposure to unavoidable shock (Wachtel, 1968) and to a simulated diving experience (Weltman, Smith & Egstrom, 1971) were impaired in detecting peripheral flashes of light. Such attentional narrowing also appears to influence the processing of perceptual objects. When processing composite forms made up global and local elements, individuals high in trait anxiety showed an attentional bias favoring the local elements, whereas those low in anxiety favored the global elements (Tyler & Tucker, 1982). In addition, anxiety may promote a restricted scope in processing semantic information. In research applying Rosch's analysis of category organization, trait anxious subjects showed a tendency to reject non-prototypical exemplars of categories, to perceive less relatedness between different categories, and to employ relatively narrow categories (Mikulincer, Kedem & Paz, 1990). Although there are studies reporting null results (e.g., Leon, 1989), the majority are consistent with Tucker's proposal that anxiety produces a state of tonic activation involving a narrow breadth of attention (Tucker & Williamson, 1984).

One complication facing the hypothesis that anxiety narrows attention arises from findings that it can also lead to increased distractibility. As described metaphorically by Wachtel (1967), anxiety narrows the beam of attention, but also causes the beam to wander over the perceptual field. Consistent with this idea, individuals high in trait anxiety (neurotic introverts) showed enhanced distraction by irrelevant letters while performing a difficult letter transformation task (Eysenck & Graydon, 1989). Distractibility is also evident in patients suffering from generalized anxiety disorder, who showed delayed choice reaction time performance when presented with irrelevant distractor words (Mathews, May, Mogg & Eysenck, 1990). Although these distraction effects appear to run counter to the narrowing effects, the two may arise from distinct adaptive mechanisms during anxious states. While the attentional narrowing may reflect dopaminergic tonic activation, the distraction may arise from an increase in general alertness, which could nonselectively facilitate attention to irrelevant stimuli. Alternatively, the distraction may reflect selective processes involved in orienting to potentially threatening stimuli, with anxious subjects tending to view distracting information as threatening to their performance on the primary task. From this perspective, anxiety involves a general narrowing of the breadth of attention, along with a tendency to be distracted by potentially threatening information. We turn next to more specific evidence regarding attention to threatening information.

2.2 Orienting to threatening information

Beginning in the 1980s, many studies using the "emotional Stroop task" provided evidence that anxiety enhances attention to threatening information. When instructed to name the ink color of a word stimulus, anxious subjects are delayed when the word's meaning is threatening, suggesting that their attention is drawn to the irrelevant but threatening information. Some of these studies suggest that the attentional bias can be quite specific. For example, patients suffering from social phobia were delayed primarily by socially-threatening words (e.g., 'rejection'; Hope, Rapee, Heimberg & Dombeck, 1990), patients with physical concerns or panic disorder were distracted by words conveying physical threat (e.g., 'injury'; Mogg, Mathews & Weinman, 1989), and spider phobics were slow in naming the color of spider-related words (e.g., 'web'; Watts, McKenna, Sharrock & Trezise, 1986). Such specificity is consistent with engagement of the defensive circuitry by highly processed semantic information, such as that delivered to the amygdala from cortical association areas. However, other studies are suggestive of more general influences. MacLeod and Rutherford (1992) presented subjects with threatening words that were either related or unrelated to an upcoming exam, and masked half the words to prevent their conscious identification. Trait anxious subjects were delayed given both types of threat words, but only given the subliminal presentations. These findings suggest that the anxiety effect involves a preconscious biasing of attention (but see Wells & Matthews (1994)), a biasing based on partially processed information that has been categorized in terms of general threat rather than a specific type of threat. Such information may be delivered directly from the thalamus to the amygdala, but the verbal nature of the stimuli is more compatible with a cortical contribution.

Although the Stroop task has provided some important findings, the contribution of attentional as opposed to response processes (i.e., response inhibition) is not always clear. A more controlled paradigm, the "dot probe task", has provided generally converging findings. This task involves the simultaneous presentation of two words, one of which is occasionally threatening in meaning, followed after 500 milliseconds by a detection target. In an initial study, MacLeod, Mathews & Tata (1986) found that patients suffering from generalized anxiety disorder were fast to detect the target when it appeared in the location of a threatening word, suggesting that their attention was preferentially allocated to the threatening location. In a subsequent study of trait anxiety, MacLeod & Mathews (1988) tested subjects under conditions involving low state anxiety (12 weeks before a major examination) and high state anxiety (1 week before the examination). Subjects high in trait anxiety showed an attentional bias favoring all threatening words during both testings, but when tested under the high state anxiety conditions, their bias was specific to examination-relevant words (e.g. 'test'). More recently, Mogg, Bradley & Williams (1995) compared anxious and depressed patients with normal controls on a dot probe task employing supraliminal and subliminal word presentations. Compared to controls, anxious subjects showed an attentional bias favoring negative words. This bias was evident under both supraliminal and subliminal presentations, and did not depend upon the semantic content (i.e., anxiety-related versus depression-related) of the negative words. In general, studies employing the dot probe task suggest that the attentional bias related to trait anxiety is not highly specific, and may be elicited by preconscious processing based on partially processed information.

A third paradigm has been developed in our laboratory in order to more precisely assess attentional operations related to anxiety. The task involves a simple game set within the spatial orienting task developed by Posner (1978). Subjects view a display consisting of a central fixation flanked by two peripheral boxes. Each trial begins with a peripheral cue that draws attention to one of the two boxes, followed by a detection target in either the cued or noncued location. One peripheral location is assigned a positive value in that 10 points can be gained if a target in that location is detected fast (i.e., faster than the subject's median reaction time). The other location is assigned a negative value in that 10 points are lost if the target is detected too slow (i.e., slower than the subject's median). It is assumed that when the positive location is cued, a motivational state related to 'reward' (i.e., potential reward) is elicited, and when the negative location is cued, a state related to threat or punishment (i.e., potential punishment) is elicited.

Several studies have found that individuals high in trait anxiety (i.e., neurotic introverts) show an attentional bias favoring the threatening location (Derryberry & Reed, 1994). This bias appears when the cue precedes to the target by only 100 milliseconds, and regardless of whether the cue is informative or noninformative concerning the actual location of the target. This suggests that the bias arises from relatively involuntary rather than intentional processes. Moreover, the negative bias does not appear in the form of faster detections of targets in negative locations following negative cues. Rather, anxious subjects are slow to detect targets in positive locations following cues in negative locations. This suggests that the bias may reflect an impairment in shifting away from negative cues, rather than a facilitation in shifting toward such cues. Relevant here is Posner's decomposition of orienting into three component operations: disengaging from the current location, moving to the next location, and engaging that location (Posner, Inhoff, Friedrich & Cohen, 1987). It may be that anxiety exerts its attentional effects on attention by directly modulating the disengage mechanism. Such an effect could be accomplished by projections from the amygdala to the cingulate cortex to the parietal cortex, a crucial region involved in disengagement. Alternatively, it is possible that anxiety functions to enhance the engage operation, which in turn leads to delayed disengagement. This type of effect could be mediated by an amygdaloid influence on the pulvinar nuclei, a set of thalamic nuclei thought to be crucial in engagement. Also, the delayed disengagement may reflect a narrowing influence arising from subcortical reticular projections. By restricting attention to the most relevant inputs, subcortical systems may limit the impact of other inputs, making it more difficult to disengage.

Regardless of the underlying neural mechanisms, these findings suggest some important characteristics of anxious states of consciousness. Rather than noticing and attending to many sources of threat, anxiety may primarily involve a tendency to lock onto a single source of threat, and to delay shifting away. Such prolonged engagement may be adaptive in promoting more extensive processing of the threatening information. During normal worrisome thought, for example, more aspects of the problem could be taken into consideration, perhaps allowing a more effective solution. This is a fine line, however, for worry can often take on a perseverative or even obsessive quality, with the individual experiencing great difficulty breaking away from the anxious content. Fortunately, there appears to be another type of attentional bias accompanying anxiety that helps counter this negative tendency.

2.3 *Orienting to relieving information*

Although most theories of anxiety emphasize threatening information, it can also be argued that anxiety facilitates information related to relief and safety. This argument is based on an adaptive view of anxiety, emphasizing that this is a defensive state designed to help the individual cope with dangerous situations. To successfully cope, it is necessary to attend not only to the threat itself, but also to environmental inputs that may help attenuate the threat. When approached by a predator, for example, the prey animal's survival depends on its capacity to attend to both the threat and the available sources of relief (e.g., escape routes, safe places) (Toates, 1986). When an infant is approached by a threatening stranger, the child is attentive to the stranger, but also seeks to regain a feeling of security by attending to the mother (Rothbart & Derryberry, 1981). In a general sense, anxiety can be viewed as a defensive state aimed designed to deal with a specific type of problem, whose solution requires attention to multiple sources of information.

To begin assessing these ideas, we have recently completed a series of studies using a modified version of the game paradigm described above. As before, subjects were presented with a peripheral cue followed by a detection target in one of two peripheral locations. In contrast to the previous studies, where it was possible to either gain or lose points on each trial, the trials were blocked to form 'appetitive' and 'defensive' games. On the appetitive games, subjects gained points for each fast detection (which constitutes 'reward' in learning theory terms; (Gray, 1987)), but did not gain any points for slow responses (which constitutes 'frustrating non-reward'). On the defensive games, subjects lost 10 points for each slow response ('punishment'), but did not lose any points if the response was fast ('relieving non-punishment'). In addition, the criterion for a 'fast' response was manipulated to create 'easy' targets (where 75% of the responses were fast) and "hard" targets (where 75% of the responses were slow). Peripheral pretarget cues signaled the probable location of the upcoming target and whether it would be hard or easy. Thus, the cue served to direct attention to peripheral location, and also to predict the probable outcome of the subject's response. On defensive games, cues in hard locations predicted probable punishment, whereas cues in easy locations predicted probable relief. On appetitive games, hard cues predicted frustration whereas easy cues predicted reward.

Three studies lead to the same pattern of results (Derryberry & Reed, 1995). On the appetitive games, low and high trait anxious subjects showed no differences in attending to cues that predicted reward or frustration. This null finding is noteworthy in that some models of anxiety (e.g., Gray) predict that anxious individuals will be sensitive to frustrating as well as threatening signals. On the defensive games, high anxious subjects showed enhanced attention (compared to low anxious subjects) to cues in potentially punishing locations. This is consistent with our earlier findings (Derryberry & Reed, 1994), and with other studies showing enhanced attention to threatening information (e.g., MacLeod & Matthews, 1988). However, anxious subjects also showed enhanced attention to the relieving cues that signaled probable non-punishment. This supports the notion that under defensive conditions, anxiety promotes attention to relieving as well as threatening sources of information.

Evidence of attention to relieving information can also be found in more complex cognitive activity. In worrisome thought, for example, the person is primarily concerned with a potential threat, but it is usually attended in relation to the various options for coping with it. Several theorists have approached worry as a form of problem solving — the individual attempts to solve a threatening problem by evaluating the relief potential of varying coping options and planning an effective strategy (Davey, 1994; Mathews, 1990; Tucker & Derryberry, 1992). More extreme examples can be found in the cognition accompanying certain anxiety disorders. In obsessive-compulsive disorder, there is often an excessive concern with the relief afforded by a particular coping behavior (Shapiro, 1965). In the many disorders that involve avoidance behavior (e.g., agoraphobia, social phobia), the tendency to avoid can be seen to be motivated by the individual's concern with the safety provided by familiar places and people.

It thus appears that anxiety is not simply a state of consciousness dominated by threatening content, but a state aimed at finding relief in relation to the threat. In some cases, both threatening and relieving content may be equally represented in consciousness, though in others either threat or relief may become dominant. Some of the factors influencing the relative distribution of attention to threat and relief have been discussed in more detail in Derryberry & Reed (1996).

2.4 *Orienting to interoceptive information*

While attention is oriented to external sources of threat and relief, a fundamental component of anxious experience involves interoceptive feelings. As discussed earlier, anxiety gives rise to numerous bodily changes which may be experienced on line or in the form of activated representations. While cortical representations are available for processing anxious feelings, the actual experience of anxiety will depend upon the manner in which the relevant representations are attended. Unfortunately, research is quite limited in this area, and only relatively general issues can be addressed.

Perhaps the most general issue concerns the relative distribution of attention to interoceptive versus exteroceptive information. In some extreme cases, attention may be so strongly directed toward to the external world that anxious feelings barely attain awareness. In fact, the nervous system is equipped with descending monoaminergic and endorphinergic pain suppression systems that can be engaged during stressful states. The resulting "stress-induced analgesia" is thought to be adaptive in suppressing pain information so that an animal can remain undistracted as it attempts to cope with the external threat (Kelly, 1986). It has been suggested that in humans such a suppressive mechanism may contribute to a 'repressive-coping style', in which the individual experiences limited subjective anxiety in spite of strong physiological reactions (Schwartz, 1990). In contrast, there appear to be other anxious states in which the individual's consciousness is more or less dominated by interoceptive feelings. Such enhanced attention to the body may be particularly likely when the sensations are themselves a source of threat, a phenomenon sometimes referred to as 'fear of fear'. Examples include panic and agoraphobic patients who fear that they may experience a heart attack or fainting in public, socially anxious individuals who fear that the exposure of their anxiety may elicit social disapproval, and test anxious individuals who fear that their anxiety may impair their exam performance (Wells & Matthews, 1994). Because they limit attention to the external world, these types of anxious states may constrain the person's ability to cope with the threatening situation (e.g. the exam or social interaction).

A second issue concerns the orientation of attention within the interoceptive domain. This is particularly important in influencing the qualitative nature of the experienced feelings. At the level of specific inputs, individuals differ in the strength of their body responses across different organ systems (Lacey,

1967), and may also differ in attending to these organs. Some individuals may experience anxiety primarily in terms of muscular tension, others in terms of respiratory symptoms, and others in terms of vestibular changes (Kenardy, Evans & Oei, 1992). As mentioned above, patients suffering from panic disorder are highly attentive to the cardiovascular sensations that may signal the feared heart attack (Ehlers & Breuer, 1992). At a more general level, individuals appear to differ in their tendencies to attend to the hedonic and energetic dimensions of affect (Feldman, 1995). Individuals with a strong 'valence focus' may experience anxiety primarily as a painful negative state, whereas those with a strong "arousal focus" may experience anxiety more in terms of its energetic properties. Also, it is worth noting that whatever information is involved in the specific and general feelings, this information is at times ambiguous. It has been suggested that the overlapping and diffuse nature of interoceptive input can generate perceptual possibilities similar to those involved in reversible visual figures, and that attention to specific aspects of the input pattern may be crucial in determining the disambiguated feeling (Cacioppo, Berntson & Klein, 1992). Although our understanding of attention within the interoceptive domain remains quite limited, these examples illustrate the potential complexity and variability of the feeling states accompanying anxiety.

A third issue concerns the role of internally-directed attention. On the surface, internally-directed attention seems unadaptive in that it exacerbates the bodily experience of anxiety, and at the same time interferes with the processing of more relevant external information (Wells & Matthews, 1994). One approach to this problem emphasizes that subjective feelings are often highly relevant (i.e., threatening) to the individual, who fears that the feelings may reflect health problems, impair their performance, and so on. More generally, however, it should be recalled that recent models propose that interoceptive information can play an important role in evaluation and behavioral guidance. By attending to hedonic information, the individual may be better able to assess the extent to which available objects support positive or negative outcomes (i.e., enhanced danger or relief). In a sense, attention to hedonic information may serve to highlight the 'somatic markers' within the representation (Damasio, 1994; Nauta, 1971), thereby promoting a more rapid and effective selection of approach and avoidance behaviors. Along similar lines, attention to energetic information may allow the individual to better assess their current energy resources, and thereby to evaluate the efficacy of different response options (Thayer, 1989). This energetic information may be crucial to the perceived controllability of the situation (Dienstbier, 1989).

A final and most difficult issue concerns the role of attention in linking interoceptive with other types of information. In some instances the relationship may be highly determined, but in others there seems to be considerable flexibility in relating different aspects of feeling to different parts of the world. When a child is punished by a parent, for example, the anxious child may attend to their interoceptive feelings, their previous behavior, the parent's behavior, and so on. If attention is directed more strongly to the parent, the anxious feelings may be linked to the external event (the punishment), but if attention favors their own behavior, the feelings may be experienced in relation to an internal source (the transgression). Such biases may contribute to the development of external and internal attributions for the negative state of affairs (Dienstbier, 1984; Hoffman, 1988). At the interoceptive level, if the child attends primarily to the hedonic information, they may feel generally distressed and unable to cope with the situation, but if attention also facilitates energetic information, the child may feel more efficacious and capable of coping (Dienstbier, 1989; Thayer, 1989). This feeling may receive additional support if the child attends to a source of potential relief, such as the options of apologizing or even avoiding the parent. Although speculative, these examples illustrate the potential flexibility with which attention can help to link various sources of information, and can thus weave many different states of anxious consciousness.

3.   Summary and conclusions

In the case of anxiety, both physiological and psychological evidence converge in suggesting that defensive motivational states influence consciousness in several crucial ways. At the perceptual level, the multidimensional bodily responses provide elaborate interoceptive input that may be experienced as the feelings central to emotion. At the attentional level, anxiety influences the general style of processing by promoting an alert and focused mode, and also the content of processing by enhancing information relevant to threat, relief, and interoceptive evaluation. Although we have a long way to go toward understanding the intricacies of emotional experience, the convergence of psychological and physiological perspectives is promising. These are complementary approaches with much to offer one another.

One of the challenges facing future research involves moving beyond anxiety to consider additional types of motivational states. As mentioned earlier, the defensive system is but one of multiple motivational systems within the brain, with others specialized to regulate processing in light of appetitive, achievement, affiliative, sexual, parental, and other types of needs. These other systems can be expected to contribute to consciousness in ways that are generally similar to anxiety; i.e., by generating bodily feelings and by regulating attention. However, the patterning of the feelings and the specific biasing of attention is likely to differ depending on the motivational state. Evidence exists, for example, that certain positive states promote an 'expansive' rather than narrow focus of attention (Isen, 1990; Tucker & Williamson, 1984), and that depressive states bias retrieval more strongly than attentional processes (Eysenck, 1992; Williams, Watts, MacLeod & Mathews, 1988). From an evolutionary perspective, each system has been shaped to deal with specific types of needs, which require distinct states within the body and the cortex.

Another challenge involves a consideration of how these parallel motivational systems work together. Environments often include signals relevant to multiple needs, promoting the coactivation of several systems. In some cases the systems may interact competitively through inhibitory interconnections, as when the defensive system suppresses appetitive motivation. However, it seems likely that the regulatory systems can also be activated simultaneously to establish cooperative interactions. Systems related to parental or achievement needs may form coalitions with systems regulating appetitive and defensive needs. If we can understand these interactions, then we can move beyond the notion of a single homunculus or executive system controlling attention and consciousness. At least at a general level, consciousness can be seen to emerge from the modulatory influence of interacting motivational systems, shaped through evolutionary history, upon the representational circuitry of the cortex.

References

Alheid, G. F. & Heimer, L. 1988. New perspectives in basal forebrain organization of special relevance for neuropsychiatric disorders: The striatopallidal, amygdaloid, and corticopetal components of substantia innominata. *Neuroscience, 27*, 1-39.

Amies, P. L., Gelder, M. G., & Shaw, P. M. 1983. Social phobia: A comparative clinical study. *British Journal of Psychiatry, 142*, 176-177.

Applegate, C. D., Kapp, B. S., Underwood, M. D., & McNall, C. L. 1983. Autonomic and somatomotor effects of amygdala central nucleus stimulation in awake rabbits. *Physiology & Behavior, 31*, 353-360.

Bandler, R. & Shipley, M. T. 1994. Columnar organization in the midbrain periaqueductal gray: modules for emotional expression? *Trends in Neurosciences, 17*, 379-389.

Bernard, J., Alden, J., & Besson, J. 1993. The organization of the efferent projections from the pontine parabrachial area to the amygdaloid complex: A phaselous vulgaris leuco-agglutinin (PHA-L) study in the rat. *Journal of Comparative Neurology, 329*, 201-229.

Bernard, J. F., Huang, G. F., & Besson, J. M. 1992. Nucleus centralis of the amygdala and the globus pallidus ventralis: Electrophysiological evidence for an involvement in pain processes. *Journal of Neurophysiology, 68*, 551-568.

Cacioppo, J. T., Berntson, G. G., & Klein, D. J. 1992. What is an emotion? The role of somatovisceral afference, with special emphasis on somatovisceral "illusions". In M. S. Clark (Ed.), *Emotion and Social Behavior*, (pp. 63-98). London: Sage Publications.

Cechetto, D. F. & Saper, C. B. 1987. Evidence for a viscerotopic sensory representation in the cortex and thalamus in the rat. *Journal of Comparative Neurology, 262*, 27-45.

Cechetto, D. F. & Saper, C. B. 1990. Role of the cerebral cortex in autonomic function. In A. D. Loewy & K. M. Spyer (Eds.), *Central regulation of autonomic functions*, (pp. 208-223). New York: Oxford University Press.

Cervero, F. & Foreman, R. D. 1990. Sensory innervation of the viscera. In A. D. Loewy & K. M. Spyer (Eds.), *Central regulation of autonomic functions*, (pp. 104-125). New York: Oxford University Press.

Clark, L. A., Watson, D., & Mineka, S. 1994. Temperament, personality, and the mood and anxiety disorders. *Journal of Abnormal Psychology, 103*, 103-116.

Damasio, A. R. 1994. *Descartes' error: Emotion, reason, and the human brain*. New York: G. P. Putnam's Sons.

Davey, G. C. L. 1994. Pathological worrying as exacerbated problem-solving. In G. C. L. Davey & F. Tallis (Eds.), *Worry: Perspectives on theory, assessment and treatment*, (pp. 35-59). New York: Wiley.

Davis, M. 1992. The role of the amygdala in fear and anxiety. *Annual Review of Neuroscience 1992, 15*, 353-375.

Davis, M., Hitchcock, J. M., & Rosen, J. B. 1987. Anxiety and the amygdala: Pharmacological and anatomical analysis of the fear-potentiated startle paradigm. In G. Bower (Ed.), *The Psychology of Learning and Motivation*, (Vol. 21, pp. 263-305). New York: Academic Press.

Derryberry, D. & Reed, M. A. 1994. Temperament and attention: Orienting toward and away from positive and negative signals. *Journal of Personality and Social Psychology, 66*, 1128-1139.

Derryberry, D. & Reed, M. A. 1995. Trait anxiety enhances attention to threatening and relieving signals. Unpublished manuscript.

Derryberry, D. & Reed, M. A. 1996. Regulatory processes and the development of cognitive representations. *Development and Psychopathology, 8*, 215-234.

Derryberry, D. & Tucker, D. M. 1991. The adaptive base of the neural hierarchy: Elementary motivational controls on network function. In R. Dienstbier (Ed.), *Nebraska symposium on motivation, Vol. 38: Perspectives on motivation*, (pp. 289-342). Lincoln, Nebraska: University of Nebraska Press.

Dienstbier, R. A. 1984. The role of emotion in moral socialization. In C. E. Izard, J. Kagan, & R. B. Zajonc (Eds.), *Emotions, cognition, and behavior*, (pp. 484-514). Cambridge: Cambridge University Press.

Dienstbier, R. A. 1989. Arousal and physiological toughness: Implications for mental and physical health. *Psychological Review, 96*, 84-100.

Dunn, A. J. & Berridge, C. W. 1990. Physiological and behavioral responses to corticotropin-releasing factor administration: Is CRF a mediator of anxiety or stress responses? *Brain Research Reviews, 15*, 71-100.

Easterbrook, J. A. 1959. The effect of emotion on cue utilization and the organization of behaviour. *Psychological Review, 66*, 183-201.

Ehlers, A. & Breuer, P. 1992. Increased cardiac awareness in panic disorder. *Journal of Abnormal Psychology, 101*, 371-382.

Eysenck, M. W. 1992. *Anxiety: The cognitive perspective.* Hillsdale, N. J.: Erlbaum.

Eysenck, M. W. & Graydon, J. 1989. Susceptibility to distraction as a function of personality. *Personality and Individual Differences, 10*, 681-687.

Feldman, L. A. 1995. Valence focus and arousal focus: Individual differences in the structure of affective experience. *Journal of Personality and Social Psychology, 69*, 153-166.

Foote, S. L., Bloom, F. E., & Aston-Jones, G. 1983. Nucleus locus coeruleus: New evidence of anatomical and physiological specificity. *Physiological Review, 63*, 844-914.

Gray, J. A. 1982. *The neuropsychology of anxiety.* London: Oxford.

Gray, J. A. 1987. *The psychology of fear and stress (Second Edition).* New York: McGraw-Hill.

Heimer, L., de Olmos, J., Alheid, G. F., & Zaborszky, L. 1991. "Perestroika" in the basal forebrain: Opening the border between neurology and psychiatry. In G. Holstege (Ed.), *Progress in brain research, Volume 87: Role of the forebrain in sensation and behavior*, (pp. 109-165). Amsterdam: Elsevier.

Hoffman, M. L. 1988. Moral development. In M. H. Bornstein & M. E. Lamb (Eds.), *Developmental psychology: An advanced textbook. Second Edition*, (pp. 497-548). Hillsdale, New Jersey: Erlbaum.

Holstege, G. 1991. Descending motor pathways and the spinal motor system: Limbic and non-limbic components. In G. Holstege (Ed.), *Progress in brain research, Volume 57. Role of the forebrain in sensation and behavior*, (pp. 307-421). New York: Elsevier.

Hope, D. A., Rapee, R. M., Heimberg, R. G., & Dombeck, M. J. 1990. Representations of the self in social phobia: Vulnerability to social threat. *Cognitive Research and Therapy, 14*, 177-190.

Isen, A. M. 1990. The influence of positive and negative affect on cognitive organization: Some implications for development. In N. Stein, B. Leventhal, & T. Trabasso (Eds.), *Psychological and biological approaches to emotion*, (pp. 75-94). Hillsdale, NJ: Erlbaum.

Iwai, E. & Yukie, M. 1987. Amygdalofugal and amygdalopetal connections with modality-specific visual cortical areas in Macaques (macaca fuscata, M. mulatta, and M. fascicularis). *Journal of Comparative Neurology, 261*, 362-387.

James, W. 1890. *Principles of psychology.* New York: Holt.

Kelly, D. D. 1986. *Stress-induced analgesia.* New York: New York Academy of Sciences.

Kenardy, J., Evans, L., & Oei, T. 1992. The latent structure of anxiety symptoms in anxiety disorders. *American Journal of Psychiatry, 149*, 1058-1061.

Krushel, L. A. & Van Der Kooy, D. 1988. Visceral cortex: Integration of the mucosal senses with limbic information in the rat agranular insular cortex. *Journal of Comparative Neurology, 270,* 39-54.

Lacey, J. I. 1967. Somatic response patterning and stress: Some revisions of activation theory. In M. J. Appley & R. Trumball (Eds.), *Psychological stress: Issues in research,* (pp. 14-37). New York: Appleton-Century-Crofts.

LeDoux, J. E. 1987. Emotion. In F. Plum (Ed.), *Handbook of physiology. Section 1: The nervous system. Volume V. Higher functions of the brain, Part 1.,* (pp. 419-459). Bethesda, MD: American Physiological Society.

LeDoux, J. E. 1995. In search of an emotional system in the brain: Leaping from fear to emotion and consciousness. In M. S. Gazzaniga (Ed.), *The cognitive neurosciences,* (pp. 1049-1062). Cambridge, MA: MIT Press.

Leon, M. R. 1989. Anxiety and the inclusiveness of information processing. *Journal of Research in Personality, 23,* 85-98.

Loewy, A. D. 1990. Central autonomic pathways. In A. D. Loewy & K. M. Spyer (Eds.), *Central regulation of autonomic functions,* (pp. 88-103). New York: Oxford University Press.

Loewy, A. D. & Spyer, K. M. Eds. (1990). *Central regulation of autonomic functions.* New York: Oxford University Press.

MacLeod, C. & Mathews, A. 1988. Anxiety and the allocation of attention to threat. *Quarterly Journal of Experimental Psychology, 40,* 653-670.

MacLeod, C., Mathews, A., & Tata, P. 1986. Attentional bias in emotional disorders. *Journal of Abnormal Psychology, 95,* 15-20.

MacLeod, C. & Rutherford, E. M. 1992. Anxiety and the selective processing of emotional information: Mediating roles of awareness, trait and state variables, and personal relevance of stimulus materials. *Behaviour Research and Therapy, 30,* 479-491.

Mathews, A. 1990. Why worry? The cognitive function of anxiety. *Behavioral Research and Therapy, 28,* 455-468.

Mathews, A., May, J., Mogg, K., & Eysenck, M. 1990. Attentional bias in anxiety: Selective search or defective filtering? *Journal of Abnormal Psychology, 99,* 166-173.

McDonald, A. J. 1991. Organization of amygdaloid projections to the prefrontal cortex and associated striatum in the rat. *Neuroscience, 44,* 1-14.

Mesulam, M. M. 1981. A cortical network for directed attention and unilateral neglect. *Annals of Neurology, 10,* 309-325.

Mikulincer, M., Kedem, P., & Paz, D. 1990. Anxiety and categorization - 1. The structure and boundaries of mental categories. *Personality and Individual Differences, 11,* 805-814.

Mogg, K., Bradley, B. P., & Williams, R. 1995. Attentional bias in anxiety and depression: The role of awareness. *British Journal of Clinical Psychology, 34,* 17-36.

Mogg, K., Mathews, A., & Weinman, J. 1989. Selective processing of threat cues in anxiety states: A replication. *Behavior Research and Therapy, 27,* 317-323.

Morgan, M. A. & LeDoux, J. E. 1995. Differential contribution of dorsal and ventral medial prefrontal cortex to the acquisition and extinction of conditioned fear. *Behavioral Neuroscience, 109,* 681-688.

Nauta, W. J. H. 1971. The problem of the frontal lobe: A reinterpretation. *Journal of Psychiatric Research, 8,* 167-187.

Ohman, A. & Soares, J. J. F. 1993. On the automatic nature of phobic fear: Conditioned electrodermal responses to masked fear-relevant stimuli. *Journal of Abnormal Psychology, 102*, 121-132.

Pascoe, J. P. & Kapp, B. S. 1993. Electrophysiology of the dorsolateral mesopontine reticular formation during pavlovian conditioning in the rabbit. *Neuroscience, 54*, 753-772.

Phillips, R. G. & LeDoux, J. E. 1995. Lesions of the fornix but not the entorhinal or perirhinal cortex interfere with contextual fear condition. *Journal of Neuroscience, 15*, 5308-5315.

Porges, S. W., Doussard-Roosevelt, J. A., & Maiti, A. K. 1994. Vagal tone and the physiological regulation of emotion. In N. A. Fox (Ed.), *The development of emotion regulation: Biological and behavioral considerations*, . Monographs of the Society for Research in Child Development, 1994, 59 (2-3, Serial No. 240), 167-186.

Posner, M. I. 1978. *Chronometric explorations of mind*. Hillsdale, New Jersey: Erlbaum.

Posner, M. I., Inhoff, A. W., Friedrich, F. J., & Cohen, A. 1987. Isolating attentional systems: A cognitive-anatomical analysis. *Psychobiology, 15*, 107-121.

Posner, M. I. & Petersen, S. E. 1990. The attention system of the human brain. *Annual Review of Neuroscience, 13*, 25-42.

Posner, M. I. & Raichle, M. E. 1994. *Images of mind*. New York: Scientific American Library.

Robbins, T. W. & Everitt, B. J. 1995. Arousal systems and attention. In M. S. Gazzaniga (Ed.), *The cognitive neurosciences*, (pp. 703-720). Cambridge, MA: MIT Press.

Rothbart, M. K. & Derryberry, D. 1981. Development of individual differences in temperament. In M. E. Lamb & A. L. Brown (Eds.), *Advances in developmental psychology, Volume I*, (pp. 37-86). Hillsdale, New Jersey: Erlbaum.

Schwartz, G. E. 1990. Psychobiology of repression and health: A systems approach. In J. L. Singer (Ed.), *Repression and dissociation: Implications for personality theory, psychopathology, and health*, (pp. 405-434). Chicago: University of Chicago Press.

Shapiro, D. 1965. *Neurotic styles*. New York: Basic Books, Inc.

Shulkin, J., McEwen, B. S., & Gold, P. W. 1994. Allostasis, amygdala, and anticipatory angst. *Neuroscience and Biobehavioral Reviews, 18*, 385-396.

Sillito, A. M. & Murphy, P. C. 1987. The cholinergic modulation of cortical function. In E. G. Jones & A. Peters (Eds.), *Cerebral cortex, Volume 6. Further aspect of cortical function, including hippocampus*, (pp. 161-185). New York: Plenum Press.

Spoont, M. R. 1992. Modulatory role of serotonin in neural information processing: Implications for human psychopathology. *Psychological Bulletin, 112*, 330-350.

Stansbury, K. & Gunnar, M. R. 1994. Adrenocortical activity and emotion regulation. In N. A. Fox (Ed.), *The development of emotion regulation: Biological and behavioral considerations*, . Monographs of the Society for Research in Child Development, 1994, 59 (2-3, Serial No. 240), 108-134.

Thayer, R. E. 1989. *The biopsychology of mood and arousal*. New York: Oxford University Press.

Toates, F. 1986. *Motivational systems*. Cambridge, England: Cambridge University Press.

Tucker, D. M. & Derryberry, D. 1992. Motivated attention: Anxiety and the frontal executive mechanisms. *Neuropsychiatry, Neuropsychology, and Behavioral Neurology, 5*, 233-252.

Tucker, D. M. & Williamson, P. A. 1984. Asymmetric neural control systems in human self-regulation. *Psychological Review, 91*, 185-215.

Tyler, S. K. & Tucker, D. M. 1982. Anxiety and perceptual structure: Individual differences in neuropsychological function. *Journal of Abnormal Psychology, 91*, 210-220.

Van Bockstaele, E. J., Pieribone, V. A., & Aston-Jones, G. 1989. Diverse afferents converge on the nucleus paragigantocellularis in the rat ventrolateral medulla: Retrograde and anterograde tracing studies. *Journal of Comparative Neurology, 290*, 561-584.

Wachtel, P. L. 1967. Conceptions of broad and narrow attention. *Psychological Bulletin, 68*, 417-429.

Wachtel, P. L. 1968. Anxiety, attention, and coping with threat. *Journal of Abnormal Psychology, 73*, 137-143.

Wallace, D. M., Magnuson, D. J., & Gray, T. S. 1992. Organization of amygdaloid projections to brainstem dopaminergic, noradrenergic, and adrenergic cell groups in the rat. *Brain Research Bulletin, 28*, 447-454.

Waterhouse, B. D., Moises, H. C., & Woodward, D. J. 1986. Interaction of serotonin with somatosensory cortical neuronal responses to afferent synaptic inputs and putative neurotransmitters. *Brain Research Bulletin, 17*, 507-518.

Watts, F. N., McKenna, F. T., Sharrock, R., & Trezise, L. 1986. Colour naming of phobic-related words. *British Journal of Psychology, 77*, 97-108.

Wells, A. & Matthews, G. 1994. *Attention and emotion: A clinical perspective.* Hillsdale, NJ: Erlbaum.

Weltman, G., Smith, J. E., & Egstrom, G. H. 1971. Perceptual narrowing during simulated pressure-chamber exposure. *Human Factors, 13*, 99-107.

Williams, M., Watts, F., MacLeod, C., & Mathews, A. 1988. *Cognitive psychology and emotional disorders.* New York: Wiley.

Section IV

Brain Evolution

Introduction

Many scientists are concerned with what consciousness does for the organism — what is its function? Attempts to answer this question are plagued with doubt: for any function that might be proposed, it may be possible to imagine that job getting done without consciousness being involved. That is, consciousness may not be *necessary* for its purported function to be achieved in some (imagined) alternative system. If this skeptical response awaits every attempt to assign function to consciousness, ought we to give up our search for consciousness in the brain? No! It is certainly the case that *the way our brains have evolved does entail consciousness*. So instead of asking "Why do we have consciousness?" it may make more sense to ask "Why do we have the neural circuits which are, in fact, involved in conscious experience?" According to current biological theory, evolution holds the key to unlocking the answer to this question.

Previous sections of this book have described several brain functions which are of critical importance to the experience of human consciousness. First we explored some of the primary boundaries of conscious content in Section I. Next we examined the ways in which environmental stimuli and ongoing behavior exert their control in selecting sensory representations in Section II. Then we discussed ways in which phenomenal experience can be influenced by neurophysiological and emotional states in Section III. These earlier portions of this book have revealed three links between mind and brain. First, *what we are conscious of* depends on patterns of neural activity in the relevant information processing circuits such as those constituting modality-specific sensory pathways in cortex. Second, conscious experience also depends on *how the mind is directed*, and this attentiveness is a function of those brain systems responsible for the selection of neurally represented information. Third, the *affective and energetic feelings* which imbue the current content of consciousness arise from neuromodulation of the information-processing pathways by transmitter-specific innervation from brainstem nuclei, and communication between these pathways and limbic circuits.

These results clearly help in answering the question "How is the brain conscious?" — so already a respectable portion of an immense and complex puzzle has been at least partly explained. This fincal Section takes a complementary approach to understanding the functional organization of the neural systems responsible for the familiar stream of consciousness.

Be warned that here we venture more *speculation* than has been offered in previous chapters on the neurobiological underpinning of subjective awareness. In Chapter 9, Phan Luu, Daniel Levitin, and John Kelley explain how the organization of cortical connectivity supports the comparison of sensory inputs to personal expectations, a process which yields conscious experience. Based on parallels between phenomenological awareness and brain evolution, Peter Grossenbacher suggests in Chapter 10 that consciousness evolved along with the coordination of communication between relatively independent brain systems. Equipped with the neurocognitive findings presented earlier in this book, you must evaluate the merit of the push beyond the envelope of current scientific consensus which constitutes this final Section.

CHAPTER 9

Consciousness

A Preparatory and comparative process

Phan Luu
Electrical Geodesics, Inc., Dept.of Psychology, Eugene, OR

John M. Kelley
Massachusetts General Hospital

Daniel J. Levitin
McGill University

The brain and its relation to consciousness has long been a topic that appeared to be intractable. Indeed, this impression ultimately led Freud (1895) to abandon his highly regarded "Project for a Scientific Psychology." However, we believe that the problem of relating consciousness to brain processes *is* tractable, as have many other scientists and philosophers (e.g., Crick, 1994; Edelman, 1989).

It is our belief that consciousness is not a uniquely human property. By analogy, vision and perception are not uniquely human attributes; they are properties that are shared by all animals with the requisite neurological hardware. But each species is endowed with slightly different 'hardware'. For example, the eyes and the (number of) identifiable cortical areas dedicated to visual processing differ from one species to another. Of course, the exact nature of the eye itself dictates the range and quality of information that can ever reach the cortex. The cortical areas dedicated to vision form a global network in which visual information is processed. To the extent that a certain cortical area and its function do differ between species, or is missing altogether in a species, the network's configuration will be altered. Thus, the network's operation will be qualitatively changed, and the experience of seeing and perceiving must be different from animal to animal. Yet, they do perceive. Similarly, we believe that some animals possess the necessary hardware for consciousness. However, to expect animals to experience consciousness *as humans do* is equivalent to expecting them to see and perceive as we do.

From this perspective, consciousness is not something mystical and ethereal that defies attempts to relate it to the brain; it is a property of all brains with the necessary hardware. The question is, what is the fundamental process of consciousness, and what are the mechanisms that give rise to our unique experience of consciousness? We believe that *being conscious reflects a preparatory and comparative process* carried out by the brain in order to ready the organism for perception. This readiness for perception, coupled with environmental input, *is* consciousness. A more difficult aspect of consciousness is the phenomenological quality of being a person with a sense of individuality and historical place. Concern for this phenomenological aspect of human consciousness has lead some people to be dissatisfied with brain models of consciousness. However, there are certain aspects of brain evolution that may allow us to adequately account for the human phenomenological experience of consciousness as well.

We mentioned that we believe consciousness to be a process, and this point is worth elaborating. Pribram (1980) made an astute comparison between gravity and consciousness. Gravity is the result of an interaction between masses, and thus it cannot be found at a particular place in an object. Pribram defined consciousness in a similar way (i.e., as an interaction between the organism and the environment). Thus, we would not expect to find a particular brain region that is responsible for the feeling of consciousness.

Again, we turn to the process of vision for an analogy. Visual perception is not a "thing" that occurs at a particular region in the brain. Rather, it is a "constructive process" that involves both global views and access to the finer details, and it is achieved in a distributed network of brain regions. The visual object-recognition pathway is organized in a hierarchical fashion: cells in the primary visual cortex respond to simple parameters (such as orientation), whereas cells further along the pathway in the inferior temporal lobe respond to complex features, such as simple geometrical designs (Tanaka, 1993). The distributed processing of object features means that there is no one place in the brain where we experience the object as a whole. Although some cells represent complex entities, such as hands and faces (Desimone, Albright, Gross & Bruce, 1984), scientists have not been able to find cells that respond exclusively to specific objects in the environment. Thus, "seeing" must involve a constructive process in which features are specified, combined, and ultimately experienced over a distributed network. To emphasize this point, it has been suggested that cognitive functions occur throughout our brain just as we live throughout our body (Herrick, 1948).

In the remaining sections of this chapter, we discuss how the process of consciousness is achieved. We also elaborate on how both cortical evolution and individual development contribute to the personal and subjective experience of consciousness. Our model contains two central ideas. First, consciousness depends upon the integration of affect and motivation with representations of sensory and motor information (i.e., sensorimotor patterns). This integration is necessary but not sufficient for consciousness. Emotions are believed to provide the guiding framework for cortical development and organization throughout the life span of the organism (Luu & Tucker, 1996; Trevarthen & Aitken, 1995; Tucker, 1992). Even in adulthood, consciousness must be mediated by brain mechanisms that have evolved to control primitive systems of affective regulation. These regulatory mechanisms are extensions of a *vertical hierarchy* of brain organization and development, and contribute directly to the character of what it feels like to be conscious. This regulation is the task of *vertical integration* (Tucker, 1993; Tucker & Derryberry, 1992) that will be discussed below. Without vertical integration, the content of our consciousness would be completely ephemeral — thoughts, images, and action plans would all pass fleetingly through our minds and would be easily displaced by the next capricious thought, image or action plan. This is not to say that the experience of consciousness is never ephemeral, such as when one is daydreaming. Rather, we want to emphasize that the forces that sustain cognition in a controlled manner are affective in nature and are represented at many levels in the neuraxis.

Our second central idea is that *consciousness is fundamentally a preparatory and comparative process*. It is *preparatory* in the sense that a frontal system dedicated to the processing of information according to internal motivational states prepares posterior, sensory regions so that they are more likely to register environmental stimuli that are congruent with the organism's affective and motivational state. Consciousness is *comparative* in that the organism is constantly comparing its internal state with representations of the external environment so that inconsistencies or mismatches can be reconciled. The cortical mantle can be divided, for explanatory purposes, into a posterior sensory system and an anterior response system (Jackson, 1931). This division is only a heuristic and should not be adhered to in the strictest sense because there are cells in the parietal lobe that are involved in movement, such as reaching and grasping (Jeannerod, Arbib, Rizzolatti, & Sakata, 1995). However, those areas contain neurons that seem to encode more global aspects of the action. Nevertheless, as we will show, the evolution of the posterior sensory system has increased the resolution and fidelity of the organism's environmental interface (Tucker, 1992). In contrast, the evolution of the anterior system has improved the representation of information regarding internal states and plans (i.e., improvements that have allowed information to be more flexibly reorganized, represented, acted upon, and perceived).

1.   Brain evolution and architecture

In this section, we lay the foundation for our model by giving a selective overview of theories of brain evolution. In addition, we describe some features of brain architecture that are relevant to our model.

1.1 *Evolution and spherical organization*

In 1948, Yakovlev proposed that the brain evolved along, and is organized as, a series of concentric spheres in which each sphere is successively and functionally linked to adjacent spheres (see Figure 9.1). At the core is the brainstem, with centers (such as the periventricular area) involved in the maintenance of homeostatic states (e.g., body temperature, blood sugar). Although this homeostatic core receives afferent signals from sensory receptors outside the brain, the core itself is also reactive to sensory stimuli. For example, heat applied to the anterior extremity of the third ventricle immediately produces changes in heat-regulating mechanisms, and CO_2 applied to the posterior brainstem produces changes in depth and rate of respiration. Moreover, these sensitive regions within the brain are modality-specific and tightly localized. From these observations Pribram (1960) argued that these homeostatic centers resemble sensory receptors.

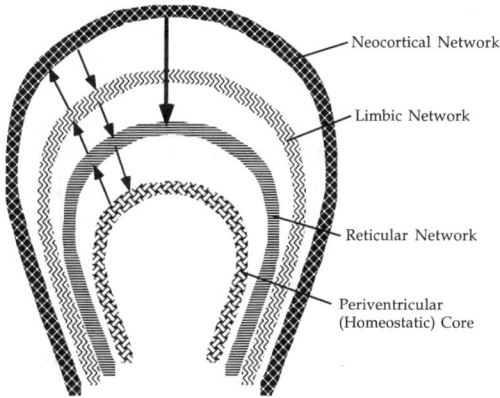

Figure 9.1. Schematic of the spherical/concentric organization of the brain as proposed by Pribram (1960) and Yakovlev (1948). Arrows indicate the bidirectional integration of information across the vertical hierarchy. The large, bold arrow indicates the frontal connections with the reticular network.

The next sphere out is the reticular network which is involved in the control of arousal. The reticular network consists of a number of nuclei located in the medulla, pons, and midbrain. We will have more to say about arousal and activation later (see also Whitehead and Schliebner, this volume). Beyond the reticular network is the limbic system, located on the medial aspects of the cerebral hemispheres. Pribram (1960) argues that the limbic network links and regulates the multiple homeostatic functions of the brain-stem with each other and with the arousal functions of the reticular activating network, thus, setting up biases that regulate behavior. The outer most shell is the neocortex.

The concentric organization of the brain reflects the organization of behavior (Yakovlev, 1948). Yakovlev argued that motility and other behavior, in relation to the world, is organized into three spheres. The simplest form of motility is that of visceration, such as respiration and blood circulation, in which changes in homeostatic states relate to the current physical environment. The next sphere of behavior involves the outward expression of internal states, that is, the expression of emotion. The final sphere involves behavior of effectuation which creates changes in the physical world. Just like the interdependence of the spheres in the cortical domain, Yakovlev states, "The behavior of the organism is total; every heart beat, every twitch of a muscle, every movement and posture is an integral part of the total behavior which evolves and proceeds as a unity in time." (p.315)

Vertical Integration. The large size of the cortical mantle (the outer most sphere) in both humans and nonhuman primates, relative to other animals, led early students of the brain to focus on its importance to behavior, and unfortunately, to overlook the importance of lower systems. Although it was known for some time that certain areas of the brainstem could be stimulated to produce alterations in the alert state, Moruzzi and Magoun (1949) were the first to systematically demonstrate that an ascending reticular activating system, arising from the brainstem, directly influenced the cortical state. Specifically, Moruzzi and Magoun found that stimulation of the brainstem reticular activating system produced changes both in the cat's ongoing EEG and in its level of alertness. This important finding demonstrated that essential aspects of cortical functioning are regulated by subcortical structures. This led to the reconceptualization of the brain as being organized vertically (Luria, 1973; Pribram, 1960), which was implicit within Yakovlev's (1948) formulation. John Hughlings Jackson (1931) proposed just such a vertical framework at the turn of the century, based on his observations of epileptic and brain-lesioned patients.

In his attempt to understand the principles underlying pathological conditions after brain damage, Jackson (1931) theorized that brain evolution has proceeded in an ascending fashion. In this theory, new structures grew outward from — and are superordinate to — older structures. The Jacksonian theory of brain evolution is that evolution moves from highly-determined and specialized brain areas to more flexible and generalized brain areas, and from a limited repertoire of automatic responses toward an increasing number of voluntary responses. In this framework, recently evolved brain structures regulate the more primitive structures which evolved earlier. Thus, the pathology observed after cerebral lesions can be understood in terms of the undamaged older structures being released from the control of the damaged, more recent structures; in this case, brain lesions lead to a dissolution (the opposite of evolution) of brain organization and function.

It is a principle of cortical organization that brain structures are *reciprocally connected*, with each area sending signals back to those structures from which it receives transmissions. Luria (1973) described a descending reticular activation network, originating in the prefrontal cortex (see Figure 9.1). In fact, it has been known for some time that stimulation applied to the ventrome-dial surface of the prefrontal cortex can elicit visceral and autonomic re-sponses such as respiratory arrest and changes in blood pressure in humans and other animals (Kaada, Pribram & Epstein, 1949; Livingston, Chapman, Livingston & Kraintz, 1947). These findings reveal connections between prefrontal cortex and lower structures involved in the regulation of those visceral and autonomic responses. Consistent with the Jacksonian principle, this descending system recruits and regulates the ascending one so that the prefrontal lobe can have broad influence over the rest of the cerebrum (Luria, 1973).

Because of the way the brain evolved and is organized, Tucker (1993) pointed out that normal cerebral function and representation inevitably face the task of vertical integration. That is, the task of integrating the multiple representations of a function across the neural hierarchy (i.e., the different concentric spheres) is to obtain coherent behavioral expression. For example, in primate vocalization it has been shown that primitive organization of phonation (e.g., respiration) takes place at the pons and medulla (Jürgens, 1979). Motivations are coupled to these elementary patterns at the level of the caudal periaqueductal gray and tegmentum. The species-specific aspect of vocalization involves contributions from the diencephalon (thalamus and hypothalamus). The voluntary aspect of uttering a call is initiated at limbic levels, and finally, voluntary call formation is organized at neocortical levels (Ploog, 1992). Thus, the final product of coherent vocalization stems from the coordination and integration of neural structures which separately compute all aspects of the call, represented at various levels of the neuraxis. As is the case for vocalization, temperature regulation (Satinoff, 1978), attention (Posner & Dehaene, 1994), and human language (Brown, 1979) also involve multiple levels of the neuraxis. For example, the superior colliculi, pulvinar, insula, cingulate, and frontal and parietal cortices are all involved in various aspects of attention.

Vertical integration is not a problem in an intact nervous system. How-ever, when the integrity of the neuraxis is damaged, the failure of integration, and thus cerebral functioning, becomes readily apparent. That is, symptoms after a lesion to the hierarchy can be taken to reflect failures of integration and as disruptions of a normally continuous process. For example, in human language, lesions at limbic levels result in semantic errors, and lesions to neocortical regions, such as Broca's area, result in difficulties with phonemic articulation, leaving semantics intact (Brown, 1979). The important point about vertical integration is that cortical functioning requires support from and continuity across the many levels of the vertical hierarchy.

1.2 Cortical evolution

Adaptive Pressure. Allman's (1990) theory of cortical evolution posits that the neocortex evolved as a consequence of emerging endothermic homeostasis in mammals (i.e., the need for warm-blooded animals to maintain their body temperatures within certain narrow limits). Internal regulation of body temperature allows the organism to function in a larger range of climates. The narrower range of internal temperature allows for finer tuning of biochemical functioning involved in metabolic processes, and thereby provides increased reliability in neuronal functioning. Unfortunately, the advantages afforded by internal regulation of thermal homeostasis are costly in terms of energy consumed. Therefore, Allman argued, the neocortex evolved to enable the animal to reliably map its food resources and plan strategies that would allow it to obtain those resources.

The environment of an organism is inherently noisy, and it is the task of the organism to extract the signal. In the mammalian brain vast regions of the neocortex are comprised of numerous topographically organized sensory maps. For example, Allman (1990) notes, through comparative studies, that in early marsupials approximately 75% of the neocortex consists of topographic sensory maps. In these topographic maps, the environmental stimuli transduced via receptor surfaces such as the retina, skin, or basilar membrane are represented at the cortical level in a spatial layout which mimics that of the receptor sheet. For example, the postcentral gyrus of each cerebral hemisphere has a map dedicated to somatosensory processing of the opposite half of the body. In this map, adjacent parts of the body surface are represented by neurons located in adjacent parts of the cortical map. It is through the correlated functioning of these sensory maps that noise is filtered and information extracted (Edelman, 1989). In other words, these sensory maps enable the recurrent images from the environment to be easily perceived.

Within this context, the evolution of the brain's sensory regions can be understood as increasing the resolution of the organism's environmental interface (Tucker, 1992). However, this enhanced representation of the external environment needs to be integrated with information from the motivational core regarding the internal environment. In this way, behavior can be modified to prioritize the organism's goals, and to achieve them in the most efficient manner possible. We will see that cortical evolution has emphasized this aspect as well.

Trends and Growth Rings. The cortex, through evolution, has substantially increased in size, accounting for more and more of the cerebral mass (Northcutt & Kaas, 1995). This expansion of the cortical surface has been argued to be a result of the extended period of cell division in the ventricular zone during neonatal development (Rakic, 1995). During embryonic development, cells created in this area migrate outwards to form columns or radial units that eventually give rise to cortical neurons. Thus, over the course of evolution, the increase in size of the cortex appears to have depended, at least in part, on the increases in the length of time that cells in the ventricular zone were allowed to divide and form cortical columns.

Along with the increase in cortical area, evolution introduced more cortical fields. For example, it is believed that most rodents have 5-8 areas dedicated to visual processing, whereas the macaque monkey, with more cortex, has approximately 30. Compare this with the belief that the *entire* isocortex (six-layer cortex) of the first mammals contained only about 20 areas, including *all* sensory modalities as well as motor areas (Northcutt & Kaas, 1995). The addition of new fields, which appears to be a hallmark of cortical evolution, has allowed for improved processing of the environmental stimuli encountered within the organism's ecological niche. Although there are conflicting views regarding the process that controls the development of new cortical fields, evidence suggests that the new areas developed from older ones (Krubitzer, 1995).

Consistent with the current proposal of new cortical fields emerging from existing cortical areas, Sanides (1970) described cortical evolution as a process of differentiation out of two primitive moieties; the paleocortex (olfactory cortex) and the archicortex (hippocampal cortex). The paleocortex gave rise to cortices on the ventral and lateral surfaces, and the archicortex gave rise to cortices on the medial and dorsal surfaces.

The cortex is a laminated structure that varies in the number of layers throughout its extent. Each layer is dominated by a certain type of cell. For example, granular layer IV, the input layer, is predominantly occupied by small stellate cells which gives it a granular appearance. The infragranular layers (below layer IV) are output layers, whose neurons project to other cortical areas and outside the cortex. The supragranular layers (above layer IV) are processing layers. Sanides (1970) used the pattern of variation in the number of layers and the cell distribution within these layers, to guide his theory of cortical evolution. He noted that, in comparison to more recently evolved tissue, primitive (i.e., limbic) cortex is characterized by sparse lamination and a laminar pattern that varies greatly between areas. With increasing differentiation from this primitive core, the cortex gained additional lamination, resulting in the six-layer isocortex. The evolutionary emphasis in cellular distribution shifted from the infragranular layers in primitive cortex[1] to the supragranular layers in isocortex, suggesting that evolution has refined and extended the processing of sensory input. The evolutionary sequence of cortical differentiation is as follows: allocortex (two layers), proceeding to periallocortex (near allocortex — the laminar pattern resembles allocortex), to proisocortex (near isocortex — the laminar pattern resembles six-layer cortex), and finally to isocortex (6 layers — homogeneous).[2] Figure 9.2 presents a schematic of this progression. Within this framework, all neocortical areas can be traced back to their roots in limbic cortex.

Pandya and associates (Pandya, Seltzer & Barbas, 1988; Pandya & Yeterian, 1990) have meticulously illustrated Sanides' theory of evolution for the frontal lobe (see Figure 9.3). The paleocortical trend begins in the temporal pole and orbitofrontal regions, and gives rise to three ventrolateral sectors in the frontal lobe: the precentral, premotor and prefrontal. In contrast, the mediodorsal aspect of the frontal cortex stems from the archicortex. The periallocortex of the anterior cingulate is the first step in the sequence away from the allocortex within this trend. It then differentiated into areas such as the premotor (including the supplementary motor area) and primary motor cortex. In the prefrontal cortex, it differentiated into cingulate areas in front of the corpus callosum, and then into areas on the dorsal surface. The principal sulcus and the inferior frontal sulcus demarcate the boundary on the dorsolateral surface of the two trends in monkeys and humans, respectively (Pandya et al., 1988; Pandya & Yeterian, 1990; Sanides, 1970). Again, this sequential differentiation has entailed an increase in the number of cells in layers III and IV.

Sanides (1970) further proposed that the evolutionary pattern of cortical differentiation proceeds in a ring-like manner — new areas form the core and contain denser thalamic input, whereas the originating cortex forms the outer ring. Sanides found that even in Brodmann's architectonic map, the ring of

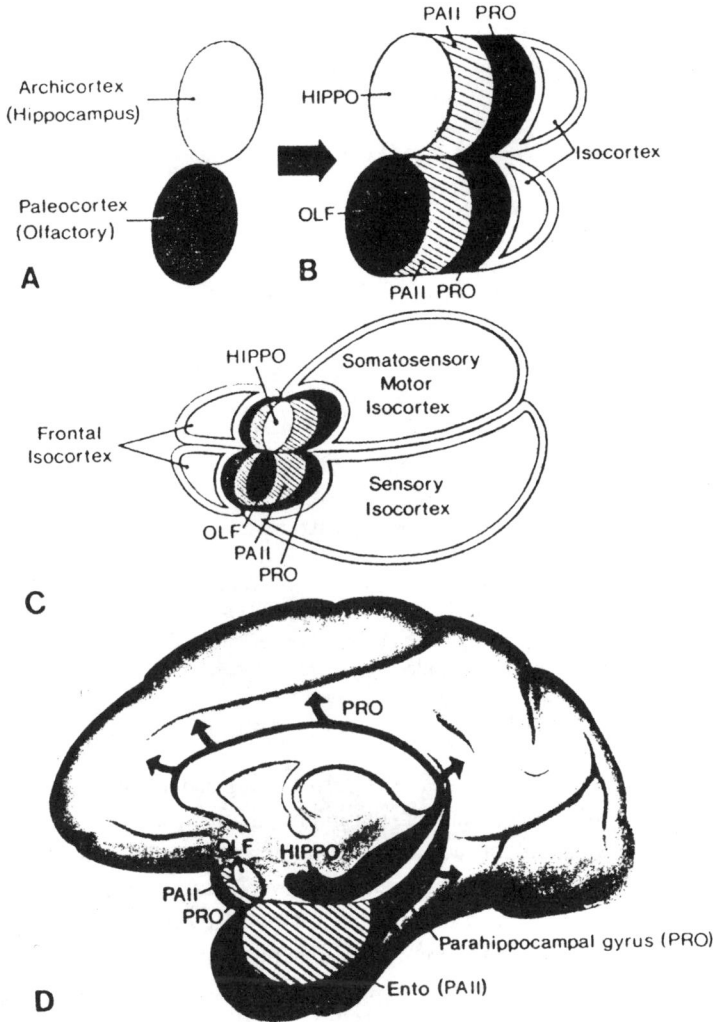

Figure 9.2, A-D. Schematic diagrams to show the progressive development of cortical areas from the two primordial moieties (archi and paleo) through successive steps: periallocortex (Pall) to proisocortex (PRO) to isocortex. (from Pandya et al., 1988; reprinted with permission).

Dorsal

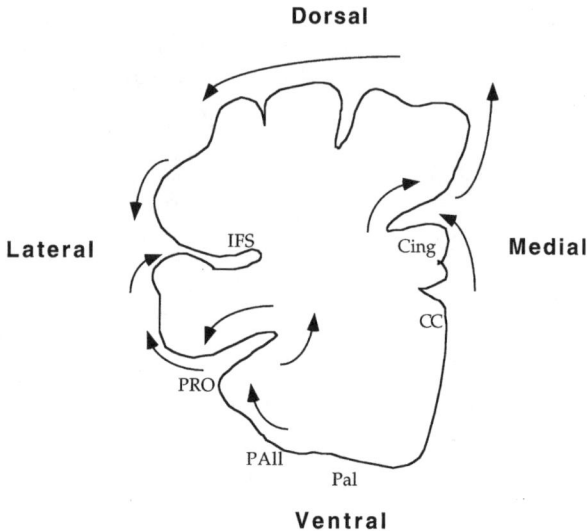

Lateral   IFS   Cing   **Medial**

CC

PRO

PAll
Pal

Ventral

Figure 9.3. Coronal diagram of the human frontal lobe. The arrows indicate the *archicortical trend*, going from medial to dorsal to lateral, and the *paleocortical trend* going from ventral to lateral. The two trends converge at the inferior frontal sulcus. CC: corpus callosum; Cing: Cingulate; IFS: inferior frontal sulcus; Pal: paleocortex; PAll: Periallocortex; PRO: Proisocortex. After Sanides (1970).

differentiation can be seen — area 18 forms a ring around area 17 and area 19 encircles area 18. In this manner, the intermediate cortex retains its connections with the other older area as it forms new connections with the newly formed field. Newly developed areas, however, retain few connections with older areas, other than with the one from which they emerged. The development of the new areas from previously existing cortex may result from new, correlated inputs invading an extant area with subsequent aggregation of similar types of inputs (Krubitzer, 1995). In other words, new inputs to an existing area grow to form modules within these extant fields. If there are modules with similar inputs they might aggregate to form a new field. This process of new field formation can then produce an island within the existing area. Thus, the evolution of the environmental sensory interface, (i.e., cortical fields involved in the processing of sensory information) can be regarded as a process of cortical differentiation into specialized new fields with an increasing emphasis on supragranular, processing layers.

1.3 *Cortical circuits organized according to evolutionary progression*

Although a variety of cortical projection patterns had been recognized for some time, it was Sanides' (1970) theory of cortical evolution that provided a truly elegant framework in which to conceptualize the organizational principle of cortical projections. Pandya and associates (Pandya et al., 1988) applied this theory to the pattern of cortical projections and found that they could be explained by the sequential stages of cortical evolution.

Pandya et al. (1988) noted that early studies of cortical projections tended to focus only on the efferents of primary areas, and that back-going projections were not given the attention that they merit. Projections from primary areas to downstream fields originate in infragranular layers and terminate in layer IV; the reciprocating back-projections from association areas to sensory and primary sensory areas originate in infragranular layers and terminate in layer I.

The forward-projections are arranged consistently with the notion of sequential processing of information, such that more abstract aspects of the information are represented and processed in areas downstream. For example, neurons in 'downstream' visual areas have larger receptive fields and respond to more complex features than neurons earlier in the system (Tanaka, 1993; Zeki & Ship, 1988). It was originally believed that the back-going projections served only to produce feedback regarding receipt of forward transmission. We now know that they do much more, and that back-going projections serve to link areas of progressive lamination (Pandya et al., 1988). Also, they may have a direct influence on the processing and representation of information in the areas upstream (Zeki & Ship, 1988). It has been suggested that cortical layer I serves as a short-term buffer, akin to short-term memory, in which sensory information is held for subsequent corticocortical integration (Vogt, 1991). If this is the case, then it can be seen that back-projections terminate in a layer that can potentially affect the processing of incoming information.

Within each cortical trend (i.e., archicortical and paleocortical), cells within a given area send axons to both a less and a more differentiated area, mainly to their evolutionary neighbors; a given area's projections to areas of distant evolutionary age is rare. Using visual cortex as an example, area 18 has many connections to areas 17 and 19, but areas 17 and 19 have fewer direct connections to each other. Presumably, these areas evolved in sequence from 19 to 17, demonstrating that projections from one area to other areas of distant evolutionary age are rare. Across the two trends, this pattern of projection indicates a similar preservation of evolutionary rank, and this pattern holds across sensory modalities as well (Pandya et al., 1988). Logically, from Sanides' (1970) growth-ring theory, this pattern of cortical projections would be expected — as more recent areas evolved, they became separated from older fields. Thus, evolution of the sensory cortex also can be regarded as a gradual transition towards sensory segregation.

2. Brain maturation

Partially paralleling the phylogenetic, vertical evolution of the brain, ontogenetic cerebral maturation proceeds in a similar manner from the central core out, at least up to the level of the cortex. Myelination and metabolism studies have shown that brainstem structures, with the exception of the reticular system, mature earlier than the striatum, and that the striatum, in turn, matures earlier than most cortical areas (Brody, Kinney, Kloman & Gilles, 1987; Chugani & Phelps, 1986; Yakovlev & Lecours, 1967).

Surprisingly, at the level of the cortex, the progression reverses. That is, the most recently evolved sensory areas are the first to mature (Brody et al., 1987; Chugani & Phelps, 1986; Gibson, 1991). The primary sensory areas are the first to myelinate and are the ones associated with the highest level of glucose metabolism at this stage of development. The maturation then proceeds outwards from the precentral and postcentral gyri towards the frontal pole and parietal/occipital association cortices (Kinney, Brody, Kloman & Gilles, 1988). The frontal and forebrain limbic cortices mature later than the posterior half of the brain and, in general, are the last cerebral structures to mature (taking up to 8 years or more to develop mature levels of myelination).

Yakovlev and Lecours (1967) observed that both progressions of maturation, from subcortical to cortical, and from primary sensory to association and limbic cortex, seem to coincide. For example, subcortical fibers mediating somatic experience myelinate before those fibers that mediate more integrative experience. This is the same pattern observed at the cortical level. Moreover, myelination of thalamic projections to association areas of the cortex is synchronized with myelination of fibers leaving these cortical areas. This overall pattern of brain maturation has been described as *a gradient of central convergence onto the prefrontal cortex and the related limbic forebrain* (Luu & Tucker, 1996).

2.1 An epicritic organizational system

Based upon these evolutionary and maturational progressions, it is possible to describe an organizational system that is specialized for the processing and representation of sensory information. As we have explained, the evolutionary trend is toward increasing differentiation and segregation. For the different sensory modalities, this trend directly influences the topographic organization, and thus representation, of sensory information. As mentioned earlier, to the extent that an organism's brain contains additional cortical fields dedicated to processing and representing information within a specific sensory modality, the more refined and accurate will be that organism's experience and representation of its environment within that modality.

The evolution of highly differentiated and articulated topographic maps has yielded posterior sensory systems equipped with highly refined localization properties. Derived from neurology, the term "epicritic," coined by Henry Head (Pribram, 1981), refers to *local signs of sensation, signs that can be localized in space and time.* For example, the pain system, as it enters the thalamus from the spinal cord, splits into *two* pathways. One of these two pain pathways leads from the ventral posterior nucleus of the thalamus and terminates in the somatosensory cortex (Vogt, Sikes, & Vogt, 1993). These cells have small receptive fields for noxious stimuli; a prick on the hand induces a sensation that is of limited duration and that can be localized to the hand. With their exquisitely precise topographic maps, the posterior sensory systems provides representation of information with high spatial and temporal resolution ideal for the extraction, processing and storage of environmental stimulus features (Damasio, 1989; Edelman, 1989). This can be contrasted with the other, more diffuse and non-localized pain pathway to be described later.

The nature of representations within this epicritic system is not only determined by the cortical fields that have evolved over the history of the species, and of the type of information that these fields receive, but it is also determined by maturationally dependent, developmental processes. Sculpting of cortical circuitry in early-maturing cortices is a result of a form of development known as "experience-expectant development" (Greenough & Black, 1992). This form of development, seen most often in posterior sensory systems (e.g., the formation of ocular dominance columns), involves an overproduction of synapses in anticipation of environmental events. Environmental stimuli then provide the input necessary for neuronal activity, and each synapse is either strengthened or eliminated depending upon its activity level and that of its neighbors. Experience-expectant development occurs only within a narrowly defined window of time (i.e., a critical period).

Greenough and Black (1992) argued that experience-expectant development mechanisms evolved under three conditions. First, information that guides the retention and elimination of synapses is common to all young members of the species. Second, the events that provide this information must occur reliably for all members of the species and must have done so throughout the evolution of the cerebral system under question. Finally, the maturational time window in which the information is provided is critical.

To summarize, the posterior epicritic system can be regarded as a sensory system that is influenced by early developmental mechanisms and that it is affected by species-specific information. Moreover, this system emphasizes temporal and spatial resolution. This system can be generally regarded as containing specialized mechanisms for the processing and representation of the environment. The epicritic system, through the correlated activity of neurons arranged in topographic maps, extracts invariances from an environment that is inherently noisy and unorganized (Edelman, 1989; Pribram, 1960). Through reentrant connections, patterns of activity in different cortical maps can become correlated so that environmental inputs become categorically grouped. Because of these properties, the structure of the categories formed and represented within the epicritic system is based upon physical stimulus features such as shape, and is analogous to basic-level and subordinate-level categories (for example, *chairs* and *rocking-chairs*, respectively). Thus, an abstraction (i.e., an average) of a category within the epicritic system still reflects the shape of individual members of the category (Rosch & Mervis, 1975; Rosch, Mervis, Gray, Johnson & Boyes-Braem, 1976).

2.2 *A protocritic recategorical system*

As noted previously, the human brain has grown in size, most notably the prefrontal lobe (Benson, 1993). During this evolution, the prefrontal lobe has also increased in its intrinsic and extrinsic connectivity (Altman, 1995; Fuster, 1989; Nauta, 1964). This suggests that the trend in prefrontal-lobe evolution is toward increasing its processing capacity and ability to organize the information provided by improved sensory systems. However, the prefrontal lobe's functions are fundamentally different from the specialized functions of the epicritic system.

The prefrontal lobe appears to be involved in learning general tasks (not limited to a single processing domain) that are not supported by specialized circuits (Gaffan, 1994). For example, learning to associate reward with a visual stimulus involves circuits connecting visual cortex with the amygdala. Lesions to these areas or to their underlying connections produce failures to overtly learn the association. Learning the association of instructional cues with an appropriate choice does not have a specialized mechanism and thus depends upon prefrontal functioning (Gaffan, 1994). Prefrontal lesions or damage to its connections with the visual cortex impair learning of these tasks. More than likely, learning to successfully navigate through a complex social world is also a task that lacks specialized mechanisms.

Because the maturation of the prefrontal and limbic cortices is protracted, their cortical wiring is subjected to a different form of cortical development than earlier maturing cortical areas. The protracted maturation of the prefrontal and limbic cortices allow for more 'experience-dependent' developmental mechanisms to sculpt their cortical wiring (Greenough & Black, 1992). This form of cortical sculpting refers to synaptic pruning that occurs later in cortical maturation. In contrast to experience-expectant mechanisms, experience-dependent development is characterized by a lack of a critical time window during maturation in which information must be present in order for functional cortical sculpting to occur. Because of this lack of a critical period, the organization of cortical circuitry through experience-dependent mechanisms is more greatly influenced by the individual experiences of the organism than on experiences that are anticipated as a result of their consistent appearance throughout the evolution of the species. For example, the location of food, warmth, shelter, and potential mates, and recognizing one's position within a social hierarchy are things in flux for a good part of (if not throughout) the life of an animal and would not benefit from predetermined cortical encoding within narrow time intervals. The synaptic changes occurring with this form of development involve massive sprouting of dendritic fields (see Greenough & Black, 1992).

The findings of experience-dependent development (Greenough & Black, 1992) and self-regulating functions of the prefrontal lobe (Tucker, Luu, & Pribram, 1995) combine to form a useful theoretical framework. From this perspective, the evidence of central convergence in cerebral maturation suggests that the representations within prefrontal and limbic networks are based upon abstractions of environmental stimuli guided by internal states (Luu &

Tucker, 1996). That is, sculpting of late-maturing cortical areas comes under the influence of early-maturing cortical sensory areas and subcortical structures. Early-developing sensory areas provide environmental information, and subcortical structures involved in affective self-regulation provide information about internal states.

Recordings from neuronal populations in the prefrontal cortex of the behaving monkey have revealed that neuronal groups undergo quick, functional reorganization as a result of stimulus context and the behavior of the animal (Aertsen et al., 1991). Although Aertsen and colleagues did not investigate the role of limbic and subcortical structures in this functional reorganization, the fact that the reorganization was dependent upon the behavior of the monkey is compelling. Similarly, single-cell recording studies have demonstrated that the correlated firing between two cells depends upon the animal attending to the stimulus (Ahissar et al., 1992). Furthermore, it has been shown that the catecholamine brainstem systems (Mattson, 1988), which have been suggested to be primitive affective/attention control systems (Tucker & Williamson, 1984), can potentially suppress or facilitate synaptic development. The information provided by internal states allows this generalized system to recategorize and represent the information extracted by the posterior sensory areas.

Thus, this generalized, recategorical system prioritizes and filters the torrent of exquisitely detailed sensory information emanating from the epicritic system, according to the relevance of the sensory information to the organism's goals. We would expect that its representations would emphasize the function and ideals of things-in-the-world: these are the features of things that are most closely linked to our motivational states and goals. These representations are extracted invariances related to the social history of the organism. Just as there are species-specific encodings of perceptual invariances, this generalized, recategorical system may be in a position to encode invariances of a personal nature. For example, in some individuals their social history has sculpted a system in which threats to the self are, more often than not, readily extracted from social situations. These representations can be regarded as personal affordances (Luu & Tucker, 1996). That is, information that is represented in this generalized, recategorical system bears a personal viewpoint upon the experience and processing of new information.

The structure of representations in this generalized, recategorical system may be similar to the structures of superordinate-level categories (e.g., tools, furniture, Rosch & Mervis, 1975), theory categories (Medin, 1989), and goal-driven categories (e.g., things to keep us warm, Barsalou, 1985). In contrast to representations in the epicritic system, in which membership in a category depends upon physical features, membership in these types of categories is defined by function, theories, and ideals. Moreover, unlike basic and subordinate categories, and in keeping with the general nature of the system, an abstraction of the members does not reflect correlation among physical features of the members.

In contrast to the local signs supported by epicritic representations, the prefrontal network's representations are non-local, being diffusely distributed across both space and time. For example, a second pain pathway (mentioned briefly above) leads from spinal cord to the medial nuclei of the thalamus and terminates in the anterior cingulate (Vogt, et al., 1993). Cells within the anterior cingulate have broad receptive fields; a single cell's receptive field for noxious stimuli can be the entire body surface. This pain pathway appears to be responding to the valence of the noxious stimuli. Pribram (1981) has designated the term "protocritic" to describe a brain system that processes and represents these non-local signs.

3.   The process of consciousness

"Emotion is assumed to be always present in ordinary consciousness, giving it a particular experiential quality and maintaining its purposeful flow." (Izard, 1980) (p. 193).

3.1 *The importance of vertical integration to the conscious process*

Tucker and Williamson (Tucker & Williamson, 1984) suggested that a unidimensional construct of arousal cannot account for an organism's complex attentional control and self-regulatory functions. They proposed that the brain has two systems pertaining to arousal, each regulated by a different brainstem neuromodulator system. These systems are believed to be inherently affective in nature, and they influence attentional, engagement, and cognitive styles. That is, these brainstem systems are not affectively neutral, but rather their activity also engenders changes in emotion, engagement tendencies, and modes of cognition.

On one hand, the *activation system* is centered upon the dopamine cells of the tegmentum and substantia nigra. Its activity produces a redundancy bias that maintains focused attention and routinizes action. It produces an analytic cognitive mode. The activation system is believed to be central to the experience of negative affect, such as anxiety. On the other hand, the *arousal system* is regulated by norepinepherine cells of the locus ceruleus. This system produces a habituation bias so that novel events capture attention. The cognitive mode of this system is holistic. Its activity is believed to be central to the experience of positive affect, such as elation (for a detailed description see Tucker and Williamson, 1984).

Psychometric studies on the structure of mood have similarly revealed that attention, engagement, and cognition are intimately entwined (Tellegen, 1985; Watson & Tellegen 1985). Traditionally, mood space has been described as consisting of a pleasant-unpleasant dimension and an engagement-disengagement dimension. However, based upon their factor analytic studies of mood descriptors, Tellegen and Watson argued that mood space can also be validly described by the dimensions of positive and negative affect. This alternative description is just a 45 degree rotation of the axes that describe the pleasant-unpleasant and engagement-disengagement dimensions. However, Watson and Tellegen argued that their rotation is preferable because their two axes conform to the natural clustering of the mood descriptors. High positive affect is characterized by words such as active, elated, and excited. High negative affect is characterized by descriptors such as nervous, jittery, and fearful. Tellegen and Watson argued that these dimensions are more than just dimensions of arousal or affective. Rather, the descriptors suggest that they are dimensions that describe mood, arousal, engagement styles, and cognitive mode. For example, high positive affect is characterized as pleasurable engagement with an orienting cognitive mode. Thus, we can argue that the association between attention, affect, engagement, and cognition as described by the activation and arousal systems and the two dimensions of affect reflect a vertical integration of cerebral functioning.

Vertical integration of valenced arousal systems serves to affectively motivate, sustain, and integrate sensorimotor patterns. Without affective and motivational input from the brainstem, sensorimotor patterns in the neocortex lose their immediacy and quickly fade (Tucker, 1992). This principle may underlie certain symptoms displayed by patients with frontal lobe and limbic lesions. For instance, Pribram (1950, 1991) argued that limbic lesions disturb complex action plans because the states and action of the homeostatic centers are no longer coordinated. In addition, it is possible to observe abnormal influences of emotion on cognition in persons with relatively intact brains. Patients with a right temporal lobe epileptic focus display exaggerated emotionality, whereas those with a left temporal lobe focus exhibit obsessive thoughts or ruminations and a catastrophic response in their self-evaluations (Bear & Fedio, 1977). Tucker (1981, 1992) suggests that the symptoms experienced by temporal-lobe epileptics reflect the exaggerated constraints applied on cognition by emotions. From this perspective, the content of the left temporal-lobe epileptic's consciousness is filled with ideational and intellectual themes (such as philosophical and religious ruminations), whereas the content of the right temporal-lobe epileptic's consciousness is affectively colored with feelings of elation. Thus, emotion, through vertical integration, is inextricably bound up with consciousness.

3.2 *The process*

A complex organism, through evolution, becomes endowed with cerebral mechanisms that allow it to go beyond simple, reflexive responses to the environment. These mechanisms allow the organism not only to respond to the environment, but also to be *aware* of it. However, the reflex-arc, a simpler conceptualization, did dominate the way in which the functioning of the nervous system was earlier construed (see Pribram, 1960). The concept of a reflex-arc in its simplest form describes a loop in which a reflex (response) is elicited by a stimulus. This loop is a closed circuit and does not account for how information picked up by sensory receptors can be internally influenced. Decades ago, Pribram noted that the available evidence suggested that such one-way construals of the function of the nervous system are missing something. The problem that must be faced in studying the functioning of the brain is to specify how efferent (back) projections influence receptor mechanisms.

Attempting to go beyond the reflex-arc to account for internal influences on information processing, Pribram (1960) proposed a cybernetic model of cortical functioning. In this model, the protocritic and epicritic systems detect discrepancies between their respective representations of the incoming information and adjust behavior accordingly so that the perturbations are reduced. In this way, the normal functioning of these systems requires a comparative process. From this perspective, the organism is no longer seen as a passive processor of information; rather, the organism is viewed as having control over the information that it receives. These systems, however, are not seen as isolated — they must function together to produce adaptive behavior.

In studying the symptoms of frontal lobe lesions, Teuber (1964) and Nauta (1971) proposed theories of cortical functioning that also emphasized the ability of the brain, especially the frontal lobe, to affect its own reception of peripheral stimuli. These theorists suggested that projections from one cortical area to another can alter the functioning of the receiving areas either by producing a corollary discharge, or by producing set-points at which ongoing actions are compared against somatic states. For Teuber (1964), the frontal lobe sends a corollary discharge to the posterior areas of the brain to prepare the sensory systems to receive information based upon the executed action. Extending this concept, Nauta (1971) proposed that the prefrontal cortex instantiates set-points that assist the organism in anticipating and integrating action patterns across time (a form of working memory for action plans).

More recently, the possible effects of back-projections on sensory processing upstream has been extended beyond the frontal cortex (see Edelman, 1989; Zeki & Ship, 1988). For example, it has been shown that projections from primary visual cortex (V1) back to the lateral geniculate nucleus (LGN) can alter the firing of cells within the LGN (Sillito, Jones, Gerstein & West, 1994). They do so by altering the firing threshold of the LGN cells.

The V1 inputs to the LGN are not strong enough to drive the LGN cells by themselves — in order for the LGN cells to depolarize they need additional retinal input. As a group, those LGN cells which receive a common input via projections from V1 will be *synchronized in their firing* when the appropriate stimulus is provided. Crick and Koch (1990) have argued that synchronization of neuronal function across separate areas of visual cortex provides a form of short-term memory that leads to visual awareness. In the study by Sillito et al., it is unlikely that visual awareness can occur in the LGN. We believe that awareness involves synchronization at a more global-network level, and that projections from the protocritic networks are required.

As explained previously, the prefrontal cortex mediates the process of recategorization and the representation of personal affordances. This protocritic system, by way of its connections to the posterior, epicritic system, facilitates the synchronization of different sensory maps within the epicritic system to form coherent percepts. However, recall the observation that cells receiving inputs from the back-projections will not depolarize, i.e., be synchronized in their functioning, without environmental input. As a result a percept is not fully achieved until environmental information provides the confirming signals. The synchronization of activity in the epicritic system as mediated by protocritic networks gives consciousness a flavor that is beyond mere stimulus awareness.

The influence of the protocritic recategorical system allows for the individual history of the organism to influence cortical binding and the experience of consciousness. Bruner (1957) has long argued for the concept of readiness during perception. That is, the act of perceiving involves the process of categorization, and the categories which are developed influence the organism's readiness and ability to perceive stimuli. To the extent that incoming information violates the expectations inherent in a preconfigured system, either the organism will attempt to correct the discrepancy (Pribram, 1960), or the information will be incorrectly perceived or not perceived at all (Bruner, 1957).

Within this framework of consciousness, the interaction between the protocritic and epicritic systems explains the personal and idiosyncratic nature of human conscious experience. The abstraction of invariances, or personal affordances, by the protocritic system preconfigures sensory systems to receive, extract, and be aware of information in idiosyncratic ways (e.g., the tendency to perceive threat in social situations). In addition, it is through the workings of this system that ongoing events may be left out of consciousness. Bruner (1957) coined the term "perceptual defense" to describe events that occur in the environment but are left out of perceptual awareness, and he suggested that this is a result of interference caused by currently active categories (for our purposes categories represented in the protocritic system). These categories distort or exclude, by way of the protocritic system's projections to the epicritic sensory system, poor-fitting environmental events such that these events are either misperceived or are not perceived at all. Thus, actual ongoing events may be left out of consciousness altogether by interference from representations in the protocritic system.

4. Summary

Cortical binding through synchronized neural firing across sensory maps has been proposed to be a mechanism for consciousness. We believe that it is a key to understanding only one aspect of consciousness and that aspect is sensory awareness. The process of consciousness that we have described extends awareness within specific sensory modalities to a global sense of awareness of the environment based upon individual experiences. This is accomplished by suggesting that consciousness is a process of preparation and comparison.

The process depends on integration of cerebral functioning across the vertical hierarchy, and also upon the interaction of two systems: the frontal protocritic system and the posterior epicritic system. Evolution has elaborated each of these two systems. The epicritic system became more differentiated and has progressed toward increasingly refined sensory processing and segregation. In following its own course of differentiation, the protocritic system has increased both in size and in the density of its intrinsic and extrinsic connections. In organisms alive today, each system differs in the rate at which it matures, and as a result each is under the influence of different developmental processes. These developmental processes determine, in part, the nature of the representations within each system.

There are certainly aspects of human consciousness that cannot be accounted for in our model. For example, cortical circuits supporting human language certainly contribute to and alter the conscious experience (see Edelman, 1989). At the present time, we do not have a clear notion of how to fit these circuits and their functions into our model. However, this should not serve as a deterrent to the claim that consciousness can be related to the brain. After all, human language is a product of the brain and its influence and contribution to consciousness must be through cerebral means.

Finally, because our model is highly influenced by theories of brain and cortical evolution, we feel obligated to take a moment to entertain the question of which non-human animals experience consciousness. It is likely that animals with a reasonably developed cortex that is able to carry out this comparative process probably do experience consciousness. But certainly the *quality* of their experience must be different from what we humans experience, because each neural mechanism involved in the process differs to some degree between species.

Notes

Phan Luu was supported by NIMH grants MH42129 and MH42669 awarded to Don M. Tucker and the James S. McDonnell Foundation to support the Center for the Cognitive Neuroscience of Attention. Daniel J. Levitin was supported by a National Defense Science and Engineering Graduate Fellowship. John M. Kelley was supported by NIMH Training Grant #5T32MH18935. The authors are indebted to the following for their helpful discussions: Gerald S. Russell, Michael I. Posner, and Don M. Tucker. We are especially grateful to Peter Grossenbacher for his extensive feedback and suggestions.

Address correspondence to Phan Luu, Electrical Geodesics, Inc., Riverfront Research Park, 1850 Millrace Drive, Eugene, OR 97403, or send electronic mail to pluu@EGI.com

1. By 'primitive,' we mean cortex with heterogeneous laminated patterns (i.e., allocortex and periallocortex, as compared to isocortex, which is homogeneous in its laminar pattern). In comparative studies of the brain, the term 'primitive' is reserved for structures and areas that are apparent between species and are thus assumed to be present in their common ancestor.

2. The term isocortex means homogeneous cortex but this is a misnomer; isocortex (neocortex) is highly heterogeneous. It was probably thought to be homogeneous in comparison to the allocortex because the term allocortex originally applied to all other cortices that were not isocortex, including cortices now referred to as periallocortex and proisocortex.

References

Aertsen, A., Vaadia, E., Abeles, M., Ahissar, E., Bergman, H., Karmon, B., Lavner, Y., Margalit, E., Nelken, I., & Rotter, S. 1991. Neural interactions in the frontal cortex of a behaving monkey: signs of dependence on stimulus context and the behavioral state. *Journal für Hirnforschung, 32*, 735-743.

Ahissar, E., Vaadia, E., Ahissar, M., Bergman, H., Arieli, A., & Abeles, M. 1992. Dependence of cortical plasticity on correlated activity of single neurons and on behavioral context. *Science, 257*, 1412-1415.

Allman, J. 1990. The origin of the neocortex. *Seminars in the Neurosciences, 2*, 257-262.

Altman, J. 1995. Deciding what to do next. *Trends in Neuroscience, 18*, 117-118.

Barsalou, L. W. 1985. Ideals, central tendency, and frequency of instantiation as determinants of graded structures in categories. *Journal of Experimental Psychology: Learning, Memory, and Cognition, 11*, 629-654.

Bear, D. M., & Fedio, P. 1977. Quantitative analysis of interictal behavior in temporal lobe epilepsy. *Archives of Neurology, 34*, 454-467.

Benson, D. F. 1993. Prefrontal abilities. *Behavioural Neurology, 6*, 75-81.

Brody, B. A., Kinney, H. C., Kloman, A. S., & Gilles, F. H. 1987. Sequence of central nervous system myelination in human infancy. I. An autopsy study myelination. *Journal of Neuropathology and Experimental Neurology, 46*, 283-301.

Brown, J. W. 1979. Language representation in the brain. In I. H. D. Steklis & M. J. Raleigh (Eds.), *Neurobiology of social communication in primates*, (pp. 133-195). New York: Academic Press.

Bruner, J. S. 1957. On perceptual readiness. *Psychological Review, 64*, 123-152.

Chugani, H. T., & Phelps, M. E. 1986. Maturational changes in cerebral function in infants determined by FDG positron emission tomography. *Science, 231*, 840-843.

Crick, F. 1994. *The astonishing hypothesis: The scientific search for the soul.* New York: Charles Scribner's Sons.

Crick, F., & Koch, C. 1990. Towards a neurobiological theory of consciousness. *Seminars in the Neurosciences, 2*, 263-275.

Damasio, A. R. 1989. Time-locked multiregional retroactivation: A systems-level proposal for the neural substrates of recall and recognition. *Cognition*(33), 25-62.

Desimone, R., Albright, T. D., Gross, C. G., & Bruce, C. 1984. Stimulus-selective properties of inferior temporal neurons in the macaque. *The Journal of Neuroscience, 4*, 2051-2062.

Edelman, G. 1989. *The Remembered Present: A Biological Theory of Consciousness.* New York: Basic Books.

Freud, S. (Ed.). 1895. *Project for a scientific psychology.* (Vol. 1). London: Hogarth Press.

Fuster, J. 1989. *The prefrontal cortex.* New York: Raven Press.

Gaffan, D. 1994. Interaction of temporal lobe and the frontal lobe in memory. In A.-M. Thierry, J. Glowinski, P. S. Goldman-Rakic, & Y. Christen (Eds.), *Motor and cognitive functions of the prefrontal cortex*, (pp. 129-138). New York: Springer-Verlag.

Gazzaniga, M. S. 1995. Consciousness and the cerebral hemispheres. In M. S. Gazzaniga (Ed.), *The cognitive neurosciences*, (pp. 1391-1400). Cambridge: MIT Press.

Gibson, K. R. 1991. Myelination and behavioral development: A comparative perspective on questions of neoteny , altriciality and intelligence. In K. R. Gibson & A. C. Petersen (Eds.), *Brain maturation and cognitive development*, (pp. 29-63). New York: Aldine de Gruyter.

Greenough, W. T., & Black, J. E. 1992. Induction of brain structure by experience: substrates for cognitive development. In M. Gunnar & C. Nelson (Eds.), *Developmental behavioral neuroscience: Minnesota symposium on child psychology*, (Vol. 24, pp. 155-200). Hillsdale: Erlbaum.

Herrick, C. J. 1948. *The brain of the tiger salamander.* Chicago: University of Chicago Press.

Izard, C. E. 1980. The emergence of emotions and the development of consciousness in infancy. In J. M. Davidson & R. J. Davidson (Eds.), *The psychobiology of consciousness*, (pp. 193-216). New York: Plenum.

Jackson, J. H. 1931. The evolution and dissolution of the nervous system, *Selected writings of John Hughlings Jackson, Vol. II*, (pp. 45-75). London: Hodder and Stoughton.

Jeannerod, M., Arbib, M. A., Rizzolatti, G., & Sakata, H. 1995. Grasping objects: the cortical mechanisms of visuomotor transformation. *Trends in Neuroscience, 18*, 314-320.

Jürgens, U. 1979. Neural control of vocalization in non human primates. In H. D. Steklis & M. J. Raleigh (Eds.), *Neurobiology of social communication in primates*, (pp. 11-44). New York: Academic Press.

Kaada, B. R., Pribram, K. H., & Epstein, J. A. 1949. Respiratory and vascular response in monkeys from temporal pole, insula, orbital surface and cingulate gyrus. *Journal of Neurophysiology, 12*, 347-356.

Kinney, H. C., Brody, B. A., Kloman, A. S., & Gilles, F. H. 1988. Sequence of central nervous system myelination in human infancy: II. patterns of myelination in autopsied infants. *Journal of Neuropathology and Experimental Neurology, 47*, 217-234.

Krubitzer, L. 1995. The organization of neocortex in mammals: are species differences really so different? *Trends in Neuroscience, 18*, 408-417.

Livingston, R. B., Chapman, W. P., Livingston, K. E., & Kraintz, L. 1947. Stimulation of the orbital surface of man prior to frontal lobotomy. In J. F. Fulton, C. D. Aring, & B. S. Wortis (Eds.), *Research publications association for research in nervous and mental disease: The frontal lobes*, (Vol. 27, pp. 421-432). Baltimore: Williams & Wilkins.

Luria, A. R. 1973. *The working brain: An introduction to neuropsychology.* New York: Basic Books.

Luu, P., & Tucker, D. M. 1996. Self-regulation and cortical development: Implications for functional studies of the brain. In R. W. Thatcher, G. R. Lyon, J. Rumsey, & N. Krasnegor (Eds.), *Developmental Neuroimaging: Mapping the development of brain and behavior,* .

Mattson, M. P. 1988. Neurotransmitters in the regulation of neuronal cytoarchitecture. *Brain Research Reviews, 13*, 179-212.

Medin, D. L. 1989. Concepts and conceptual structure. *American Psychologist, 44*(12), 1469-1481.

Moruzzi, G., & Magoun, H. W. 1949. Brain stem reticular formation and activation of the EEG. *Electroencephalography and Clinical Neurophysiology, 1*, 445-473.

Nauta, W. J. H. 1964. Some efferent connections of the prefrontal cortex in the monkey. In J. M. Warren & K. Akert (Eds.), *The frontal granular cortex and behavior,* (pp. 397-409). New York: McGraw Hill.

Nauta, W. J. H. 1971. The problem of the frontal lobe: A reinterpretation. *Journal of Psychiatric Research, 8*, 167-187.

Northcutt, G. R., & Kaas, J. H. 1995. The emergence and evolution of the mammalian neocortex. *Trends in Neuroscience, 18*, 373-379.

Pandya, D. N., Seltzer, B., & Barbas, H. 1988. Input-output organization of the primate cerebral cortex. *Comparative Primate Biology, 4*, 39-80.

Pandya, D. N., & Yeterian, E. H. 1990. Prefrontal cortex in relation to other cortical areas in rhesus monkey: Architecture and connections. *Progress in Brain Research, 85*, 63-94.

Ploog, D. W. 1992. Neuroethological perspectives on the human brain: From the expression of emotions to intentional signing and speech. In A. Harrington (Ed.), *So Human a Brain: Knowledge and values in the neurosciences,* (pp. 3-13). Boston: Birkhauser.

Posner, M. I. 1994. Attention: The mechanisms of consciousness. *Proceedings of the National Academy of Science, USA, 91*.

Posner, M. I., & Dehaene, S. 1994. Attentional networks. *Trends in Neuroscience, 17*, 75-79.

Pribram, K. H. 1950. Psychosurgery in midcentury. *Surgery, Gynecology, and Obstetrics, 91*, 346-367.

Pribram, K. H. 1960. A review of theory in physiological psychology. *Annual Review of Psychology, 11*, 1-40.

Pribram, K. H. 1980. Mind, brain, and consciousness: The organization of competence and conduct. In J. M. Davidson & R. J. Davidson (Eds.), *The psychobiology of consciousness,* (pp. 47-63). New York: Plenum.

Pribram, K. H. 1981. Emotions. In S. K. Filskov & T. J. Boll (Eds.), *Handbook of clinical neuropsychology*, (pp. 102-134). New York: Wiley.

Pribram, K. H. 1991. *Brain and perception: Holonomy and structure in figural processing.* Hillsdale: Earlbaum.

Rakic, P. 1995. A small step for the cell, a giant leap for mankind: a hypothesis for neocortical expansion during evolution. *Trends in Neuroscience, 18*, 383-388.

Rosch, E., & Mervis, C. 1975. Family resemblances: Studies in the internal structure of categories. *Cognitive Psychology, 7*, 573-605.

Rosch, E., Mervis, C. B., Gray, W. D., Johnson, D. M., & Boyes-Braem, P. 1976. Basic objects in natural categories. *Cognitive Psychology*(8), 382-439.

Sanides, F. 1970. Functional architecture of motor and sensory cortices in primates in the light of a new concept of neocortex evolution. In C. R. Noback & W. Montagna (Eds.), *The primate brain: Advances in primatology*, (Vol. 1, pp. 137-208). New York: Appleton-Century-Crofts.

Satinoff, E. 1978. Neural organization and evolution of thermal regulation in mammals. *Science, 201*, 16-22.

Shepard, R. N. 1984. Ecological constraints on internal representation: Resonant kinematics of perceiving, imagining, thinking, and dreaming. *Psychological Review, 91*(4), 417-447.

Sillito, A. M., Jones, H. E., Gerstein, G. L., & West, D. C. 1994. Feature-linked synchronization of thalamic relay cell firing induced by feedback from the visual cortex. *Nature, 369*, 479-482.

Singer, W. 1993. Synchronization of cortical activity and its putative role in information processing and learning. *Annual Review of Physiology, 55*, 349-374.

Squire, L. R. 1992. Memory and the hippocampus: A synthesis from findings with rats, monkeys, and humans. *Psychological Review, 99*, 195-231.

Tanaka, K. 1993. Neuronal mechanisms of object recognition. *Science, 262*, 685-688.

Tellegen, A. 1985. Structures of mood and personality and their relevance to assessing anxiety, with and emphasis on self-report. In A. H. Tuma & J. D. Maser (Eds.), *Anxiety and the anxiety related disorders*, (pp. 681-706). Hillsdale: Erlbaum.

Teuber, H. L. 1964. The riddle of the frontal lobe function in man. In J. M. Warren & K. Akert (Eds.), *The frontal granular cortex and behavior*, (pp. 410-444). New York: McGraw Hill.

Tranel, D. 1993. The covert learning of affective valence does not require structures in hippocampal system or amygdala. *Journal of Cognitive Neuroscience, 5*, 79-88.

Trevarthen, C., & Aitken, K. J. 1995. Brain development, infant communication and empathy disorders: Intrinsic factors in child mental health. *Development and Psychopathology, 6*(597-633).

Tucker, D. M. 1981. Lateral brain function, emotion, and conceptualization. *Psychological Bulletin, 89*, 19-46.

Tucker, D. M. 1992. Developing emotions and cortical networks. In M. Gunnar & C. Nelson (Eds.), *Developmental behavioral neuroscience: Minnesota symposium on child psychology*, (Vol. 24, pp. 75-127). Hillsdale: Erlbaum.

Tucker, D. M. 1993. Emotional experience and the problem of vertical integration: Discussion of the special section on emotion. *Neuropsychology, 7*, 500-509.

Tucker, D. M., & Derryberry, D. 1992. Motivated attention: Anxiety and the frontal executive functions. *Neuropsychiatry, Neuropsychology, and Behavioral Neurology, 5,* 233-252.

Tucker, D. M., Luu, P., & Pribram, K. H. 1995. Social and emotional self-regulation. In J. Grafman, K. J. Holyoak, & F. Boller (Eds.), *Annals of the New York Academy of Sciences,* (Vol. 769, pp. 213-239). New York: New York Academy of Sciences.

Tucker, D. M., & Williamson, P. A. 1984. Asymmetric neural control systems in human self-regulation. *Psychological Review, 91,* 185-215.

Vogt, B. A. 1991. The role of layer I in cortical function. In A. Peters & E. G. Jones (Eds.), *Cerebral cortex,* (pp. 49-80). New York: Plenum.

Vogt, B. A., Sikes, R. W., & Vogt, L. J. 1993. Anterior cingulate cortex and the medial pain system. In B. A. Vogt & M. Gabriel (Eds.), *Neurobiology of the cingulate cortex and limbic thalamus,* (pp. 314-344). Boston: Birkhäuser.

Watson, D., & Tellegen, A. 1985. Toward a consensual structure of mood. *Psychological Bulletin, 98*(2), 219-235.

Yakovlev, P. I. 1948. Motility, behavior and the brain. *The Journal of Nervous and Mental Disease, 107,* 313-335.

Yakovlev, P. I., & Lecours, A. R. 1967. The myelinogenetic cycles of regional maturation of the brain. In A. Minkowski (Ed.), *regional development of the brain in early life,* (pp. 3-70). Oxford: Blackwell.

Zeki, S., & Ship, S. 1988. The functional logic of cortical connections. *Nature, 335,* 311-817.

CHAPTER 10

Multisensory Coordination and the Evolution of Consciousness

Peter G. Grossenbacher
National Institute of Mental Health

Percepts, the phenomenal contents of perception, are available during any waking state, presumably in every human being and in other species too. Perception entails becoming conscious of physical events, enabling you to sense bodily events such as breathing or limb movement, as well as objects and events in the environment surrounding your body. Some kinds of perceptual information *predominate* over other kinds in perceptual experience. That is, some sorts of sensory information appear more salient in the content of consciousness than do other sorts. This phenomenal preponderance is subjectively observable: When you perceive something, what aspects tend to most fully occupy your conscious awareness?[1]

||••||

Fortunately, we find that human nervous systems are capable of reflective self-examination. By repeating this exercise in the subjective appraisal of your experience, you can accumulate introspective data regarding your own perceptual habits. You are clearly in a unique position for observing the functioning of your own nervous system *from the inside.*

Now make use of your powers of observation: Choose something which you can perceive right now, and compare the relative salience of *where* that something is and *what* it is.

||••||

I can only guess at your experience, and your experience depends on the circumstances of your current situation... Did an object's location in space serve as background information, with other aspects such as color and shape more directly occupying your conscious awareness? For many people, certain visual aspects such as color tend to appear more salient than other visual information such as location. Certainly no two human brains are the same, and each of us has our own individual arrangement of biases among the possible contents of consciousness. However, despite these individual differences, it is the general trends which provide some common ground to human experience.

The experience of visual attributes often predominates over non-visual sensations, though exceptions to this rule are numerous: conscious awareness can certainly be absorbed by music, or the fragrance of a flower. But in these cases many people *close their eyes*, thereby blocking visual input, in order to best appreciate an orchestral melody or the scent of a rose.

What determines the relative impact on awareness brought by different kinds of perceptual information? As you now know from having read the previous chapters in this book, various answers to this question have emphasized either the role of physical stimulation (see chapter by Price, this volume), arousal (see chapter by Whitehead & Schliebner, this volume), emotion (see chapter by Derryberry), attention (chapter by Stein & Wallace), or ongoing behavior (see chapter by Rossetti). An additional approach stems from an understanding of brain evolution as a framework which reveals important aspects of mental organization (see chapter by Luu, Levitin & Kelley). This line of thought starts with the recognition that brain evolution supplies a coherent framework for understanding the communication channels which connect functionally related parts of the brain to each other. But beyond this important insight, the evolutionary past of the human brain holds further clues for understanding the nature of conscious experience.

This chapter explores the evolutionary origin of the areas in mammalian neocortex that are involved in conscious experience in an alternative approach to that pursued by Luu, Levitin & Kelley. We will start by noting the extent to which the different senses are kept departmentally segregated from one another at several stages of cortical processing. Because human brains have such a high level of modality segregation among cortical areas, the demands of multisensory integration require a sophisticated neural circuitry for coordinating the flow and processing of information arising in the multiple sensory systems. I suggest a role for content-general brain areas, which are not restricted to a single sense, in this multisensory coordination of the processing which occurs in content-specific brain areas. Direct connections between content-specific areas are contrasted with an indirect circuitry by which content-general areas can mediate the communication among content-specific areas. Some of these indirect circuits are thought to mediate recipient brain areas access to the outputs of content-specific areas. The content-specific areas, by dint of supplying the signals for access mediation, thereby supply the contents of conscious experience. This arrangement of inter-network communication is to be offered as a model which relates phenomenal awareness to the evolution of cortical neural networks.[2]

1. Anatomical segregation of sense modalities in primate cortex

Most mammals make good use of the sense modalities of vision, hearing, touch, smell, and taste. But mammalian species differ widely in the detailed circuitry of the neural networks in the brain which process the sensory information picked up by each sense. By studying the relations between brains of different species, the field of comparative neuroanatomy offers hints regarding the evolutionary path which has produced this variation. For example, during the evolution of our pre-human ancestors, *visual* brain systems became greatly elaborated, whereas *olfactory* systems (which process smells) diminished in proportion (Stephan & Andy, 1970). In contrast, the superior sense of smell enjoyed by dogs illustrates that evolution has not wrought change in this same direction for all mammals.

If the content of conscious experience is influenced by the anatomical extent of modality-specific brain systems, then visual percepts should occur more frequently than smells in the content of human consciousness. Is this the case for you?

‖••‖

In addition to inter-species differences in how much brain tissue is devoted to each sense modality, there is another important way in which the living neural machinery which processes sensory information differs among species. Comparison of brain organization in the rat and the monkey illustrates that there has been an evolutionary trend toward greater segregation of sense modalities in primates compared to other mammalian families[3]. Rat brain, as typical of rodents, contains a highly interconnected cortex with substantial crosstalk between sense modalities (Paperna & Malach, 1991; Reep, Corwin, Hashimoto & Watson, 1984). In the rat, the cortical areas most specialized for the sense modality of vision in fact receive neural signals from other modalities as well (Miller & Vogt, 1984). This direct intersensory crosstalk is also found in the cortex of many other mammals such as cat (Bental, Dafny & Feldman, 1968; Fishman & Michael, 1973; Horn, 1965; Murata, Cramer & Bach-y-Rita, 1965; Spinelli, Starr & Barrett, 1968), and rabbit (Beteleva, 1975; Chow, Masland & Stewart, 1971; Voronin & Skrebitsky, 1965). Unlike these mammals, the cortex of primates has fully independent primary receiving areas which do not directly interconnect. So the entrance of sensory information into primate cortex may be uniquely fragmented, compared to other animals, as it maintains the strict segregation of sense modalities already established before reaching the cortex (see Figure 10.1).

Since the content of consciousness is largely supplied by cortical representations, the organization of sensory information in cortex may influence the psychological structure of perceptual awareness. Consider the psychological ramifications of the increase in cortical segregation among sense modalities over the course of human evolution. Certainly, under most conditions we encounter today, sensory information is readily tagged as heard, seen, touched, smelled, or tasted. As a greater proportion of cortical circuits have become increasingly devoted to a single sense modality, the route by which sensory information enters the nervous system may now be more readily identified in conscious awareness than in our evolutionary past. Is this true for you? If so, then you should find it effortless to know the sense modality in which a sensory experience arises. Try it now: notice which sense is providing the content of your phenomenal awareness.

||••||

Since our senses are so easily distinguished from each other, it is possible for us to separately consider what each sense modality contributes to the ever-changing amalgam of conscious contents. Indeed, humans are quite capable of comparing the impressions provided by two distinct senses, such as vision and touch, in order to verify that the impressions are consistent with each other. Checking for mutual consistency between sense modalities provides a useful

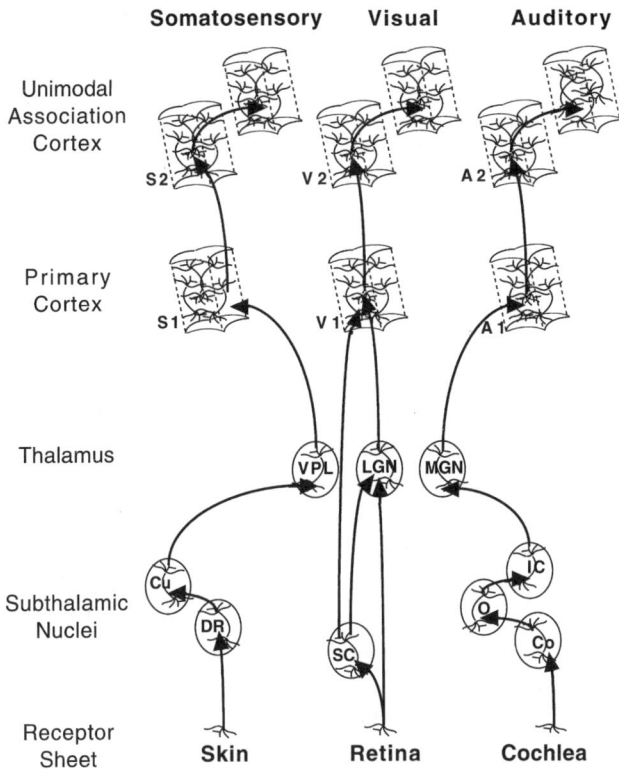

Figure 10.1. Each sense modality converts perceptible changes in physical energy into neural signals which are passed form one neuron to the next in a sensory pathway. In all mammals, these great sensory pathways start at the periphery of the nervous system with anatomically isolated sheets of receptor cells. These pathways first proceed through nuclei below the level of the thalamus (see a neuroscience text for details). Only in primates does this anatomical segregation of sense modalities persist past the thalamus and into the brain's cortex. Arrows indicate axonal projections which convey feedforward flow of neural signals (feedbackward connections which reciprocate these are omitted from this figure for simplicity). In the brains of primates (including humans), cortical pathways specific to each sense modality are anatomically segregated from each other as indicated by the lack of arrows interconnecting somatosensory, visual, and auditory areas. The primary sensory cortical areas of somatosensation, vision, and audition (S1, V1, and A1) each process elemental features of stimulus attributes in their respective modality, and their outputs are directed to other modality-specific parts of cortex, so-called "unimodal association areas," which appear responsible for computing more sophisticated representations (see Kosslyn, chapter 4 of this book).

means for noticing and correcting a sensory illusion or perceptual error: If an approaching animal first looks like a waddling *duck* but it sounds like a hissing *snake*, it might be worth a second look! In the course of normal development, everyday experience with crossmodal checking throughout childhood and adulthood may reinforce the conception of sensory modalities as separate departments of mind. Although the critical developmental factors could be debated, it is self-evident that sense modalities constitute a primary basis for categorization of conscious human sensory experience.

A role for the senses in categorization is evident in conversations which mention the *scent* of a rose, the *sound* of an engine, or the *look* of a stranger. Distinct sense modalities are easily discernible because each encompasses unique qualities not shared with other modalities. The temperature of a thing is given only by touch, and its color only by sight. It may surprise no one to point out that each sense modality tends to fail under circumstances which do not impede other senses. For example, whether the eyes become injured or darkness falls, it is a simple matter to know that visual percepts have diminished without similar changes in other senses. Daily life experience provides repeated encounters with this kind of independence between modalities, making it easy to recognize multiple differences among our senses. In addition, many behavioral acts can be organized according to the principle of separately identifiable senses: when about to cross a road, children are admonished to "stop, look, and listen" in order to emphasize the equal relevance of both vision and hearing to avoiding the danger of automotive traffic.

2.   Multisensory processing requires intersensory coordination

Along with the evolution of increased modality segregation, there has been an accompanying increase in the number of modality-specific association areas in primate cortex (Carlson, Huerta, Cusick & Kaas, 1986; Preuss & Goldman-Rakic, 1991). Indeed, as brain size has increased, the number of separately identifiable cortical areas has grown substantially. This proliferation of cortical areas is important because each area may support the neural computations needed to sustain one or more qualitatively distinct mental representations. Certainly, a large number of sensory dimensions are represented in the human brain, and with greater anatomical specificity than in the brains of other species which have less modality segregation. It is my impression that the numerous cortical areas account for the impressive variety and complexity available in human sensory experience. Because there are many areas of cortex in the human brain whose function has yet to be identified, there is as yet only partial evidence for this presumed correspondence between the number of cortical areas and the complexity of sensory dimensions. Nonetheless, I believe that recent proliferation of sensory cortical areas during evolution has greatly enriched the variety of sensory dimensions which are available for both perception and imagery — mental functions which depend on much the same brain areas (Zatorre & Halpern, 1993) (and see chapter by Kosslyn, this volume).

Despite its highly segregated input pathways and its cortical areas devoted to modality-specific sensory processing, the human brain does manage to bring together some of the information being processed in multiple sense modalities. Multisensory integration is achieved by virtue of connections between neurons which receive inputs from each of the modality-specific neural pathways. The subjective consequences of these intersensory connections are evident in conscious human experience, in which the feel and the sight of a fruit can be recognized as pertaining to the same object. That is, a hand-held apple *feels* round and smooth, and it also *looks* round and smooth. Humans are not unique in this ability for crossmodal recognition; other primates also demonstrate crossmodal recognition between touch and vision (Ettlinger & Wilson, 1990). Scientists do not yet know exactly how our primate brains, which lack direct connections between primary sensory areas in cortex, accomplish such crossmodal recognition. But it appears that the brain represents multisensory properties of objects, such as shape and smoothness, which can be sensed in more than one modality.

In primate brains, the combination of information across modalities requires integration among anatomically segregated sensory pathways. Despite undisputed separation between modalities in some parts of the brain, many existing mental functions in fact require some form of neural communication between sense modalities. Indeed, various forms of collaboration between the senses appear to involve several multisensory circuits. For example, the superior colliculus is a sub-cortical structure which plays a critically important role in the processing of multisensory spatial information. That is, where something is (its location in space) can be determined through a combination of sound, sight, and touch (see Stein and Wallace, chapter 5 of this book). It is also clear that some areas in parietal cortex are also involved in coordinating spatial information which comes from multiple sense modalities (Graziano & Gross, 1993). Indeed, at least some of these parietal areas interconnect with the superior colliculus, to provide the anatomical basis for coherent processing of location information across brain areas.

Whereas perception and imagery *of space* depends on multisensory cortical areas in the parietal lobe, perception and imagery *of objects* relies on multisensory cortical areas in the temporal lobe (Roland, Skinhoj & Lassen, 1981; Ungerleider, 1995). It may be these temporal areas of cortex which store learned intersensory combinations of the qualitatively distinct sensations coming from our different senses. When a person encounters objects or environments for the first time, new associations between sense modalities are learned: a fruit which looks *like so,* tastes *like so.*

In the human brain, many neural circuits in frontal, temporal and parietal lobes of cortex receive information from the different senses which has already been pre-processed by modality-specific cortical areas. Abstract reasoning skills such as analogy and metaphor probably depend on multisensory neural circuits such as these. One well-known area of cortex called 'Wernicke's area,' considered necessary for meaningful language, provides a good example. Wernicke's area resides in a multisensory region of the temporal lobe of cortex, and it receives input from several unisensory areas (Galaburda & Pandya, 1982; Galaburda & Pandya, 1983; Geschwind, 1965). It should not surprise us that the brain area thought to be most devoted to the processing of semantics or meaning is not restricted to a single sensory modality: the word 'dog' refers to a dog regardless of whether it is conjured up by an experience of smelling, touching, seeing, or hearing the presence of a canine. Likewise, the concept of *sharp teeth* may hold the same meaning whether a dog's teeth are touched or seen (or imagined). Let us consider the kind of meaning that pertains to information spanning more than one sense modality. Given its multisensory inputs, Wernicke's area ought to be especially important for representing this kind of meaning.

How do neurons subserving distinct sensory modalities interact? Let us distinguish between two kinds of circuitry which allow distinct parts of the brain to communicate with each other: direct and indirect. *Direct* communication is supported by neural connections which convey signals directly (possibly monosynaptically) from one brain area to another (see Figure 10.2 for two kinds of direct communication). If all communication between different parts of the brain were direct, then as new brain areas appear in the course of evolution, they might each require direct connectivity with a large number of other brain areas. Each new area would add to the possibilities for direct connections linking every already existing brain area. In this scenario, the potential demands for increasing connectivity could lead to very expansive growth in the mass of fiber tracts needed to conduct signals from one part of the brain to the rest. In other words, there would be a potentially high anatomical cost for the evolution of a new brain area if inter-area communication depended solely on direct connections.

Feedforward Connections Only

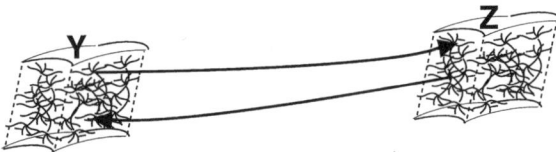

Feedforward and Reciprocating
Feedbackward Connections

Figure 10.2. In order to illustrate how groups of neurons which are located in different parts of the brain physically connect to each other to form communicating circuits, let us oversimplify the matter by considering only two brain areas at once. One part of the brain (W) contains neurons whose axons project directly to synapse with neurons located in another part of the brain (X). The activity of the receiving neurons in X is influenced by the axonally conducted signals sent from W. The top panel shows a feedforward circuit in which signals only pass from W to X. Note that in this circuit, neurons in X cannot influence the activity of neurons in W. The bottom panel shows a feedback circuit in which signals pass both in a forward direction from Y to Z and also in a backward direction from Z to Y. In this circuit, Y neurons can influence neural activity in Z, *and* neurons in Z can influence the activity of neurons in Y. Feedback circuits are found in many neural pathways in the brain, and they provide rapid, secure transfer of information.

However, the mammalian brain is not limited to direct communication; it also employs *indirect* circuits, and these can take any of several forms. Indirect connectivity between two brain areas requires mediation by at least one other brain area. Figure 10.3 shows a simple example of one such circuit.

Indirect circuits provide the benefit of greater flexibility compared to direct circuits. An indirect circuit can support a greater variety of neural computations in comparison to a direct circuit. Another advantage of the indirect circuit is the relatively low anatomical cost of evolving new brain areas which can become integrated into such a network. That is, the only additional fiber tracts required to put a new area Z in communication with the areas X and Y depicted in Figure 10.3 would be direct connections between Z and A. This same fixed cost applies no matter how many content-specific areas are already communicating with each other via A. One caveat on these benefits of indirect circuitry concerns the speed of signal transmission: communication within a more complex circuit would tend to be a bit slower than monosynaptic (direct) connections. Another potential limitation on the utility

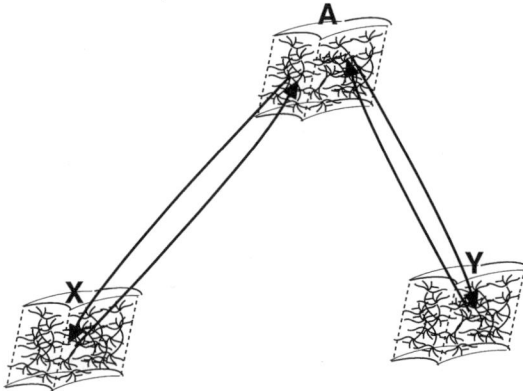

Communication between
X and Y mediated by A

Figure 10.3 Brain areas X and Y are not directly connected with each other, whereas brain area A has bi-directional connections with both X and Y as indicated by the arrows. This indirect circuit allows A to receive signals from and transmit signals to X and also to and from Y, so as to *mediate* communication between two parts of the brain which are not directly connected to each other. The connectivity between X and Y is indirect because they are not directly interconnected as indicated by the lack of arrows between X and Y.

of indirect circuits regards the opportunity for inappropriate scrambling of messages generated from disparate brain areas if they are routed through a common mediating circuit.

Modality-specific neural pathways make use of direct circuits to convey signals between the linked brain areas which constitute these pathways (Gattas, Sousa, Mishkin & Ungerleider, 1997). We would expect indirect circuits to play a role in multisensory coordination because of the exorbitant anatomical cost (explained above) for the large number of direct connections which would be required to handle the same linkages which can be achieved with a smaller number of indirect circuits. Fortunately, this theoretic argument is bolstered by neuroanatomical findings of a role for indirect circuits in multisensory coordination (as discussed in the following section of this chapter). The central role for indirect cortical circuits in multisensory coordination might not pertain to non-primate species having brains in which primary sensory areas of cortex are directly interconnected. In contrast, the need for coordinating inter-system communication among sensory circuits has probably grown during our own evolution toward anatomical separation of modality-specific representations in neocortex.

3.   Coordination among networks in the brain

The brain's repertoire of mental skills depends on the neural computations performed by a large number of content-specific networks, many of which are located in the brain's cortex. One localized cortical network, for example, specializes in the processing of visual information relevant to motion (Krubitzer & Kaas, 1990; Newsome, Britten & Movshon, 1989). Similarly, another network specializes in the perception of faces (Clark et al., 1995; Heywood & Cowey, 1992). These functional subsystems of the brain are not mere blocks of tissue, but in the living organism they behave as sensitive, responsive cellular networks which communicate with each other. It is this communication between one part of the brain and another which governs the moment-to-moment changes in what goes on in the mind. For example, upon glimpsing a ferocious predator, it is imperative that those neural signals which relate to this visual experience communicate with other parts of the brain which control whether and how the body moves. It appears that the relevant brain areas succeed in getting their signals delivered correctly to other parts of the brain often enough for adequate mental functioning.[4]

It is clear that the efficiency with which a cortical network computes and communicates its output can change. Indeed, this efficiency can be subject to voluntary control! In our Arbitrary Priority study (Grossenbacher, Posner, Compton & Tucker, 1991; Posner & Raichle, 1994), we instructed subjects to prioritize one of two simultaneous decisions over the other. By comparing the time it took subjects to respond under different conditions (in a speeded, conjoint decision task), we found that each component decision delivered its output most quickly when it had been given priority, compared to when the other decision was given priority. At the time, we considered this result to support the idea that each component decision was completed at a rate which depended on its priority. But in that study we were unable to distinguish between the intra-network computation which produced the result, and the communicating of this computational result to other parts of the brain. It now seems just as plausible that we influenced the efficiency of inter-network communication instead of (or in addition to) changing the efficiency of computation within a cortical area.

What do we know about how the brain goes about coordinating its numerous activities? In the Arbitrary Priority study we also recorded brain electrical activity with sixty-three electrodes placed on the scalp. The electrical recordings we obtained revealed event-related brain potentials with effects of priority evident as early as three tenths of a second after onset of the visual stimulus. This indicates that the brain does selectively influence the efficiency of neural processing in different parts of cortex. Indeed, our instructions for arbitrarily prioritized coordination of mental activity resulted in a pattern of localized activations and deactivations distributed throughout the brain. The effects of this large-scale coordination are visible in the images of brain activations obtained using measures of blood flow (Posner & Raichle, 1994).

The neural mechanisms which control inter-area communication are not well understood, but this scale of inter-network communication may depend on two types of neural circuits. Using a distinction established in the previous section, we may consider the control exerted by neural signals carried by both *direct* connections between two brain areas, and a more remote form of control exerted by signals carried by *indirect* circuits involving additional brain areas specialized for the purpose of coordinating communication between brain areas.

In this sense, the coordination among cortical functions may occur at two anatomic levels. At the more local anatomic level, a brain area communicates directly with some of its neighbors, possibly with signals traveling in both directions, to and fro. If the bidirectional circuit is organized for mutual inhibition, then only one of two reciprocally connected areas could be highly active at a time. Or, if the circuit is organized for mutual excitation, then both would tend to be active at once. If such local mechanisms were sufficient, then large-scale coordination of brain activity would be nothing more than the product of numerous, local effects, just as the V formed by flying geese emerges from each bird determining its location in the formation relative to its nearest neighbor.

It would be difficult, if we were limited to local mechanisms which rely only on direct connections between content-specific areas, to account for intentional (top-down) coordination exemplified by the Arbitrary Priority study. Could it really be the case that the neural underpinnings of any two mental representations *must* be directly connected to each other in order for both representations to be brought into a single decision? That seems incredible given the geometrically vast number of possible combinations that appear amenable to conjoint evaluation. It seems more likely that the impressive coordination thought to govern communication among the large number of content-specific systems in the brain depends on indirect circuitry mediated by areas of the brain which are not content-specific, but are instead systems of a more general nature. At this more global level of coordination, those content-specific processes which are currently active may require discourse with content-general circuits, while for those content-specific processes which are not currently engaged, the need for communication via content-general circuits is minimized.

4.   Inter-network coordination, cingulate cortex, and awareness

By surveying the neural connections between different parts of the brain, we can determine which brain structures have the connections necessary for mediating communication between other parts of the brain. One brain area does stand out as especially well connected in a manner which could support the mediation of communication between a large number of content-specific cortical areas. The *cingulate gyrus* receives inputs from most every sense modality, and also sends output signals to a large number of brain areas (Pandya, Van Hoesen & Mesulam, 1981). Figure 10.4 shows some of the connections which uniquely position this portion of cortex with respect to communication among other brain areas.

Some of the parietal areas which trade signals back and forth with cingulate cortex are multisensory association areas which process information deriving from (and may exert influence upon) many earlier stages of sensory processing. Likewise, the frontal areas which send signals to and receive signals from cingulate cortex are also positioned high in the cortical hierarchies of sensory and motor processing, apparently contributing information to

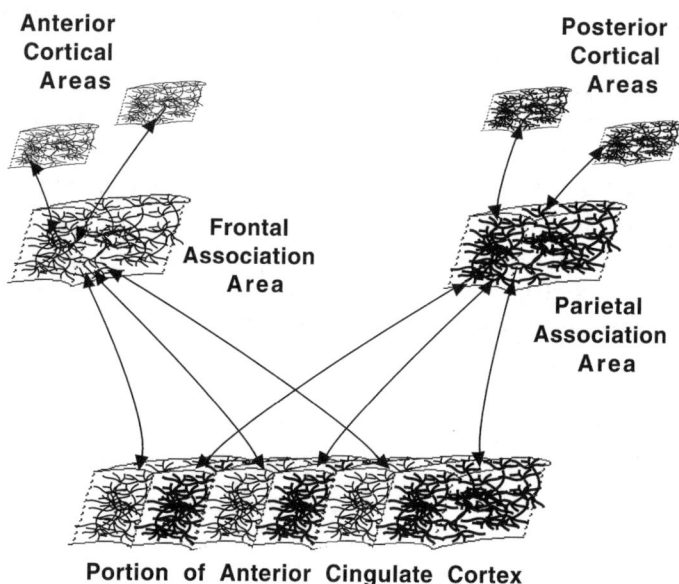

Portion of Anterior Cingulate Cortex

Figure 10.4. The cingulate gyrus is a large fold of cortex located just above the corpus callosum (the largest fiber bundle which connects left and right cortical hemispheres). Visible at the base of each hemisphere's inner surface, and elongated in both anterior and posterior directions, the cingulate occupies a very central locus in the web of communicating connections which link brain areas together. Cingulate cortex connects with frontal and parietal association areas as shown by the arrows (the connections are in fact bidirectional). Note the impressive pattern of interdigitation in which alternating bands of cingulate tissue communicate with frontal and parietal lobes (Goldman-Rakic, 1988). This part of the brain plays a critical role in conscious awareness (see text).

the current and pending contents of consciousness. This means that anterior cingulate itself connects indirectly with a large proportion of the entire cortex in addition to the areas with which it connects directly. Having indirect connections with so many areas of cortex suggests that cingulate cortex may be critically important for facilitating inter-network communication.[5]

Within the tissue that comprises the anterior portion of the cingulate gyrus, the neurons which communicate with parietal areas are clumped together into bands which are positioned in alternation with the remaining bands of neurons which communicate with frontal areas (see Figure 10.4). These two types of cingulate cells do not merely cluster together by type, but are precisely arrayed in a pattern of interdigitation so that each band concerned with parietal information is bordered on both sides by bands which communicate with frontal cortex. To the extent that adjacent bands communicate with each other, this architecture could mediate coordination at a scale which spans a majority of the entire cortical terrain.

Given its connections with a large number of cortical areas, anterior cingulate is well positioned for providing content-general circuits which mediate conscious awareness. Consistent with this possible role are findings that sensory awareness appears to depend on neural networks residing within the anterior cingulate gyrus (Posner & Petersen, 1990). The anterior cingulate is metabolically active during many different kinds of perception and cognition (Pardo, Pardo, Janer & Raichle, 1990). Positive correlation between its level of metabolic activity and the number of noteworthy stimuli perceived during one minute suggests a consistent relation between neural activity and conscious content (Grasby et al., 1993; Posner, Petersen, Fox & Raichle, 1988). Thus consciousness, of sensory information at least, in this way depends on neural circuitry in anterior cingulate cortex (Posner & Rothbart, 1991). In addition, lesions in this part of the brain result in complete lack of self-initiated movement (Damasio & Van Hoesen, 1983), indicating a possible role in volition as well as perception. Together with more superior portions of medial frontal cortex, activity in the anterior cingulate correlates with the frequency of thoughts which crop up on their own (McGuire, Paulesu, Frackowiak & Frith, 1996). Though other brain areas may also be important for mediating conscious awareness, the anterior cingulate gyrus provides a prime example of a content-general part of the brain which has been empirically linked to conscious awareness. Evidence for this comes from *activations* associated with awareness of sensory content, *deactivations* associated with lack of conscious content, and loss of consciousness consequent to lesion.

The outputs of content-specific networks in cortex appear to be coordinated by networks such as those in anterior cingulate cortex which mediate conscious awareness. Put another way, content-specific sensory processes share their outputs in a functional foreground coordinated by content-general brain systems. Could it be that it is *only* these shared outputs which can be experienced in conscious awareness?

5. Access mediation and consciousness: A new model

Several of the points raised in previous sections of this chapter can now be brought together to explain the neural processes by which sensory representations contribute to the contents of consciousness. The Access Mediation Model (AMM, pronounced "ahm") asserts: *The content-general neural circuits which thread output from any of several cortical areas as input into other cortical areas are the very same circuits which sustain conscious experience.*[6]

According to AMM, subjective experience arises by virtue of indirect neural communication between the numerous brain areas devoted to sensory, motor, and other domains of cortical processing.[7] In this view, consciousness depends on the sharing of information between one part of the brain and another, excluding the cases in which the sharing is entirely conducted by direct connections between content-specific neural networks. With respect to computational function, consciousness reflects the operation of the mediating content-general circuits which coordinate the outputs of content-specific cortical networks. Because AMM posits appearance in the contents of consciousness *only* if there is adequate indirect connectivity, mediation is understood to be logically *necessary* for consciousness. This formal relation can be expressed in two equivalent ways:

1. Any sensory conscious content must pertain to cortically represented information which is communicated to other parts of the brain via mediation by content-general circuits.

2. If a content-specific cortical area *lacks* mediation for its outputs via the content-general circuits which mediate inter-cortical communication, then the information as processed within this cortical area is *inaccessible* to consciousness.

AMM does not claim that access mediation is *sufficient* for consciousness. If mediation were sufficient for appearance in conscious experience, that would mean that every cortical area having outputs which reach the requisite mediating circuits *must* have corresponding content in consciousness. Put another way, it would also follow that any cortically processed information which never appears in consciousness *must* lack access mediation. Given the plurality of cortical representations that are available to conscious experience, competition among the many cortical areas may allow the outputs of relatively weakly connected cortical areas to be overshadowed by activity of better connected areas. At any rate, regarding the question of sufficiency, AMM takes a conservative approach and does not claim that anatomical connections suffice for generating conscious contents.

In addition to determining whether a neural representation ever could contribute to conscious mental content, the mechanism of content-general mediation may also influence *quantitative* aspects of phenomenal experience. The explanatory framework provided by AMM naturally leads to inferences regarding subjectively observed biases among conscious contents.[8] One inference holds that *the phenomenal **salience** of conscious content depends on the bandwidth provided by mediating circuits: the more output signals from a content-specific cortical area to reach anterior cingulate cortex, the more strong and clear will be the conscious content sustained by these signals.* Other things being equal, a conscious content will tend to appear more salient to the extent that the output signals originating from the relevant content-specific area of cortex are conveyed to (and through) content-general areas such as anterior cingulate cortex. This inferred relation between quantity of connections and phenomenal salience provides a testable claim which should prove tractable in coming years, although at this time insufficient data area available to determine its veracity.

In addition to salience, AMM supports a second quantitative inference: that *the **frequency** with which a specific type of information appears in the content of consciousness is in part determined by the relative quantity of its connections with the content-general circuits which coordinate inter-network communication.* As discussed above, imbalances in salience and frequency arise from a variety of sources including evolution, ontogenetic development, arousal and emotion, and selective attention. Some of these factors, such as evolution, influence whether the indirect connections necessary for access mediation exist at all, and in what number they exist. Other factors, such as attention, influence the degree to which existing indirect circuits come into play for mediating access at different times. These more dynamic influences will tend to make frequency less reliably biased than salience as a quantitative consequence of the mediating bandwidth. None the less, I suggest that it is the quantity of the indirect neural connections appropriate for mediated access which explains any anatomically determined tendencies in the frequency with which specific types of information appear in the contents of awareness.

In terms of neuroanatomy, the access mediation which sustains conscious experience is probably controlled, at least in part, by neurons located in anterior cingulate cortex. As presented above, evidence regarding anterior cingulate's connections with a host of other brain areas supports a possible role for this brain structure in inter-network communication, and physiological data indicate its metabolic involvement in conscious awareness. Moreover, a role for anterior cingulate in mediating conscious access is consistent with the reliable finding that executive cognitive control depends on this part of the brain (LaBerge, 1990; Pardo et al., 1990; Vogt, Finch & Olson, 1992). Although it is easy, from a psychological perspective, to consider conscious awareness and executive control to be possibly distinct components of mind, neurologically they may be inseparable. Suppose that AMM is correct in that the contents of conscious awareness really do arise by virtue of the content-general circuitry in anterior cingulate cortex governing which cortical areas' output is to get routed as input to other cortical areas. AMM thus depicts conscious experience as depending on the executive coordination which determines the representations threaded as input to other cortical areas.

6. Discussion

According to AMM, subjective experience arises by virtue of indirect neural communication between numerous brain areas devoted to sensory, motor, and other domains of cortical processing. One corollary to AMM is that *direct* communication between content-specific areas does not suffice to produce conscious content. We can distinguish between three possible ways in which communication among a given set of brain areas could lack mediated access, thereby ensuring that the outputs of these areas do not appear in the contents of conscious awareness. First, it could be the case that indirect circuitry simply has never evolved to provide mediated access to the given set of brain areas. That is, perhaps our ancestral lineage includes no instances of the given set of cortical areas participating in consciousness. A second possibility consists in currently existing indirect connections becoming less involved, temporarily making them functionally inconsequential as the communication among a set of brain areas comes to rely more exclusively on direct connections. Finally, if the impact of indirect connections can be dynamically reduced during on-line processing, perhaps their numerosity can diminish over the much longer time-scale of evolutionary development. These three scenarios are discussed in turn below. The subsequent portion of this discussion will focus on the evolution-ary development of the content-general areas which mediate cortico-cortical communication, and repercussions of this development for the evolution of content-specific cortical areas.

6.1 Circuits which lack access mediation cannot contribute content to consciousness

Some brain areas which connect directly with their recipient areas may never have had much mediation by indirect connections at any stage in their evolu-tion. This could be one evolutionary route by which some circuits found in modern brains do not exhibit mediated communication. Likely as a general rule for subcortical structures, it could also apply to at least some brain areas which comprise cortical pathways, such as adjacent areas which are bidirec-tionally connected with both feedforward and feedback projections only to their nearest cortical neighbors.

In an earlier chapter, Kosslyn suggests that only those representations which must be recoded in order to be communicated from one cortical area to another are available as conscious content. This Recoding Theory and the Access Mediation Model both emphasize the communication between distinct parts of the brain which differ in their representation of information. But in other respects these two theories do not resemble each other. Whereas Kosslyn focuses on directly interconnected cortical areas within a (visual) pathway, AMM encompasses indirect communication between neural repre-sentations regardless of sense modality. For two areas which are directly interconnected, AMM asserts that any appearance in conscious content will only be a function of additional indirect connectivity between the source and other cortical areas. The two theories have opposite predictions for the case, if it can be found, in which two areas of cortex are directly interconnected, are internally organized so differently from each other that information must be remapped between them, and they are not linked indirectly via a content-general circuit. In this case, Recoding Theory predicts that the information would appear in the contents of conscious experience, whereas AMM asserts that the information could not appear in conscious awareness.

6.2 *Learned automaticity bypasses indirect connections*

The content of consciousness can include some aspects of the voluntary control of bodily acts. Consider what happens when you learn a complex skill. Let us take automobile driving as a familiar example, although the logic applies equally well to other tasks such as making stone tools or climbing a tree. The novice is all too aware of each component action involved in shifting gears, steering, braking, etc. With many hours of practice, the disturbingly detailed effort of conscious control gradually gives way to less conscious involvement in performance. Extended practice can naturally lead to *learned automatization* of the behavior — when you can perform a familiar task without much thinking about what you are doing. Automatic behaviors have been described as efficient yet inflexible compared to similar behaviors guided by conscious control (Schneider & Shiffrin, 1977).

What happens in the brain of an individual during acquisition of learned automatization? Presumably, activity in circuits mediating conscious control mechanisms gradually becomes less important for performance. If neural signaling in these circuits decrease as automaticity is learned, then the remaining circuits (i.e., direct communication between content-specific areas) must change to embody the learned skill. Indeed, recent blood flow studies show exactly this decrease in activation during extended practice[9] (Friston, Frith, Passingham, Liddle & Frackowiak, 1992; Karni et al., 1995). Such learned changes could happen in brain circuits simply by modification of the synapses which subserve communication between interconnected neurons (see Figure 10.5). In terms of neural network modeling, it is the relative weights of connections (between neurons) which change. In this way, directly connected circuits involving content-specific cortical areas can perform the learned task, leaving behind the slower and more complicated mediated access which was initially required.

Before:
Primarily Indirect
Communication

After:
Primariy Direct
Communication

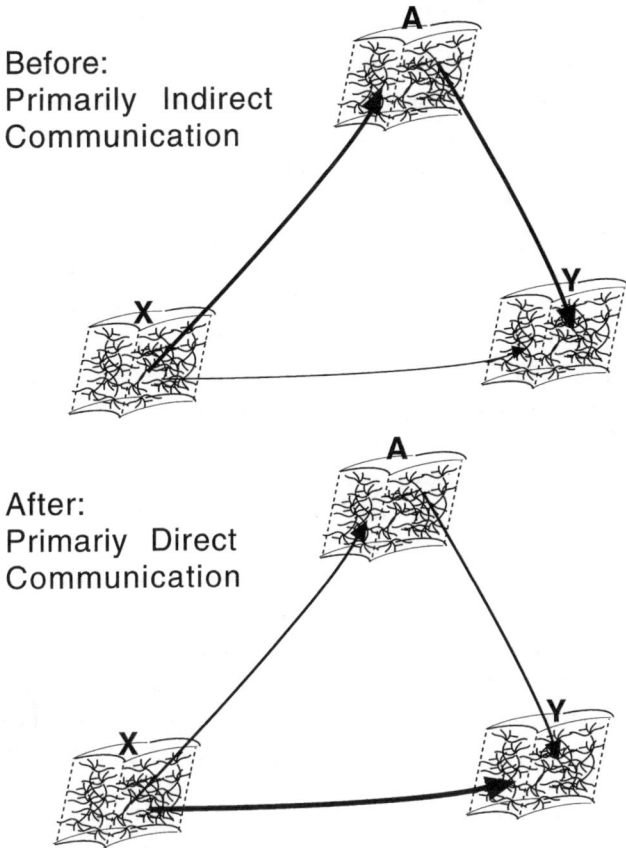

Figure 10.5. During the learning of a cognitive or motor skill, the neural route for communication between the involved cortical areas may undergo changes over the course of repeated practice. If the direct circuits which interconnect two areas X and Y are initially able to convey only relatively weak signaling (designated by a thin line in the upper panel), the strengths of synapses within Y can change so as to boost this line of communication between X and Y (indicated by the thick line in lower panel). Of course, it is not the axonal diameter that changes, but merely the efficacy of synapses from the neurons in X that project their axons into Y. Although the precise scenario of change depicted in this figure is probably not the only mechanism by which automatization can be learned, change in synaptic weighting is well accepted as a learning mechanism, and the depicted transition from primarily indirect to primarily direct communication is pedagogically useful.

6.3 *Evolved automatization eliminates indirect connections*

Access mediation is a critical function under conditions of evolving an increasing diversity of functional brain areas, clearly the case for many primates. But reliance on mediation, once established, may not persist immutably in future generations as a *permanent* feature of a particular cortical function. On the evolutionary time scale, dependence of a given sensory representation on conscious mediation could wane. It may be easiest to understand how mediated access for a given content-specific neural network could become lost during evolution by first drawing an analogy to learned automaticity.

Once a skilled behavior has been practiced many times, through learned automaticity it can place reduced demands on conscious control mechanisms. Perhaps there is a similar transition in evolution such that a previously controlled process can evolve to proceed with less conscious control. A content-specific system which can operate with some independence from consciously coordinated brain systems could become increasingly automated during evolutionary development.[10]

Human somatosensory networks (which process information pertaining to bodily sensations) may have relatively *less* direct connectivity with content-general circuits involved in conscious awareness (in proportion to the total set of content-specific systems subject to conscious mediation) *now* compared to eons ago. Decreasing connectivity between a content-specific network and content-general awareness circuits would presumably be accompanied by reduction in the conscious content which depends specifically on that content-specific network. Evolved automaticity entailing a decrease in conscious awareness of bodily states during our evolutionary development could account for this neuroanatomical change. Why might this loss of conscious sensitivity come about? I submit that evolved automaticity would be most likely to develop for those processes having inputs and computational demands which change little over large numbers of generations. This stability constraint may be more likely to be met in perception of internal environments (interoception) than for perception of stimuli outside the organism (exteroception) because the internal environment is subject to genetic controls which have much less influence on the external environment.

Of course, there must be important differences between the neural changes underlying *evolved* automaticity and *learned* automaticity. Rather than modification of synaptic strength (the probable mechanism for learned automaticity), evolved automaticity could be accomplished by an increase in the number of axonal connections linking a content-specific system to other, functionally related brain systems (see Figure 10.6). Indeed, many cortical areas do project directly to several remote brain areas, delivering their outputs rapidly and securely, an efficient solution for frequently followed threads of

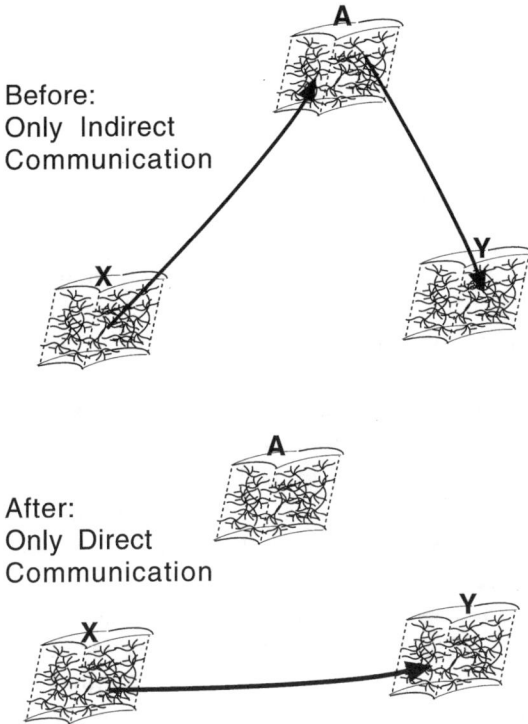

Before:
Only Indirect
Communication

After:
Only Direct
Communication

Figure 10.6. After a newly evolved cortical area has already proven useful via mediated communication with other brain areas, there is the possibility of it developing direct connections with those areas which benefit most from receiving its outputs. During the evolution of direct communication between two brain areas (X and Y), the mediation of a third brain area (A) can help. This figure depicts the evolutionary development from one extreme (initial lack of any direct connectivity between X and Y) to another (complete loss of mediating connections involving area A). This complete reversal makes for a clear example but is not strictly necessary for Evolved Automatization. Regardless of whether the original mediating connections involving area A are preserved, Evolved Automatization ensures that the primary route for communication between area X and area Y is direct, and no longer involves area A.

inter-process communication. Because it is difficult to imagine how *all* such direct connections could already be coexisting with the earliest appearance of a new cortical area, it seems reasonable to suppose that they could more readily appear subsequent to indirect connectivity. Assuming that neural connections are genetically variable, and assuming that the anatomical cost of installing direct connectivity with another brain area is not too great, then new connections will tend to evolve when there is most benefit to the organism and its offspring (adaptive value). In sum, connections from a given cortical area to coordinating circuits may diminish over time, under the pressure of competition for connectivity with limited coordinating circuits, as other, newer cortical areas assert greater need for access mediation.

6.4 *Evolved enrichment increases content capacity*

Humans may demonstrate another important adaptation to an increasing number of content-specific systems if content-general neural circuitry has *increased* in generality (the number of areas which it coordinates) during our evolution. It makes sense that brain mechanisms involved in conscious awareness have accumulated an increasing capacity for coordination as the complete pool of content-specific processes has evolved into an increasingly large repertoire. This growth in content-general capacity enriches the pool of accessible content-specific systems with which a given brain area might (indirectly) communicate, as shown in Figure 10.7. By bolstering the number of potential content-specific recipients of signals, this scenario of *evolved enrichment* satisfies the requirement to coordinate among a larger and more diverse assemblage of specialized systems. In this way, the evolutionary development of each new content-specific cortical area could contribute to an increase in the extent of connectivity between content-general mediating circuitry and content-specific areas.

Some thinkers claim that conscious experience must be unique to humans and is not shared by other animals. Due to the discrepant frameworks of subjectivity and objectivity, it can be extraordinarily difficult even to prove the existence of conscious awareness in another person, and this problem worsens when we consider species which do not use a human language to communicate. However, if the anterior cingulate does play the role posited by AMM in the coordination of inter-area communication in cortex, then there is some evidence that does bear on the question of whether consciousness

Less Capacity:

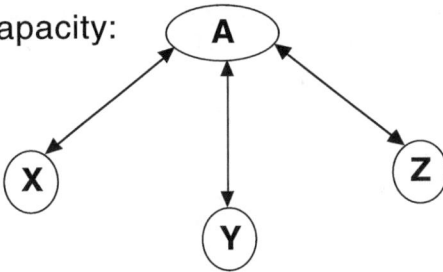

More Capacity:

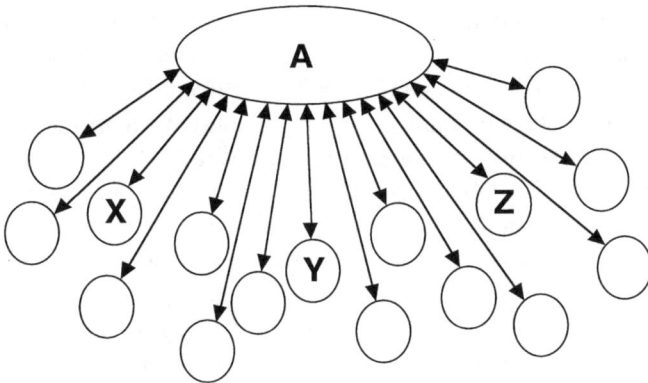

Figure 10.7. The ellipse labeled "A" represents the content-general system which mediates access among content-specific brain areas. The size of this ellipse indicates the capacity of A for mediating access among multiple content-specific cortical areas. Ellipses "X" through "Z" represent content-specific areas which both send and receive signals via A, as indicated by the two-headed arrows. The top panel depicts an early stage of human brain evolution, with a later stage shown in the bottom panel. As the circuitry of content-general access mediation evolves to handle the communication needs of a larger number of content-specific cortical areas, its capacity for providing useful connections to cortical areas also grows. To the extent that new cortical areas are more likely to evolve under conditions of higher capacity for access mediation, this increase in capacity for inter-network communication could explain the extremely rapid rate at which the human cortex has increased in size during our recent evolution (see text).

evolved recently. Conscious experience might have to have evolved recently if its underlying neural systems themselves evolved only recently. But in point of fact, the cingulate cortex is composed of an ancient cytoarchitecture, suggesting that its coordinating role is not a new one. This possibility is further supported by the similarity in patterns of cortical connectivity to and from anterior cingulate cortex found both inside and outside the primate family (Pandya et al., 1981). Widespread distribution of such connectivity across mammals reveals that cingulate cortex has probably played a central role in cortical communication for eons (Vogt & Miller, 1983).

6.5 *Rapidly increasing size of human brains*

Since the time when common ancestors of humans and non-human primates were living, human brains have dramatically expanded in size at a rate unequaled by any other animal (Passingham, 1975; Pilbeam & Gould, 1974). How has our heritage of explosively rapid brain growth afforded the mental abilities and subjective experiences now enjoyed by modern humans? The paired tracks of rapid brain growth and intellectual gains are historically inseparable, and no doubt depend on one another.

Perhaps the organization of neural connections within the human brain holds clues for unraveling the mystery of unusually rapid growth of neocortex in human evolution. Recall that in the brains of primates, sense modalities are more highly segregated than in any other mammalian family, and among primates, it is we humans who have the largest mass of neural circuitry positioned to make sense of these parallel, independent input streams. Might the rapid increase in human brain size somehow be a consequence of our increasingly segregated sensory systems? For example, there may have been evolutionary pressure on the content-general circuitry which coordinates the activity of multiple cortical pathways to expand in size and capacity in order to mediate communication between areas which lack direct interconnectivity. This notion appears to be supported by the fact that human brains have extensive portions of cortex which process information from more than one sense modality (Augustine, 1996; Hikosaka, Iwai, Saito & Tanaka, 1988; Molchan, Sunderland, McIntosh, Herscovitch & Schreurs, 1994; Rolls, 1996). However, we should not assume that these multisensory cortical areas all need to be content-general. Indeed, many of the multisensory areas of cortex are probably content-specific, concerned with integrating particular features of

objects which are available to more than one sense modality. (For example,
visual and tactile information regarding the surface roughness of an object
becomes integrated in this way (Grossenbacher, 1998 submitted).) In fact,
there is no clear evidence for the idea that content-general circuitry occupies
an especially voluminous proportion of brain tissue. Indeed, the primary
benefit of *mediated* communication between cortical networks is anatomical
efficiency.[11] So if the prodigious size of human brains cannot be accounted for
merely in terms of massive mediating circuitry, why are they so big? The rest
of this section speculates on a more likely role for the neural circuitry of
consciousness in the accelerated evolution of large human brains.

　　In order to discuss the possible influence of consciousness on brain
evolution, a brief reminder of relevant genetic and evolutionary principles is in
order. *Genetic inheritance* refers to the passing of shared traits from one
generation to the next via shared genetic material. The idea of inheritance
presupposes the possibility of alternative traits appearing in some individuals
within a population (variation among phenotypes). The so-called "gene pool"
of a species constitutes all the varieties of genetic material carried by living
members of the population (variation among genotypes). New heritable traits
appear in a population for the first time upon the occurrence of a novel genotype,
and recurring appearance of novel genetic variants is critical to evolutionary
change. Of course, in order for a new trait to survive into subsequent generations
and to successfully compete with genetic alternatives, it has either to be unduly
favored by chance alone, which is considered very unlikely, or it must have
adaptive value. Any trait which increases the likelihood of surviving to
reproduce and raise offspring has adaptive value. To the extent that human
intelligence has contributed to our thriving as a species despite the host of
predators, diseases, and climactic challenges which have accompanied our
evolution, higher brain functions carry adaptive value. So we are justified in
contemplating the genetic inheritance of all human brain systems, both content-
specific and content-general. In both of these kinds of systems, a random
perturbation in the genotype (DNA) can lead to the expression of a new
phenotype (e.g., an adult organism). Whether this new trait survives to infuse
later generations depends on its adaptive value. In the case of content-general
circuits which mediate conscious access, we should take an interest in the
evolutionary pressures which have made mediated access so adaptive.

　　It need not be mysterious that the access mediation responsible for
consciousness can also affect brain evolution. The evolutionary argument

rests on brain circuitry influencing fitness (the likelihood of successful survival, reproduction, and rearing of offspring). It has already been demonstrated that behavior contributes to fitness (Bateson, 1988).[12] I assume that
some of the behaviors which contribute to fitness depend on content-specific
brain systems sharing their outputs. Because behavior contributes to the
fitness of an individual, it would be wrong to suppose that access mediation
has evolved without also itself influencing brain evolution at the same time.[13]

Let us attempt to envision the way that adult brain organization has
evolved over time. The grim reality of living in a harsh world where mishaps
in neural communication can cause death exerts great pressure on brain
evolution for ensuring reliability. Against this severe backdrop, a light sprin-
kling of occasional and small changes in brain organization counters the
natural inertia toward conservation of the status quo. Due to genetic variation,
the brains of a small proportion of individuals within a population do manifest
uniquely novel perturbations which depart from the status quo. For many of
these mutants, their idiosyncratic neuroanatomy is maladaptive and leads to
premature death, or failure to reproduce, or inability to successfully rear
offspring. But for others, their deviated brains do not disrupt fitness, and may
even enhance their viability. These are the individuals who pave the way
toward evolutionary changes in human nature.

Let us now narrow our focus to consider how cortico-cortical access via
conscious mediation could influence the evolutionary development of content-specific cortical systems. As the number of distinct content-specific
cortical areas has increased during human evolution, new areas have appeared
within the cortical terrain during this period. As yet we may only guess at the
process by which this evolutionary step has been taken. As discussed above,
the capacity for inter-network communication can be inherited, and this
capacity for mediating inter-network communication could affect the viability
of content-specific areas (especially those having little or no direct connectivity with other content-specific areas).

During the process of evolution, some nascent cortical areas could disappear from a population, while others will catch on and persist into future
generations. For a new brain area to increase the fitness of an organism, its
output signals must reach an adequate number of other brain areas. (Otherwise
the information processed within the new brain area could have no impact.) If
the outputs of a new area have access to a greater number of brain areas, it
stands a greater chance of contributing to the viability of the central nervous

system and organism as a whole (other things being equal). For this reason, ready mediation of such access by content-general systems could amplify the contribution of potentially useful brain areas. As long as the outputs of a new area reach systems geared for access mediation, then the mechanism of mediated access could contribute to the perpetuity of recently evolved cortical areas. In this way, innovative mental functions enabled by small genetic perturbations in content-specific circuits can become consolidated in subsequent generations through natural selection, thanks to the catalysis of access mediation. To summarize this line of speculation, the neural mechanisms involved in conscious awareness support the functioning of recently evolved (content-specific) capacities which might otherwise come to an evolutionary dead-end without access mediation.

As explained earlier in this chapter, the evolution of increasing modality segregation has been accompanied by a corresponding increase in the (conscious) mediation of communication between content-specific sensory representations. Over the course of our evolution, an increasing capacity in the content-general circuitry which mediates inter-network communication and consciousness has improved the efficiency with which nascent content-specific areas could benefit from mediated communication with other cortical areas. This improved efficiency with which a nascent content-specific area benefits from mediated communication with other cortical areas could augment the adaptive value of this new cortical area. Neural mechanisms underlying consciousness support new, tentative mental capacities which might otherwise fail to function sufficiently well to provide adaptive advantage because their outputs could reach a larger number of cortical areas, thereby raising the chances of these outputs being of greater use to the organism. In other words, a larger capacity for inter-network facilitation should increase the likelihood that a nascent neural system will contribute to fitness. For this reason, conscious mediation of cortico-cortical communication may facilitate the evolutionary development of additional content-specific neural networks.

If increased mediating capacity does facilitate the evolutionary emergence of new content-specific cortical areas, this could help explain the positively accelerated rate of brain growth evident in the recent evolution of our own species. In this way, *consciousness might guide the leading edge of evolutionary development by facilitating communication between a newly evolved mental function and other mental functions.* The net effect could be an increase in the speed with which new mental capacities evolve. Conscious mediation of inter-cortical access provides a likely neurobiological basis for the very rapid increase in cognitive abilities realized by our unique (hominid) ancestors.

6.6 Cortical recency and phenomenal salience

An intriguing prediction derives from AMM which more deeply explains why the contents of consciousness are distributed over processing domains in unequal proportions. AMM asserts that those content-specific cortical areas which compute the representation which contribute most to the contents of consciousness are those with the preponderance of connections to the content-general coordinating circuits which mediate access to the input of other cortical areas. However, this begs the question of *why* it is these cortical areas and not others that have greatest amounts of mediation. Why, for example, is *color* apparently so well connected? One possibility is that cortical areas which have recently evolved tend to enjoy greater connectivity with coordinating circuits than do comparable cortical areas which have been on the scene much longer (Grossenbacher, 1996). The reasoning behind this proposition proceeds as follows.

Contents of consciousness range widely over perceptual, motor and semantic processes. The richness of conscious experience must somehow depend on the complex web of interconnected cortical systems involved in perceptual, semantic and motor processing. As already mentioned, some phylogenetically older cortical functions may become automated during evolutionary development (Evolved Automatization), thereby requiring less conscious guidance. In the course of evolution, conscious control capacities originally devoted to these older functions may become available for new functions.

Consider the case where mediating circuits do become available for coordinating the outputs of cortical areas. I will argue that newly evolved cortical areas are more likely than previously established areas to maximize their output projections to the circuits which handle access mediation. As a consequence of this trend, *more recently evolved cortical functions have greater access to consciousness than do older brain systems*. This Cortical Recency principle provides a new theoretical framework for explaining tendencies in human phenomenology (Grossenbacher, 1996).

According to the Cortical Recency principle, the information which is more salient in most people's conscious experience is represented (computed and sustained) by cortical circuits which evolved more recently than other, phylogenetically older cortical circuits. Apparent correlation between phylogenetic recency and phenomenology tentatively affirms that representations computed in recently evolved cortical systems are prioritized in consciousness over those involving cortical systems which evolved earlier (Grossenbacher, 1996). Two complementary evolutionary processes could synergistically produce these effects of cortical recency on bias in consciousness. First, the connections which route outputs from a new cortical area to coordinating circuits may evolve prior to the appearance of direct links between this new area and other content-specific areas of the brain. Indeed, direct connections, even to distant areas of cortex, could evolve most easily *subsequent* to indirect connections via mediating circuits. Second, the disparity in conscious mediation among cortical systems as a function of evolutionary recency could also arise because the older cortical functions have for a longer time been subject to pressures toward evolved automatization. For these reasons, compared to similar cortical areas which have benefited from mediated access for many more generations, newer cortical areas having outputs which reach distant areas will tend to depend more on mediated access to these areas.

Strong evidence for Cortical Recency may not already be in hand, but Cortical Recency does generate clear predictions for the cortical systems which represent conscious contents. The domains of language and visual perception provide clear testing grounds which should soon generate solid evidence for or against Cortical Recency. Everyone who understands the words in this sentence is performing the mundane yet phenomenal task of comprehending language. Human language makes use of elaborated sensory representations and ability to image (as well as the ability to plan and sequence complex motor activity). To the extent that glottological thought ("mental speech") does dominate consciousness, then conscious verbal streams should depend on recently evolved cortical functions according to the Cortical Recency principle. We do know that human language does depend on systems located in multimodal association areas of cortex (Galaburda, 1982). As yet, it remains to be determined whether or not *conscious* aspects of language (e.g., semantics) depend on recently evolved cortical systems more than do *unconscious* aspects (e.g., syntax). This clear prediction of Cortical Recency makes it falsifiable, i.e., subject to disproof, and the determination will come once the evolutionary ordering among the relevant cortical systems becomes known.

Human brain systems appear to be biased so that *color* occupies conscious awareness more than do other perceptible attributes such as *location* of objects in visual space. Color salience may generalize across many fruit-eating primates, as fructivorous diet offers a plausible adaptive impetus for color salience in primates (Mollon, 1989). The always-important location of objects has direct access to brain areas which control bodily movement such as reaching (see chapter by Rossetti), and the mediation of content-general circuitry is not usually required. I suggest that information relevant to visual object recognition may dominate conscious awareness in perception because the brain systems involved in processing this information are recently evolved, perhaps more recently than the neural mechanisms required for other perceptual functions such as spatial localization (Komatsu, Ideura, Kaji & Yamane, 1992; Preuss & Goldman-Rakic, 1991). This constitutes another testable prediction of Cortical Recency, one for which preliminary findings do show support.

The Cortical Recency principle has found tentative support in the above observations regarding the cortical basis for visual perception. But given only weak supporting evidence, it makes sense to muster any countervailing observations. Possibly all potential counter-examples to Cortical Recency share this common aspect: the capacity to reliably usurp conscious control on a short-term basis, apparently for purposes of survival of self or offspring. The examples of pain, nausea, orgasm, fatigue, rage, and fear all demonstrate that voluntary control does not always have a sure grip on behavior. Urges toward fight or flight, mediated by brain structures outside neocortex, can take over control when great need arises. Nonetheless, within neocortex, it appears that the *customary* content of conscious awareness is supplied by the more recently evolved circuitry, except for these occasional interruptions from stereotyped processes with strong links to issues of survival.

7. Summary

A complete cognitive neuroscience of consciousness need not be limited to the (incredibly productive) set of approaches which study the on-line functioning of healthy and damaged nervous systems. Evolution offers a unique and important additional basis for explaining psychological and neural mechanisms, and it is helpful to view current functional human brain anatomy within the context of its evolutionary development. In accounting for conscious human mental function in terms of brain evolution, this chapter speculates on the neural underpinnings of consciousness, and examines possible relations between the content of consciousness and cortical circuitry.

The Access Mediation Model (AMM) constrains the content of consciousness to just those neural representations which have sufficient connections to the content-general coordinating network located in anterior cingulate cortex. And further, the salience of a cortical representation in conscious awareness depends directly on the degree to which its outputs are available to mediated communication via this content-general network. In actual fact, mediated access and direct connections need not be mutually exclusive. Communication among many brain areas no doubt involves *both* indirect and direct connections; it becomes a matter of degree as to which route dominates. This balance between (conscious) mediated access and (unconscious) direct communication can change with conditions, including effects of attention.

In the course of evolution, consciousness need not be an ever-expanding enterprise. The principle of Evolved Automatization explains how a content-specific cortical area can evolve to contribute *less* to the contents of consciousness than it had in earlier generations. As a complement to Evolved Automatization, the principle of Evolved Enrichment accounts for there being more conscious processing at later stages of evolution, and explains how humans have recently come to have such big brains.

An intriguing extension of the Access Mediation Model claims that the salience of a cortical representation in conscious awareness depends indirectly on the recency of evolution of that neural representation. The Cortical Recency principle suggests that a newly evolved cortical representation is likely to depend on mediated communication. Cortical recency facilitates salience in conscious awareness by virtue of a new cortical area communicating with other cortical areas through the mediation of coordinating networks. This Cortical Recency principle derives from the role that evolution is assumed to play in shaping the organization of consciousness in the brain. However, it is possible that this principle will need to be qualified as more data become available. For instance, it might be the case that some cortical representations arise with only very local, direct connections with other cortical areas.[14]

Although brain evolution holds great explanatory potential for consciousness, it cannot tell the whole story. Each person's life experience guides how their brain develops. Beyond the profound constraints imposed by evolution, the largest impact on functional brain organization no doubt stems from neonatal and childhood experience. But even in adulthood, learning and experience influence the functional weighting of connections between neurons (for example, Learned Automatization). Moreover, one's current frame of mind determines *how much* fits into the content of consciousness in any one moment by limiting the number of distinct brain areas whose outputs can be mediated without confusion. Both the quantity and the specific qualities of conscious contents are subject to numerous and shifting constraints. Upon directing your mind to focus on, for example, the sounds that you can hear right now, you can boost the presence and salience of auditory information in the contents of your consciousness. This act of selective attention probably depends on the same circuitry which mediates access of cortical auditory representations to other brain areas.

Notes

1. As explained in the chapter which introduces this book, the "‖••‖" symbol instructs you to participate in an exercise involving subjective observation.

2. This chapter builds on the ideas developed in the preceding portions of this book, and is not intended to be read in isolation from them.

3. It would be a mistake to equate differences among living species with evolutionary history: monkeys did not evolve from rats. According to the theory of evolution, rats and monkeys share a common ancestral species from which both lines diverged. Because brains are composed of soft tissue which does not preserve in the fossil record, brain evolution can only be deduced from comparative neuroanatomy of existing species, as considered with respect to their *phylogeny*, the inferred branching tree of ancestral divergence which estimates the relatedness between species.

4. Perhaps of equal importance, the outputs of disengaged processes, stemming from brain areas which may have been recently important for previously ongoing mental activity but are no longer pertinent, are successfully ignored.

5. It may be most accurate to consider some of the frontal and parietal areas which are reciprocally connected with anterior cingulate as parts of a distributed content-general system for mediating communication between other parts of cortex. But for sake of simplicity, in this chapter I will treat this system as primarily localized to anterior cingulate cortex.

6. Granted, consciousness may not be a necessary attribute of any system imaginable which *could* mediate inter-network communication in a brain. The claim here is more empirical, and asserts that the human nervous system happened to evolve such that conscious experience *is* made possible by the neural circuits which evolved to mediate the coordination of a variety of content-specific networks in cortex.

7. This model was developed on the basis of observations on how sensory information appears in consciousness, and on how the sensory cortical circuits are organized. It seems

likely that identical or similar mechanisms of mediated access may also facilitate the communication of non-sensory information.

8. Because each individual brain and its corresponding phenomenal experience is unique, AMM should not be misconstrued as a model which is expressed identically in all human brains. Rather, the references to conscious content and neural connection are meant to be considered within the context of variation across individual people. Just as the subjective data regarding conscious content must be observed on an individual basis, so too ought the idiosyncrasies of neural connectivity to be considered. General principles can only emerge inductively, from many people sharing contents of conscious experience, and from their brains having patterns of neural connections in common as well.

9. In this study, an initial decrease in activation due to practice was countered by a subsequent increase in activation, possibly due to an increase in the perceived importance of the task when performance was required to extend over multiple days.

10. This form of automaticity constitutes an evolutionary analog to learned automaticity which, though analogous at one level, must occur within the brain in an entirely different way.

11. Mediated neural communication avoids the vast number of *direct* connections which would be required to support the great flexibility in cortical communication provided by indirect circuits.

12. The claim is *not* that mental skills acquired during an individual's life are passed on to its offspring!

13. The useful term "co-evolution" was coined to refer to feedback interaction between brain evolution and the evolution of mental activity (Deacon, 1992).

14. If this is ever found to be the case, it might constitute a relatively brief evolutionary stage which progresses toward mediated communication with more brain areas.

References

Augustine, J. R. 1996. Circuitry and functional aspects of the insular lobe in primates including humans. *Brain Research Reviews, 22(3)*, 229-44.

Bateson, P. 1988. The active role of behaviour in evolution. In M.-W. Ho & S. W. Fox (Eds.), *Evolutionary Processes and Metaphors*, (pp. 191-207). Chichester: John Wiley & Sons.

Bental, E., Dafny, N., & Feldman, S. 1968. Convergence of auditory and visual stimuli on single cells in the primary visual cortex of unanesthetized unrestrained cats. *Experimental Neurology, 20*, 341-351.

Beteleva, T. G. 1975. Responses of the rabbit's lateral-geniculate body to sound stimuli and to electrical stimulation of the reticular formation of the brain stem. In E. N. Sokolov & O. S. Vinogradova (Eds.), *Neuronal Mechanisms of the Orienting Reflex*, (pp. 170-177). Hillsdale, NJ: Lawrence Erlbaum.

Carlson, M., Huerta, M. F., Cusick, C. G., & Kaas, J. H. 1986. Studies on the evolution of multiple somatosensory representations in primates: the organization of anterior parietal

cortex in the New World Callitrichid, Saguinus. *Journal of Comparative Neurology,*
246(3), 409-26.

Chow, K. L., Masland, R. H., & Stewart, D. L. 1971. Receptive field characteristics of
striate cortical neurons in the rabbit. *Brain Research, 33(2),* 337-52.

Clark, V. P., Keil, K., Maisog, J. M., Courtney, S., Ungerleider, L. G., & Haxby, J. V. 1995.
Functional magnetic resonance imaging of human visual cortex during face matching: a
comparison with positron emission tomography. *NeuroImage, 4,* 1-15.

Damasio, A. R., & Van Hoesen, G. W. 1983. Emotional disturbances associated with focal
lesions of the limbic frontal lobe. In K. Heilman & P. Satz (Eds.), *Neuropsychology of*
Human Emotion, (pp. 85-99). New York: Guilford.

Deacon, T. W. 1992. Brain-language coevolution. In J. A. Hawkins & M. Gell-Mann
(Eds.), *The Evolution of Human Languages,* (pp. 49-83). Redwood City, CA: Addison-
Wesley.

Ettlinger, G., & Wilson, W. A. 1990. Cross-modal performance: behavioural processes,
phylogenetic considerations and neural mechanisms. *Behavioural Brain Research,*
40(3), 169-92.

Fishman, M. C., & Michael, C. R. 1973. Integration of auditory information in the cat's
visual cortex. *Vision Research, 13,* 1415-1419.

Friston, K. J., Frith, C. D., Passingham, R. E., Liddle, P. F., & Frackowiak, R. S. 1992.
Motor practice and neurophysiological adaptation in the cerebellum: a positron tomography study. *Proceedings of the Royal Society of London [Biology], 248*(1323), 223-8.

Galaburda, A. M. 1982. Histology, architectonics, and asymmetry of language areas,
Neural Models of Language Processes, (pp. 435-445): Academic Press.

Galaburda, A. M., & Pandya, D. N. 1982. Role of architectonics and connections in the
study of primate brain evolution. In E. Armstrong & D. Falk (Eds.), *Primate Brain*
Evolution, (pp. 203-216). New York: Plenum Press.

Galaburda, A. M., & Pandya, D. N. 1983. The intrinsic architectonic and connectional
organization of the superior temporal region in the rhesus monkey. *Journal of Com-*
parative Neurology, 221, 169-184.

Gattas, R., Sousa, A. P., Mishkin, M., & Ungerleider, L. G. 1997. Cortical projections of
area V2 in the macaque. *Cerebral Cortex, 7(2),* 110-29.

Geschwind, N. 1965. Disconnexion syndromes in animals and man. *Brain, 88,* 237-294;
585-644.

Goldman-Rakic, P. S. 1988. Topography of cognition: Parallel distributed networks in
primate association cortex. *Annual Review of Neuroscience, 11,* 137-156.

Grasby, P. M., Frith, C. D., Friston, K. J., Bench, C., Frackowiak, R. S., & Dolan, R. J.
1993. Functional mapping of brain areas implicated in auditory—verbal memory function. *Brain, 116* (Pt 1), 1-20.

Graziano, M. S. A., & Gross, C. G. 1993. A bimodal map of space - Somatosensory
receptive-fields in the macaque putamen with corresponding visual receptive-fields.
Experimental Brain Research, 97, 96-109.

Grossenbacher, P., Posner, M. I., Compton, P., & Tucker, D. 1991. *Combining isolable*
physical and semantic codes. Paper presented at the 32nd Annual Meeting of the
Psychonomic Society, San Francisco, CA.

Grossenbacher, P. G. 1996. Consciousness and evolution in neocortex. In P. A. Mellars &

K. R. Gibson (Eds.), *Modelling the Early Human Mind*, (pp. 119-130). Cambridge, England: McDonald Institute for Archaelogical Research.

Grossenbacher, P. G. 1999. Correspondence between tactile temporal frequency and visual spatial frequency. In P.R. Killeen & W.R. Uttal (Eds.), *Fechner Day 99: The End of 20th Century Psychophysics. Proceedings of the 15th Annual Meeting of the International Society for Psychophysics*, (pp. 56-61). Tempe, AZ, USA: The International Society for Psychophysics.

Heywood, C. A., & Cowey, A. 1992. The role of the 'face-cell' area in the discrimination and recognition of faces by monkeys. *Philosophical Transaction of the Royal Society of London B, 335(1273)*, 31-7; discussion 37-8.

Hikosaka, K., Iwai, E., Saito, H., & Tanaka, K. 1988. Polysensory properties of neurons in the anterior bank of the caudal superior temporal sulcus of the macaque monkey. *Journal of Neurophysiology, 60(5)*, 1615-37.

Horn, G. 1965. The effect of somaesthetic and photic stimuuli on the activity of units in the striate cortex of unanaesthetized, unrestrained cats. *Journal of Physiology, 179*, 263-277.

Karni, A., Meyer, G., Jezzard, P., Adams, M. M., Turner, R., & Ungerleider, L. G. 1995. Functional MRI evidence for adult motor cortex plasticity during motor skill learning. *Nature, 377(6545)*, 155-8.

Komatsu, H., Ideura, Y., Kaji, S., & Yamane, S. 1992. Color selectivity of neurons in the inferior temporal cortex of the awake Macaque monkey. *Journal of Neuroscience, 12*, 408-424.

Krubitzer, L. A., & Kaas, J. H. 1990. Cortical connections of MT in four species of primates: areal, modular, and retinotopic patterns. *Visual Neuroscience, 5(2)*, 165-204.

LaBerge, D. 1990. Thalamic and cortical mechanisms of attention suggested by recent positron emission tomographic experiments. *Journal of Cognitive Neuroscience, 2*, 358-372.

McGuire, P. K., Paulesu, E., Frackowiak, R. S., & Frith, C. D. 1996. Brain activity during stimulus independent thought. *Neuroreport, 7(13)*, 2095-9.

Miller, M. W., & Vogt, B. A. 1984. Direct connections of rat visual cortex with sensory, motor, and association cortices. *Journal of Comparative Neurology, 226*, 184-202.

Molchan, S. E., Sunderland, T., McIntosh, A. R., Herscovitch, P., & Schreurs, B. G. 1994. A functional anatomical study of associative learning in humans. *Proceedings of the National Academy of Sciences of the United States of America, 91(17)*, 8122-6.

Mollon, J. D. 1989. "Tho' she kneel'd in that place where they grew...": The uses and origins of primate colour vision. *Journal of Experimental Biology, 146*, 21-38.

Murata, K., Cramer, H., & Bach-y-Rita, P. 1965. Neuronal convergence of noxious, acoustic and visual stimuli in the visual cortex of the cat. *Journal of Neurophysiology, 28*, 1223-1239.

Newsome, W. T., Britten, K. H., & Movshon, J. A. 1989. Neuronal correlates of a perceptual decision. *Nature, 341(6237)*, 52-4.

Pandya, D. N., Van Hoesen, G. W., & Mesulam, M. M. 1981. Efferent connections of the cingulate gyrus in the rhesus monkey. *Experimental Brain Research, 42(3-4)*, 319-30.

Paperna, T., & Malach, R. 1991. Patterns of sensory intermodality relationships in the cerebral cortex of the rat. *Journal of Comparative Neurology, 308*, 432-456.

314 PETER G. GROSSENBACHER

Pardo, J. V., Pardo, P. J., Janer, K. W., & Raichle, M. E. 1990. The anterior cingulate cortex mediates processing selection in the Stroop attentional conflict paradigm. *Proceedings of the National Academy of Sciences of the United States of America, 87(1)*, 256-9.

Passingham, R. E. 1975. Changes in the size and organisation of the brain in man and his ancestors. *Brain, Behavior and Evolution, 11*, 73-90.

Pilbeam, D., & Gould, S. J. 1974. Size and scaling in human evolution. *Science, 186*, 892-901.

Posner, M. I., & Petersen, S. E. 1990. The attention system of the human brain. *Annual Review of Neuroscience, 13*, 25-42.

Posner, M. I., Petersen, S. E., Fox, P. T., & Raichle, M. E. 1988. Localization of cognitive operations in the human brain. *Science, 240*, 1627-1631.

Posner, M. I., & Raichle, M. E. 1994. *Images of Mind.* New York: Scientific American Press.

Posner, M. I., & Rothbart, M. K. 1991. Attentional mechanisms and conscious experience. In D. Milner & M. Rugg (Eds.), *The Neuropsychology of Consciousness*, (pp. 91-111). New York: Academic Press.

Preuss, T. M., & Goldman-Rakic, P. S. 1991. Architectonics of the parietal and temporal association cortex in the strepsirhine primate Galago compared to the anthropoid primate Macaca. *Journal of Comparative Neurology, 310(4)*, 475-506.

Reep, R. L., Corwin, J. V., Hashimoto, A., & Watson, R. T. 1984. Afferent connections of medial precentral cortex in the rat. *Neuroscience Letters, 44(3)*, 247-52.

Roland, P. E., Skinhoj, E., & Lassen, N. A. 1981. Focal activations of human cerebral cortex during auditory discrimination. *Journal of Neurophysiology, 45(6)*, 1139-51.

Rolls, E. T. 1996. The orbitofrontal cortex. *Philosophical Transaction of the Royal Society of London B, 351(1346)*, 1433-43; discussion 1443-4.

Schneider, W., & Shiffrin, R. M. 1977. Controlled and automatic human information processing: 1. Detection, search and attention. *Psychological Review, 84*, 1-66.

Spinelli, D. N., Starr, A., & Barrett, T. W. 1968. Auditory specificity in unit recordings from cat's visual cortex. *Experimental Neurology, 22*, 75-84.

Stephan, H., & Andy, O. J. 1970. The allocortex in primates. In C. R. Noback & W. Montagna (Eds.), *The Primate Brain*, (pp. 109-135). New York: Appleton-Century-Crofts.

Ungerleider, L. G. 1995. Functional brain imaging studies of cortical mechanisms for memory. *Science, 270(5237)*, 769-75.

Vogt, B. A., Finch, D. M., & Olson, C. R. 1992. Functional heterogeneity in cingulate cortex: the anterior executive and posterior evaluative regions. *Cerebral Cortex, 2(6)*, 435-43.

Vogt, B. A., & Miller, M. W. 1983. Cortical connections between rat cingulate cortex and visual, motor, and postsubicular cortices. *Journal of Comparative Neurology, 216*, 192-210.

Voronin, L. L., & Skrebitsky, V. G. 1965. Spontaneous and induced potential of the cortex neurons in non-anaesthetized rabbits. *Electroencephalography and Clinical Neurophysiology, Abstract of paper presented at 6th International Congress in Vienna*, 79.

Zatorre, R. J., & Halpern, A. R. 1993. Effect of unilateral temporal-lobe excision on perception and imagery of songs. *Neuropsychologia, 31*, 221-232.

Subject Index

Name Index

In the series ADVANCES IN CONSCIOUSNESS RESEARCH (AiCR) the following titles have been published thus far or are scheduled for publication:

1. GLOBUS, Gordon G.: *The Postmodern Brain*. 1995.
2. ELLIS, Ralph D.: *Questioning Consciousness. The interplay of imagery, cognition, and emotion in the human brain*. 1995.
3. JIBU, Mari and Kunio YASUE: *Quantum Brain Dynamics and Consciousness. An introduction*. 1995.
4. HARDCASTLE, Valerie Gray: *Locating Consciousness*. 1995.
5. STUBENBERG, Leopold: *Consciousness and Qualia*. 1998.
6. GENNARO, Rocco J.: *Consciousness and Self-Consciousness. A defense of the higher-order thought theory of consciousness*. 1996.
7. MAC CORMAC, Earl and Maxim I. STAMENOV (eds): *Fractals of Brain, Fractals of Mind. In search of a symmetry bond*. 1996.
8. GROSSENBACHER, Peter G. (ed.): *Finding Consciousness in the Brain. A neuro-cognitive approach*. 2001.
9. Ó NUALLÁIN, Seán, Paul MC KEVITT and Eoghan MAC AOGÁIN (eds): *Two Sciences of Mind. Readings in cognitive science and consciousness*. 1997.
10. NEWTON, Natika: *Foundations of Understanding*. 1996.
11. PYLKKÖ, Pauli: *The Aconceptual Mind. Heideggerian themes in holistic naturalism*. 1998.
12. STAMENOV, Maxim I. (ed.): *Language Structure, Discourse and the Access to Consciousness*. 1997.
13. VELMANS, Max (ed.): *Investigating Phenomenal Consciousness. Methodologies and Maps*. 2000.
14. SHEETS-JOHNSTONE, Maxine: *The Primacy of Movement*. 1999.
15. CHALLIS, Bradford H. and Boris M. VELICHKOVSKY (eds.): *Stratification in Cognition and Consciousness*. 1999.
16. ELLIS, Ralph D. and Natika NEWTON (eds.): *The Caldron of Consciousness. Motivation, affect and self-organization – An anthology*. 2000.
17. HUTTO, Daniel D.: *The Presence of Mind*. 1999.
18. PALMER, Gary B. and Debra J. OCCHI (eds.): *Languages of Sentiment. Cultural constructions of emotional substrates*. 1999.
19. DAUTENHAHN, Kerstin (ed.): *Human Cognition and Social Agent Technology*. 2000.
20. KUNZENDORF, Robert G. and Benjamin WALLACE (eds.): *Individual Differences in Conscious Experience*. 2000.
21. HUTTO, Daniel D.: *Beyond Physicalism*. 2000.
22. ROSSETTI, Yves and Antti REVONSUO (eds.): *Beyond Dissociation. Interaction between dissociated implicit and explicit processing*. 2000.
23. ZAHAVI, Dan (ed.): *Exploring the Self. Philosophical and psychopathological perspectives on self-experience*. 2000.
24. ROVEE-COLLIER, Carolyn, Harlene HAYNE and Michael COLOMBO: *The Development of Implicit and Explicit Memory*. 2000.
25. BACHMANN, Talis: *Microgenetic Approach to the Conscious Mind*. 2000.
26. Ó NUALLÁIN, Seán (ed.): *Spatial Cognition. Selected papers from Mind III, Annual Conference of the Cognitive Science Society of Ireland, 1998*. 2000.

27. McMILLAN, John and Grant R. GILLETT: *Consciousness and Intentionality.* 2001.
28. ZACHAR, Peter: *Psychological Concepts and Biological Psychiatry. A philosophical analysis.* 2000.
29. VAN LOOCKE, Philip (ed.): *The Physical Nature of Consciousness.* 2001.
30. BROOK, Andrew and Richard C. DeVIDI (eds.): *Self-awareness and Self-reference.* n.y.p.
31. RAKOVER, Sam S. and Baruch CAHLON: *Face Recognition. Cognitive and computational processes.* n.y.p.
32. VITIELLO, Giuseppe: *My Double Unveiled. The dissipative quantum model of the brain.* n.y.p.